116 Advances in Polymer Science

Atomistic Modeling of Physical Properties

Editors: L. Monnerie, U. W. Suter

With contributions by
D. B. Adolf, I. Bahar, R. H. Boyd, J. G. Curro,
J. J. de Pablo, L. R. Dodd, M. D. Ediger,
B. Erman, A. A. Gusev, M. Laso, E. Leontidis,
G. L. Liang, L. Monnerie, F. Müller-Plathe,
D. W. Noid, R.-J. Roe, K. S. Schweizer,
B. G. Sumpter, U. W. Suter, D. N. Theodorou,
W. F. van Gunsteren, B. Wunderlich

With 145 Figures and 19 Tables

Springer-Verlag
Berlin Heidelberg GmbH

Volume Editors:

Prof. Dr. Lucien Monnerie
Ecole Superieure De Physique et de Chimie Industrielles
ESPCI, Laboratoire, de Physico-Chimie Structurale et
Macromoleculaire
10, Rue Vauquelin, 75231 Paris Cedex 05, France

Prof. Dr. U.W. Suter
ETH Zürich, Institut für Polymere, Makromolekulare Chemie
ETH Zentrum, 8092 Zürich, Switzerland

ISBN 978-3-662-14913-3   ISBN 978-3-540-48352-6 (eBook)
DOI 10.1007/978-3-540-48352-6

© Springer-Verlag Berlin Heidelberg 1994
Originally published by Springer-Verlag Berlin Heidelberg New York in 1994
Softcover reprint of the hardcover 1st edition 1994

Library of Congress Catalog Card Number 61-642

Typesetting: Macmillan India Ltd., Bangalore-25

SPIN: 10126785   02/3020 - 5 4 3 2 1 0 - Printed on acid-free paper

Editors

Preface

The understanding of structure-properties relationships is one of the main goals of research in polymer physical chemistry. For a long time, this approach has been based on experimental studies which require the synthesis of specific structures and the preparation of large quantities of materials for mechanical property tests.

With the considerable increase in computer capacity and speed during the last decade, a new approach has developed, involving computer simulations. This research domain is progressing very rapidly and the purpose of this issue is to describe the present state of the art in the atomistic modeling of polymer properties.

The two first papers (Boyd, Wunderlich et al.) concern polymer crystal structures, their static and dynamic properties respectively.

The following three papers focus on local dynamics of either a single chain (Ediger, Bahar et al.) or a polymer melt (Roe).

After a critical discussion of novel algorithms for condensed polymer phases (Leontidis et al.), the structural and equation of state properties of polymers (Theodorou et al.), as well as the dynamics of small molecules in bulk polymers (Gusev et al.) are presented.

The last paper, by Schweizer and Curro, introduces a new theoretical approach, which leads to the prediction of thermodynamic properties of pure polymers and polymer blends.

L. Monnerie
U. W. Suter

Table of Contents

Prediction of Polymer Crystal Structures and Properties

R. H. Boyd
Department of Materials Science and Engineering and Department of Chemical
Engineering, University of Utah, Salt Lake City, UT 84112, USA

The simulation of the packing of polymer molecules in crystals is, in a sense, computationally simple. The number of degrees of freedom to be considered is quite limited; they are comprised only of the atom coordinates of the repeat unit and the unit cell parameters. Energy minimization serves to define the structure and often yields very useful results. Computer strategies can be designed to allow simultaneous optimization of the internal molecular geometry and the crystal packing. A number of useful crystal properties can be calculated in addition to finding the structure. However, because the molecules can usually pack in a number of stable forms, the application actually can be far from simple. These various forms have to be discovered. The evaluation of their relative stability puts great demands on the quality of the empirical force field parameters.

Advances in Polymer Science, Vol. 116
© Springer-Verlag Berlin Heidelberg 1994

1 Introduction

Polymers that have regularly repeating structural units are capable of producing periodic structures when packed together in bulk. Therefore such polymers can and, unless kinetic factors intervene, presumably will form crystals. Many of the technologically important polymeric materials are used in a state that is at least in part crystalline. If polymeric materials are to be modelled at the molecular level it is therefore required that crystals be addressed.

In amorphous materials there is but one "structure" or packing, even though it is a highly complicated and disordered one. This is not true of crystals. The same polymer molecule, either by virtue of adopting multiple chain conformations or multiple packings of the same conformation, or both, can often achieve a number of stable crystal arrangements. Thus a central problem in modelling polymer crystals is the discovery, and the assessment of the relative stability, of these arrangements. Beyond this, another important problem to be addressed is the determination of the physical properties of a given crystal structure. The present review discusses the methodology, and some results, connected with the use of atomistic modelling in determining crystal structures and the subsequent calculation of the properties of the crystals. It is concerned only with the concept of perfect crystals in the crystallographic sense. That is, crystals are infinite periodic arrays of objects. These objects, in addition to the translations inherent in the periodic array, may also be related by various symmetry operations. It is this picture that is pursued here.

2 Basic Assumptions

The first assumption is that the methods of molecular mechanics can be applied. Empirically calibrated potential functions that describe the internal structure of the polymer molecule and the intermolecular interactions are invoked. In addition, most of the work that has pursued structure and packing has been done by energy minimization. Thus the calculations are appropriate only for zero K. To some degree this condition could be removed by using molecular dynamics. However there are some non-trivial complications in this approach and these will be discussed in a separate section.

The polymer chain itself is to be described as a helix. The chain building block, which is a chemical repeat unit or a small group of them, generates an entire chain by the helix operation. Rotation about the helix axis through a given angle is accompanied by advancement along the helix axis by a given distance.

Next, it must be decided how general the approach should be. Ideally the question should be framed as: given the building block of the polymer chain, i.e., in crystallographic terms the asymmetric unit, what it the resulting crystal structure? Thus no symmetry relations, beyond the helix operation, are assumed. They are to be inferred from the resulting stable packing arrangements. However, this plan results in discovering the primitive unit cell. Very often it is a convenience to impose some symmetry operations so that a symmetry-centered cell results directly. It is quite feasible to write a computer code that allows these operations to be imposed or not, according to the problem studied.

A simplification that has often been made is to assume that the polymer chain conformation and the internal geometry may be determined in a separate step as an isolated chain. Then the crystal packing is studied under the assumption of an internally rigid molecule. Much of the packing energeties analysis of polymer crystal has used this assumption [1]. It is adequate in some cases and in others it is not. Computer codes can be formulated that incorporate this feature or not, as is desired.

3 Description of a General Procedure

The considerations of the preceding section are illustrated in more detail here by describing a procedure that has been implemented for studying polymer crystal packing [2]. It allows the simultaneous optimization of the internal atom coordinates of the chain unit and the helix parameters along with the cell packing parameters.

First, a definition is made. The crystal is generated by propagating or replicating the asymmetric unit. In the case of a polymer molecule this could well be the chemical repeat unit of the chain. However there is no guarantee that the actual helix adopted by the chain will have just one chemical repeat unit as the basis for the helix operation. The torsional angles in two adjacent chemical repeat units could, in principle, be different. Thus we beg the question by defining the "CRU". It is the conformational repeat unit of the helix; the unit in which the internal structure and conformation are allowed to adjust in the minimization. The infinite chain is generated by propagating the CRU by the helix operation.

Crystal structures are to be determined by finding local minima in the total energy. The procedure below illustrates an iterative Newton-Raphson method for accomplishing this. Being a 'second derivative' method, it is efficient in terms of the number of iterations required. The number of degrees of freedom in crystal packing simulations are rather small since the lattice is generated by replicating the CRU. Thus it is not compromised by the 'N^3' nature of the Newton-Raphson equation solving. It also has a number of additional advantages since the converged Hessian second derivative matrix can be used to calculate a number of properties.

The energy is considered to be a sum of two terms. One is the intramolecular energy and is a function of valence coordinates and the intramolecular non-bonded distances (or Coulombic charge interaction distances). The other is the intermolecular nonbonded energy. The valence coordinates are bond distances, R_{ij}, valence angles, θ_{ijk} and torsional angles, ϕ_{ijkl}. Thus the total valence inter-action energy, U_{intra}, is

$$U_{intra} = \Sigma_{Rij} u_R(R_{ij}) + \Sigma_{\theta ijk} u_\theta(\theta_{ijk}) + \Sigma_{\phi ijkl} u_\phi(\phi_{ijkl}) + \Sigma_{Rij} u_{nb}(R_{ij}) \qquad (1)$$

and the sums are over the bond stretching, valence angle bending, torsions and intramolecular nonbonded interactions present in the CRU. The total inter-molecular energy, U_{inter}, is a function only of intermolecular $R_{i'j'}$ nonbonded (or Coulombic) distances

$$U_{inter} = \Sigma_{Ri'j'} u_{nb}(R_{i'j'}) . \qquad (2)$$

There are several coordinate systems that have to be dealt with. Ultimately, in order to carry out the minimization process, the total energy is best expressed in terms of Cartesian coordinates. However, a general unit cell or lattice is characterized by non-orthogonal basis vectors. A cylindrical coordinate system is used to represent the molecular helix. The intramolecular energy is expressed in terms of valence coordinates. Thus transformations must be set up that relate the Cartesian coordinates to the helix parameters, the unit cell parameters and the valence coordinates. The helix operations and the unit cell parameters are considered first.

The CRU rotates through a helical advance angle, σ, and a translational advance, d, in replicating itself (see Fig. 1). The helix is arbitrarily oriented with respect to a Cartesian coordinate system such that the helix axis is along the z axis and the first atom of the repeat unit is on the x axis at z = 0. The Cartesian

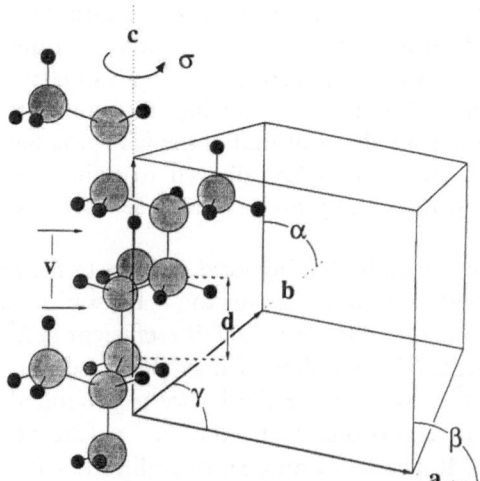

Fig. 1. A schematic depiction of the parameters involved in simultaneous inter and intra molecular energy minimization of a polymer crystal. The repeat unit internal coordinates are designated by v, the helix advance angle and distance by σ and d, the chain setting angle by θ_x (not shown) and the lattice parameters by a, b, α, β, γ

coordinates of the CRU in this orientation may be designated the "unrotated coordinates". They are represented by $\mathbf{v}$, a 3N dimensional vector made up of vectors, $\mathbf{v}(i)$, $i = 1, N$ with components $v(i)_x$, $v(i)_y$, $v(i)_z$ that are the coordinates of the N atoms of the helical repeat unit. The crystal lattice is represented by nonorthogonal basis vectors $\mathbf{a}, \mathbf{b}, \mathbf{c}$ that are oriented with $\mathbf{c}$ along the Cartesian z axis and with the projection of $\mathbf{a}$ in the x, y plane along x. The position vector, $\mathbf{x}_n(i)$, of the i-th atom in the n-th repeat unit along a central chain located at $x, y = 0$ is expressible in terms of $\mathbf{v}(i)$, the helical advance angle, σ, and translation, d, as

$$\mathbf{x}_n(i) = \mathbf{R}\{\sigma, \theta_x\}\, \mathbf{v}(i) + \mathbf{T} \tag{3}$$

where $\mathbf{R}\{\sigma, \theta_x\}$ is a rotation matrix,

$$\mathbf{R}\{\sigma, \theta_x\} = \begin{bmatrix} \cos((n-1)\sigma + \theta_x) & -\sin((n-1)\sigma + \theta_x) & 0 \\ \sin((n-1)\sigma + \theta_x) & \cos((n-1)\sigma + \theta_x) & 0 \\ 0 & 0 & 0 \end{bmatrix}$$

that is a function of a possible chain setting angle, θ_x, as well as the advance angle, σ, and where $\mathbf{T}$ is the translation vector,

$$\mathbf{T} = [0, 0, (n-1)d]^t\,.$$

The position vector, $\mathbf{x}_{n,k}(i)$, of the i-th atom in the n-th repeat unit in a distant chain, k, located at a', b', $(c = 0)$, (where the primes on a, b indicate the possibility of fractional values of a, b in lattices with more than one chain per cell) can be expressed as

$$\mathbf{x}_{n,k}(i) = \mathbf{R}\{\sigma, \theta_{x(k)}\}\, \mathbf{v}(i) + \mathbf{R}\{\alpha, \beta, \gamma\}\, \mathbf{T}_k \tag{4}$$

using the above helix rotation matrix (but with a k subscript on σ_x to indicate the possibilty of a $\pm$ dependence on k) and with the aid of a lattice rotation matrix $\mathbf{R}\{\alpha, \beta, \gamma\}$,

$$\mathbf{R}\{\alpha, \beta, \gamma\} = \begin{bmatrix} \sin\beta & C_2/\sin\beta & 0 \\ 0 & C_1/\sin\beta & 0 \\ \cos\beta & \cos\alpha & 1 \end{bmatrix}$$

in which α, β, γ are the unit cell angles (γ between $\mathbf{a}, \mathbf{b}$ etc.) and

$$C_1 = (1 - \cos^2\alpha - \cos^2\beta - \cos^2\gamma + 2\cos\alpha\,\cos\beta\,\cos\gamma)^{1/2}$$

$$C_2 = \cos\gamma - \cos\alpha\,\cos\beta$$

and a translation vector $\mathbf{T}_k$,

$$\mathbf{T}_k = [A_k a, B_k b, (n-1)d]^t\,,$$

in which A_k, B_k are appropriate multiples of a and b required to place the chain at a', b'. The helix d translation in Eq. (3) is now absorbed into T_k. A chain translational offset along c equal to $Z_{0(k)}$, a possibility when more than one chain per cell is present, can be incorporated as a parameter into each $v_z(i)$ as $v_z(i) + Z_{0(k)}$.

Since the crystal energy is expressed as a function of the position vectors $x_{n,k}(i)$ it follows that the independent basis parameters to be optimized in energy minimization are the unrotated CRU coordinates, v, the helix parameters, σ and d, a chain setting angle, θ_x, chain c axis offset, Z_0 and the lattice parameters, a, b, α, β, γ. The basis parameters are collected into a vector, **P**,

$$\mathbf{P} = [\mathbf{v}, \sigma, d, \theta_x, Z_0, a, b, \alpha, \beta, \gamma]^t . \tag{5}$$

The intramolecular energy is now expressed in terms of the unrotated CRU Cartesian coordinates, v, as follows. In the Newton-Raphson method the energy is required as a quadratic function of the parameter displacements. The valence coordinate energy functions are easily expanded as quadratics in r, θ, ϕ, but then transformations to the basis parameters must be provided. Let q, q' be generalized valence coordinates (any of R, θ, or ϕ) expressing an interaction between a set of atoms i, j, k, ... and a set i', j', k', ... Then the explicit interaction energy expression, U(q, q'), may be expanded,

$$U(q, q') = U^0(q, q') + \partial U/\partial q \Delta q + \partial U/\partial q' \Delta q'$$
$$+ 1/2 \Sigma \partial^2 U/\partial q \partial q' \Delta q \Delta q' + \cdots .$$
$$\tag{6}$$

As an example, in the case of a simple bond stretching function, where $U = 1/2k(r_{ij} - R_{ij}^0)^2$, q, q' = $r_{i,j}$ and

$$U_r = U^0(r_{ij}^0) + k(r_{ij}^0 - R_{ij}^0)\Delta r_{ij} + 1/2k\Delta r_{ij}^2 .$$

The coordinate q is then transformed to the Cartesian basis, **x**. Let

$$\mathbf{D}_1(s) = [\partial q/\partial x_s, \partial q/\partial y_s, \partial q/\partial z_s]^t, \quad s = i, j, k, \ldots$$

be the vector containing the first derivatives of q with respect to the Cartesian coordinate, **x**(p), let $\mathbf{D}_2(s, s')$ be the 3×3 matrix of second derivatives of q with respect to x_s, y_s, z_s x_s', y_s', z_s' and let $\Delta \mathbf{x}(s)$ be the vector = $[\Delta x_s, \Delta y_s, \Delta z_s]^t$. Then

$$\Delta q = \Sigma_{s=i,j,k,\ldots} \mathbf{D}_1(s)\Delta\mathbf{x}(s) + 1/2\Sigma_{s,s'} \Delta\mathbf{x}(s)^t \mathbf{D}_2(s, s') \Delta\mathbf{x}(s') + \cdots . \tag{7}$$

Use of Eq. (7) in Eq. (5) then expresses a potential energy interaction term as a quadratic in Cartesian coordinates, $\Delta\mathbf{x}$. It remains to transform the $\Delta\mathbf{x}$ coordinates to the basis parameter set as

$$\Delta x^\alpha(s) = \mathbf{B}_1 \Delta\mathbf{P} + \Delta\mathbf{P}^t \mathbf{B}_2 \Delta\mathbf{P} + \cdots \tag{8}$$

where $\Delta x^\alpha(s)$ is a component ($\alpha = x, y, z$) of $\Delta\mathbf{x}(s)$, $\mathbf{B}_1$ is the vector of first derivatives of x with respect to the basis parameters in $\mathbf{P}$ (Eq. (5)). $\mathbf{B_2}$ is the matrix of second derivatives and from Eq. (5),

$$\Delta \mathbf{P} = [\Delta v, \Delta \sigma, \Delta d, \Delta \theta_x, \Delta Z_0, \Delta a, \Delta b, \Delta \alpha, \Delta \beta, \Delta \gamma]^t .$$

The above relations result in a quadratic expansion in terms of the independent basis parameters for a single potential energy interaction expression. When this process is repeated for all the potential energy terms considered and the results accumulated, a quadratic expression for the total energy is obtained that may be expressed as

$$U = U^0 + \mathbf{A} \Delta \mathbf{P} + 1/2 \Delta \mathbf{P}^t \mathbf{C} \Delta \mathbf{P} + \cdots \qquad (9)$$

where all of the first order terms are accumulated into $\mathbf{A}$ and the second order ones into $\mathbf{C}$. The condition for stationary energy, $\delta U = 0$, since the elements of $\Delta \mathbf{P}$ are independent, leads, on differentiation of U with respect to each of the parameters in $\Delta \mathbf{P}$ and setting the result equal to zero, to a set of linear algebraic equations,

$$\mathbf{C} \Delta \mathbf{P} = -\mathbf{A} . \qquad (10)$$

On solving for $\Delta \mathbf{P}$ the parameters may be updated $\mathbf{P}' = \mathbf{P}' + \Delta \mathbf{P}$ and the process repeated until convergence, $\mathbf{A} \to 0$, $\Delta \mathbf{P} \to 0$.

The final parameter vector, $\mathbf{P}$, contains all of the information about the structure including the unit cell parameters and the atom coordinates of the CRU.

4 Polarity and Electrostatic Energy

In the development above it is considered that polar interactions are of the Coulombic type where partial charges $\delta(i)$, $\delta(j)$, residing on atom centers i, j interact (in electrostatic units) with energy, $u_{es}(i, j)$, according to

$$u_{se}(i, j) = \delta(i) \delta(j) / R_{ij} . \qquad (11)$$

This supposes that the charges are fixed and are properties of the atom centers. Under these conditions the electrostatic interaction is simply a function of the coordinate R_{ij} and minimization is effected accordingly. Attention must be paid to proper convergence when summations over the centers are carried out. If integral numbers of complete, electrically neutral, unit cells are always considered in summation this causes no difficulty.

The above recipe is useful but does not embrace an important physical reality. The charges are actually not fixed. The atom centers containing them are polarizable. The field at any center causes the induction of a local moment, μ (induced), in addition to the permanent charge, or, μ (induced) = α E, where α is the polarizability at the center and E is the electric field at the center. The fields from these induced moments induce further moments at other centers. Thus the internal or local field problem arises. Since the induced moments are linear in the local field the whole problem is linear. The local field at each center may be found from solving a set of linear algebraic equations. In a representation where permanent bond moments, $\mu^0(i)$ are invoked rather than permanent atom based charges, $\delta(i)$, the total electrostatic energy is given by

$$U_{es} = -1/2\Sigma_i\Sigma_p\mu_p^0(i)E_p(i) \tag{12}$$

where $p = x, y, z$ refers to the components of the permanent moment $\mu^0(i)$ and the local field, $E(i)$ at each bond, i, in a cell.

Thus, including polarization or mutual induction effects or properly calculating the local field, whatever words are used to describe the problem, is entirely practical. It has been implemented [3] and applied to several problems [3–8]. Parameters are available for a number of polar bonds [3, 6, 7].

The above procedure does have the problem that it is not amenable to direct inclusion in the energy minimization scheme. Thus the expedient has been adopted of using effective charges and Eq. (11) in the minimization, but using Eq. (12) as applied in a post minimization step, for representing the electrostatic energy. This is justified by the fact that the electrostatic forces are very 'soft', i.e., very slowly varying with distance. Molecular geometry (atom coordinates) tends to be determined by the 'harder' steric repulsion terms and the bonded R, θ, φ terms in the conformational energy, Eqs. (1) and (2).

The covalent shell model (CSM) [9] allows introduction of polarization effects but permits the simplicity of Eq. (11) to be preserved. Instead of writing the induced moment as the product of the field and local polarizability, μ (induced) = α E, each center is represented by two opposite charges connected by an isotropic spring. One of the charges is fixed to the atom center (the core) and the other is the shell. The induced moment results from displacement of the shell away from the core. The effective polarizability $\alpha_{csm} = Z^2/k$ where Z is the shell charge and k is the force constant of the spring. At the expense of increasing the number of computational centers, a shell at each atom center, the method allows the minimization to be carried out in terms of R coordinates, through Eq. (11).

Since the above procedures are non-trivial there has been emphasis on simpler approximations. The simplest is to include an effective dielectric constant in the denominator of Eq. (11). The dielectric constant is, of course, a macroscopic concept that results from the microscopic treatment of polarization above. Invoking it microscopically does not recognize the many local environments in which a polar bond or atom may find itself. However, some theoretical justification has been presented [10].

5 Crystal Properties

The final coefficient matrix, **C**, in Eqs. (9) and (10) can be used for further property prediction. It can be regarded as the dynamical matrix in analysis of crystal vibrations by the Born-Huang [11] formalism. Frequencies from the resulting vibrational dispersion curves can be used to compute the vibrational partition function. The latter leads to the vibrational free energy and heat capacity. The free energy may be useful in assessing relative stabilities of various crystal forms at finite temperatures.

The elastic constants are the second derivatives of the energy with respect to the crystal distortions. Since the coefficient matrix, **C**, contains this information, it may also be used to compute the elastic constants.

5.1 Vibrational Analysis and Spectra

On convergence of the Newton-Raphson iteration process, Eq. (7) is substituted into Eq. (6). The result for the total energy/reference unit in Cartesian basis is given by

$$U = 1/2\Sigma_{i=1,N}\Sigma_j\Sigma_{\alpha,\beta}\Delta x^\alpha(i)g^{\alpha,\beta}(i,j)\Delta x(j)^\beta \, , \qquad (13)$$

where $\Delta x^\alpha(i)$, $\Delta x(j)^\beta$ are the α, $\beta = x, y, z$ components of (vibrational) displacements of atoms i, j, and $g^{\alpha,\beta}(i,j)$ are the elements of the matrix resulting from the above substitution. The summation over i is over the N atoms in the reference unit and that over j is unrestricted. The equation of motion of the atom i in the CRU is

$$- m_i\Delta\ddot{x}^\alpha(i) = \Sigma_j\Sigma_\beta g^{\alpha,\beta}(i,j)\Delta x(j)^\beta \, . \qquad (14)$$

The vibrations are analyzed in terms of dispersion curves by letting the displacement of atom j differ from its counterpart in the reference cell, j', by a cell dependent phase and writing solutions to Eq. (14) as

$$\Delta x^\alpha(j) = \Delta x_0^\alpha(j')e^{i*(\omega t - \mathbf{k}\cdot\delta)} \qquad (15)$$

where $\Delta x_0^\alpha(j)$ is a displacement amplitude, **k**, is a wave vector

$$\mathbf{k} = k_a\vec{i} + k_b\vec{j} + k_c\vec{k}$$

and δ is the phase shift

$$\delta = \delta_a\vec{i} + \delta_b\vec{j} + \delta_c\vec{k} \, .$$

The Eq. (14) set is solved, with the displacements $\Delta x^\alpha(j)$ for atoms, j, outside the CRU eliminated by means of Eq. (15), for various values of the phases δ_a, δ_b, δ_c to generate the dispersion curves.

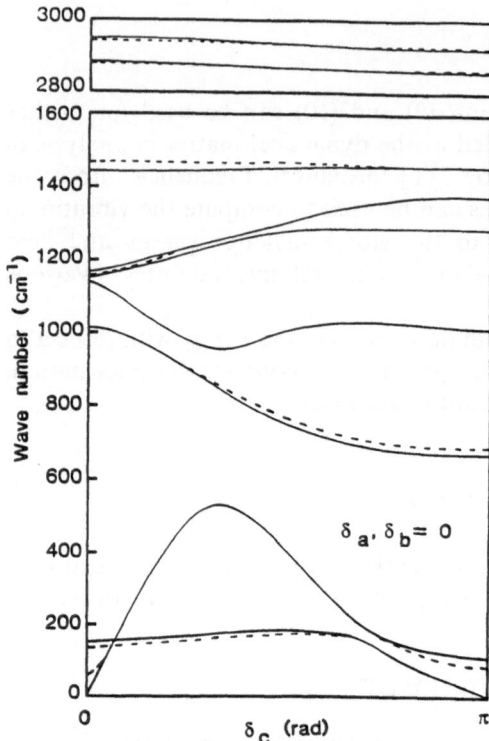

Fig. 2. Vibrational dispersion curves for polyethylene plotted along $\delta_c = 0, \pi$ for $\delta_a, \delta_b = 0$ [6]

The vibrations that are active under infra-red and Raman vibrational spectroscopy are determined from the dispersion curves at phase points (values of δ) that are appropriate for the helix present. In the case of PE in Fig. 2 these are at the values $\delta_c = 0, 180°$.

5.2 Elastic Constants

The elastic constants are the second derivatives of the energy with respect to the strains. The converged matrix, **C**, in Eq. (9) contains the second derivatives with respect to the unit cell parameters, a, b, d, α, β, γ and others parameters such as σ and the atom coordinates, **v**. The cell parameters are strains only in crystal systems with orthogonal basis vectors (cubic, tetragonal and orthorhombic). Thus two further operations are required, elimination of the extra parameters and transformation to Cartesian basis. The elimination of the extra parameters to find the elastic constants has been described [2] as has their elimination to find the compliance matrix [12]. The transformation to Cartesian basis has also been described [2].

Probably of most interest has been the compliance constant along the chain axis, S_{33}. The reciprocal of the latter is the tensile modulus in the chain direction. It is of interest in assessing the maximum achievable stiffness of highly drawn crystalline fibers. For polyethylene, S_{33}^{-1} has been calculated to be 340 GPa [6]. This is a high value and it is the basis for allowing utilization of polyethylene as an ultra-high modulus fiber. Such estimates of modulus are useful in that they form upper limits to the values achievable by drawing. Thus they are an index of the effectiveness of the latter. The question arises as to whether other polymers chains are suitable as well. In polyethylene the high value is to a large degree the result of the planar zig-zag conformation in the crystal and the fact that no torsional angle distortion accompanies tensile distortion. In polymers with truly helical conformations it is of interest to know to what degree the high modulus possibility is compromised.

Figure 3 shows the results of experimentally determining the tensile modulus of polyoxymethylene by X-ray diffraction [13]. The X-ray method determines the crystal strain from the unit cell distortion but the stress is assumed to be the macroscopic stress. This is not generally valid because of the heterogeneous nature of the material, i.e., due the existence of both crystalline and amorphous fractions. However, as temperature is lowered, the difference in the phase stiffnesses becomes less and the method becomes more accurate. Thus in Fig. 3 the apparent tensile modulus at low temperature, below the γ-relaxation, becomes more reflective of the crystal modulus. A calculated value of the modulus is also shown. It may be seen that the latter is only marginally higher than the low temperature, high draw ratio experimental values. This indicates that further processing would probably be ineffective in achieving higher modulus

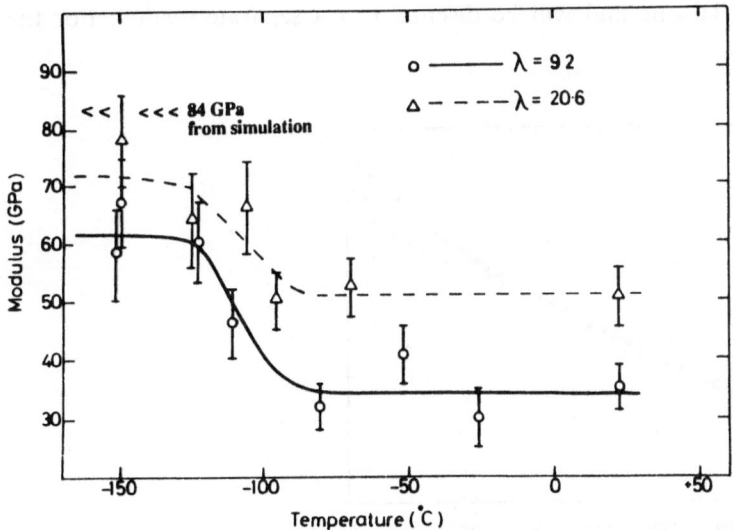

Fig. 3. Tensile or fiber modulus (S_{33}^{-1}) for polyoxymethylene as determined from X-ray diffraction experiments [13] and compared with simulation [6]. The two curves (*solid and dashed*) go through experimental points determined at two draw ratios, λ, as a function of temperature. The value from simulation is appropriate for zero K and is indicated by the < < < line

and that the theoretical limit is being approached. The limit of 84 GPa is much lower than the corresponding value for polyethylene and is the result of torsional angle distortions in tension. The torsional potential is much softer than for bond stretching and bond bending.

5.3 Heat Capacity

In the vibrational analysis of Sect. 5.1, the crystal consists of a set of harmonic oscillators whose frequencies are expressed in terms of the dispersion curves. Thus the thermodynamic properties follow from the partition functions of this set of harmonic oscillators. In practice, the dispersion curves are sampled at enough points along $\delta_a, \delta_b, \delta_c$ to give a representative population of oscillators. This procedure however leads to the constant volume capacity. The constant pressure heat capacity may be computed from the standard relation $C_p - C_v = V\alpha_T^2 T/\beta_c$, where α_T is the coefficient of thermal expansion and β_c is the compressibility. Since the elastic and compliance constants can be found as above, the compressibility may be found from the relation

$$\beta_c = -(\Delta V/\Delta P)/V = -(\varepsilon_{11} + \varepsilon_{22} + + \varepsilon_{33})/\Delta P = \Sigma_{i,j=1.3} S_{ij} \qquad (16)$$

where $\varepsilon_{11}, \varepsilon_{22}, \varepsilon_{33}$ are the tensile strains, and S_{ij} are the components of the compliance matrix for $i, j = 1, 3$. The thermal expansion, α_T, is not available. This is obviously the result of a method that finds zero temperature structures and only harmonic vibrations. The question of determining thermal expansion, α_T, is a non-trivial one and will be discussed in a separate section. For the

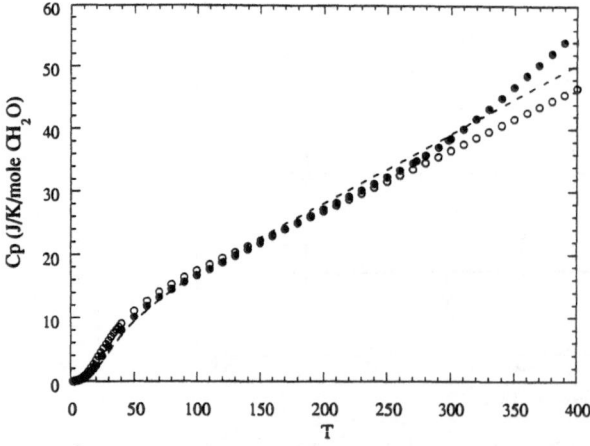

Fig. 4. The heat capacity of hexagonal polyoxymethylene from simulation [6] (*open circles*) compared with experiment [14] (*filled circles*). The *dashed line* is for orthorhombic polyoxymethylene from simulation

purposes of the present section it is assumed that it is known with sufficient precision from experiments or estimates to be able to make the $C_p - C_v$ correction.

The heat capacity of polyoxymethylene, determined as above [6], is shown in Fig. 4 along with experimental values [14]. It is worth noting that the heat capacity experiments must be performed on specimens that are semicrystalline. Extrapolations are made vs crystallinity to arrive at the values tabulated and shown as experimental. Thus there can be uncertainties due to the extrapolation process and therefore the calculated values form a valuable check on the experimental ones.

It may seen in Fig. 4 that the calculations and experiments agree well, up to approximately room temperature. Above this, they diverge due to the onset of significant anharmonicity in the vibrations, an effect not incorporated in the calculated values.

6 Thermal Effects

As emphasized above, most of the work devoted to investigating packing in crystals has utilized energy minimization. Thus the results are appropriate only for zero K. Some comments on extensions are made below.

6.1 Application of Molecular Dynamics (MD) to Structure Determination

It is an attractive prospect to replace the procedure of Sect. 3, which is coupled with energy minimization, by one coupled to MD. Then the determination of structures at finite temperatures could be contemplated. There are some problems with this that need to be addressed. The first one is fundamental and others are of a more practical nature.

MD, as normally implemented in polymer simulations, is based on classical mechanics. Thus any considerations about stabilities of structures will be based on free energies in a classical system. Since the vibrations are far from classical at most temperatures of interest, there is no guarantee that the classical free energy will correctly represent stabilities.

In order to implement MD, the simple plan of Sect. 3 has to be greatly modified. That is, the CRU, consisting of the limited number of atoms in the repeat unit, was taken there as the unit considered and thus as the number of degrees of freedom (less the lattice parameters). All other centers are generated by helix operations and translations. Because, in general, the vibrations are not in phase, this means that the system in MD has to consist of an array of chains large enough to mimic a macroscopic system. Presumably, this is alleviated by invoking periodic boundary conditions on a box consisting of a large array of chains. Thus a crystal is no more simple in MD than a liquid or amorphous system. This should not be a deterrent however. It does require that the periodic box be a general triclinic one. The periodic boundary conditions have to be formulated for arbitrary a, b, c and α, β, γ. Nearest image selection becomes more complicated [15]. Further, the dynamics have to be carried out so that these parameters are free to adjust under isostatic pressure conditions. Such general MD methods have been formulated for mechanical property simulation [16]. However, so far, they do not seem to have been applied to the study of crystal structures.

6.2 Thermal Expansion, the Quasi-Harmonic Approach

In the past, most attention has centered on the use of the quasi-approximation. No matter what the size or shape of the unit cell the vibrations may be calculated in the harmonic approximation. It is assumed that the vibrational free energy difference between two cells of different size and shape are sufficiently accurately represented by the difference in the free energies calculated in the harmonic approximation. In general, when the cell expands the packing energy will increase. However the vibrational frequencies will decrease, causing a decrease in vibrational free energy. At a given temperature, the cell dimensions will adjust to an optimum where the two effects are balanced in minimizing the free energy. This approach, where the changes in vibrational frequencies with cell dimension are calculated has been formulated in an approximate way for *n*-alkanes and polyethylene. The thermal expansion of crystals of the dimension of folded polyethylene chains was successfully accounted for [17].

7 Some Examples of Crystal Structure Prediction

Some typical examples of the application of energy minimization to polymeric crystal structures are given below. By no means all of them, or even most of them, follow the complete prescription given in Sect. 3 above where the internal geometry is adjusted along with the packing parameters. Nevertheless a great deal of insight is gained into the systems studied.

7.1 Polyethylene

This polymer has been the archetypal one for applications of molecular mechanics to lattice packing. It is found for well-calibrated energy function parameter sets, i.e., 'force fields', that the crystal structure is reproduced well. Table 1 shows a comparison for lattice parameters between a calculation [6] and experiment

[18]. The latter is for data extrapolated to zero K and thus is one of the few cases where direct comparison with energy minimization is entirely and directly appropriate. The room temperature cell parameters are also shown [19]. Thus an indication of the effect of comparison of energy minimized structures with finite temperature experimental ones is gained.

It appears that the monoclinic form of polyethylene has a slightly lower energy by calculation than the orthorhombic form [20, 21].

7.2 Polymorphs of Polyoxymethylene

Polyoxymethylene is an interesting example for computation [6]. It exhibits polymorphism, one of two structures is hexagonal and the other is ortho-rhombic, Figs. 5 and 6. In both of them, the chain conformation is basically an all gauche sequence. However, the values of the torsional angles are significantly different in the two forms. Thus it is an example where the necessity for including internal molecular flexibility in the modelling is apparent. Both forms result as energy minima in the modelling. It is found that the torsional angle in the free chain is intermediate between the values found in the two crystal forms (see Table 1). The crystal packing forces cause a compression of the helix to form the orthorhombic phase and an extension to form the hexagonal one.

The stability, as judged from the molecular mechanics calculations appropri-ate to zero K, is greater for the orthorhombic form in comparison with the hexagonal one (Table 2). Experimentally, the hexagonal form is the commonly observed one and is the stable form at room temperature. However, the transition from orthorhombic to hexagonal is found experimentally [22] to be endothermic, indicating, as found from calculation, that the orthorhombic form is energetically more stable. Thus entropy effects are responsible for stability of

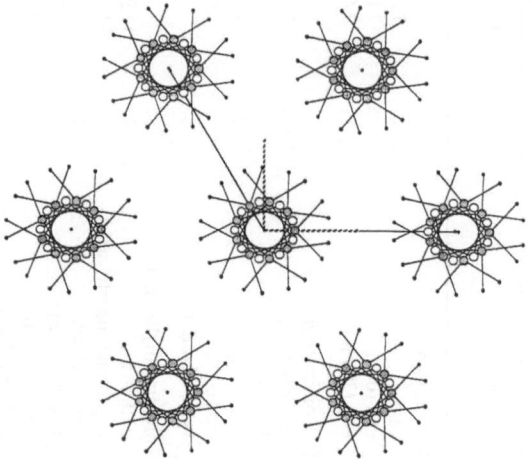

Fig. 5. Hexagonal (trigonal) packing of polyoxymethylene [6]

Table 1. Calculated and experimental lattice parameters[a]

Polymer	N	a	b	c	α	β	γ	θ_x	Helix/ conformation	Chain torsion	Ref
Polyethylene											
calc	2	7.05	4.94	2.544	90.0	90.06	90.0	46.9	2/1	180	[6]
exp (0 K)		7.15	4.91	2.546	90	90	90		2/1	180	[18]
exp (300 K)		7.42	4.95	2.547	90	90	90	~45			[19]
Polyoxymethylene											
hex calc[b]	(2)	7.77	4.50	17.1	90.2	90.1	90.3	16	9/5	78	[6]
hex exp[b]		7.74	4.47	17.4	90	90	90		9/5	77	[6]
ortho calc	2	7.47	4.81	3.52	90.0	89.9	90.2	47.5	2/1	64.6	[6]
ortho exp		7.65	4.77	3.56	90	90	90		2/1	63	[6]
isolated chain										71.6	[6]
Poly(dimethylsilane)											
ortho calc	2	11.9	7.8		90	90	90	42	2/1	180	[23]
exp(mono, nearly ortho)	1	11.18	8.00	3.88	90	90	91			180	[24]
hex (2/1) calc		7.6					0		2/1	180	[23]
hex (15/7) calc		7.8					0		15/7	165	[23]
hex (2/1) exp		7.79								180	[24]
isolated chain									15/7	165	[23]

	N	a	b	c	α	β	γ	θ_x	conformation	ref
Aliphatic polyesters										
2–4 kink	4	10.24	7.62	8.52	90.0	90.1	90.0	0.0	1/1	[8]
planar		5.57	7.11	9.63	82.9	115.4	93.8	153		[8]
exp	4	10.75	7.60·	8.33	90	90	90			[8]
2–6 kink	2	5.46	6.96	11.62	90.0	113.2	90.0	124.7	1/1	[8]
planar		5.50	7.21	12.18	102.2	106.2	90.7	160.8		[8]
exp		5.47	7.23	11.72	90	113.5	90			[8]
2–8 kink	2	5.49	6.98	14.16	90.0	114.1	90.0	134.0	1/1	[8]
planar		5.50	7.31	14.72	99.5	114.1	84.9	146.3	1/1	[8]
exp		5.51	7.25	14.28	90	114.5	90		1/1	[8]
Isotactic Polypropylene										
α form calc	4	6.51	21.07		88.9	96.0	90.2	43.2	3/1	[27]
exp		6.65	20.96	6.50	90	99.3	90	45	3/1	[19]
Polyvinylidine fluoride										
I (or β) calc	2	8.61	4.72	2.56	90.0	90.0	90.0		T	[9]
exp		8.58	4.91	2.56	90	90	90			[9]
II (or α) nonpolar calc	2	5.07	9.47	4.59	90.1	92.0	90.0		TG^+TG^-	[9]
exp		4.96	9.64	4.62	90	90	90			[9]
III (or γ) calc	2	5.02	9.53	9.14	90.0	97.7	90.0		$T_3G^+T_3G^-$	[9]
exp		4.96	9.67	9.20	90	93.0	90			[9]
IV (or δ) polar calc	4	5.08	9.32	4.58	90.1	90.0	90.0		TG^+TG^-	[9]
exp		4.96	9.64	4.62	90	90	90			[9]

[a] N = chains per cell, a, b, c = unit cell dimensions, Å, α, β, γ unit cell angles (in degrees), θ_x = chain setting angle

[b] a, b are values for structure overlaid on orthorhombic grid, $a = \sqrt{3}b$, a(hex) = b

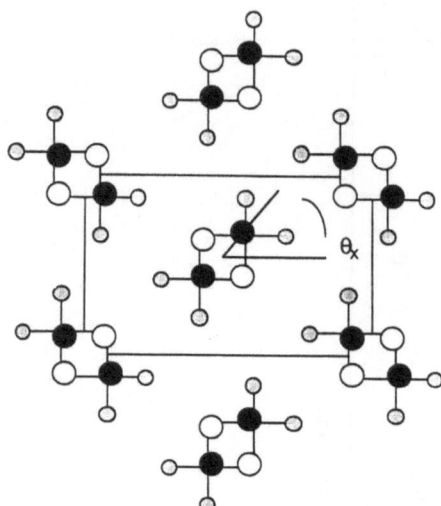

Fig. 6. Orthorhombic packing of polyoxy-methylene [6]

the hexagonal phase. The vibrational thermodynamic functions give an indication of this. In fact, the calculated vibrational free energy is lower for the hexagonal phase [6]. However, the transition temperature inferred from the calculated vibrational free energy is higher than the observed one at 342 K. The measured transition enthalpy is ~ 0.6 kJ/mol CH_2O whereas that from calculation is 1.3 kJ/mol (the zero K value of Table 2 corrected to 342 K with the calculated vibrational enthalpy). These numbers illustrate the difficulty in making predictions about phase stabilities. The reliability of results from the transferable force fields used in molecular mechanics is in general, at best, of the order of 0.5 kJ/mol per chain bond.

7.3 Poly (di-n-alkylsilanes)

The polysilanes, $-[Si(R)_2]_n-$, are interesting polymers. The Si–Si bond is significantly longer than the C–C bond (2.34 vs 1.54 Å). This allows the chain to carry substituents on every Si atom, a situation that would be too sterically crowded in the case of carbon. Many of the polymers show thermochromic transitions. This is apparently the result of σ-electron delocalization along the backbone. In the present context they are examples where the chain conformation appears to be significantly distorted by crystal packing forces [23]. The perferred conformation, with respect to electronic energy or an 'intrinsic' barrier about the Si–Si bond appears to be trans. However, in energy calculations on the isolated chain, the steric interactions between the substituents drive the chain torsional angles away from exact trans (180°). In the poly(di-methylsilane) example (Tables 1 and 2), the minimum energy torsional angle is about 165°, resulting in a helical

Table 2. Packing energetics[a]

Polymer	Intra-molecular		Inter-molec. nonbond	Electro-static	Total	Lattice[b]
	Bonded	Nonbonded				
Polyethylene						
ortho			− 16.2			16.2
exp						15.4
Polyoxymethylene						
hex	7.4	3.7	− 14.1	26.0	23.0	15.1
ortho	6.4	3.9	− 14.5	25.7	21.5	16.6
exp						16.2
isolated chain	6.5	3.6		28.0	38.1	
Poly(dimethylsilane)						
ortho	0.50	− 9.2	− 28.9		− 37.7	
hex (2/1)	0.50	− 9.2	− 24.0		− 32.7	
hex (15/7)	0.16	− 11.2	− 20.6		− 31.7	
Aliphatic polyesters						
2–4 kink		19.0	− 64.3	19.6	− 25.7	
planar		21.5	− 64.1	19.4	− 23.2	
2–6 kink		25.8	− 88.8	19.6	− 43.4	
planar		24.4	− 72.6	21.2	− 27.0	
2–8 kink		28.5	− 105.0	19.3	− 57.2	
planar		28.6	− 92.6	20.1	− 43.9	

Isotactic polypropylene (energy/trimer)	intra-Molecular	monolayer	bilayer	between bilayers	total	
α C2/c	45.15	− 15.90	− 21.46	− 20.20	− 12.22	
Cc	45.32	− 15.98	− 22.05	− 19.62	− 12.34	
P2$_1$/c	45.32	− 15.94	− 22.13	− 19.96	− 12.72	
γ Fddd	45.03	− 16.57	− 21.26	− 19.87	− 12.68	
Fd2d	44.98	− 16.65	− 21.46	− 19.79	− 12.93	
F2dd	45.07	− 16.44	− 22.22	− 19.29	− 12.89	

Polyvinylidine fluoride						
I (or β)					− 509.19	
II (or α) nonpolar					− 505.92	
III (or γ)					− 507.93	
IV (or δ) polar					− 506.30	

[a] kJ/mole per chemical repeat unit; (−CH$_2$–CH$_2$− for PE). See Table 1 for references
[b] Lattice energy is for the process of separating the packed molecules at zero K. It is listed only where experimental values are available

conformation (∼ 15/7). The experimental crystal structure is essentially ortho-rhombic with the chains in the trans conformation [24]. Energy calculations on the crystal show that the 15/7 helix packs in a metrically hexagonal pattern [23]. Crystal calculations were also made with a (rigid) trans planar zig-zag chain. It packs in the polyethylene-like orthorhombic cell (Fig. 7) and this latter arrangement is of higher overall stability than the hexagonal pattern of the 15/7 chain. Interestingly, this polymer undergoes two thermal transitions. One, at 160°C, is to a metrically hexagonal structure that is all trans. At 220° another transition to

R.H. Boyd

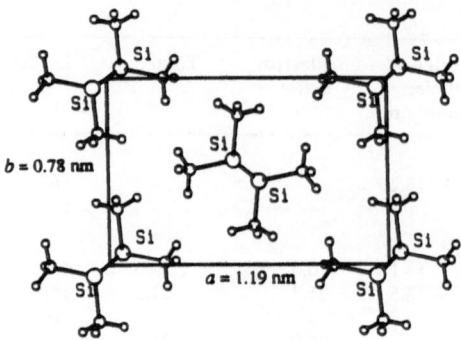

Fig. 7. Orthorhombic packing of poly(dimethylsilane) [23]

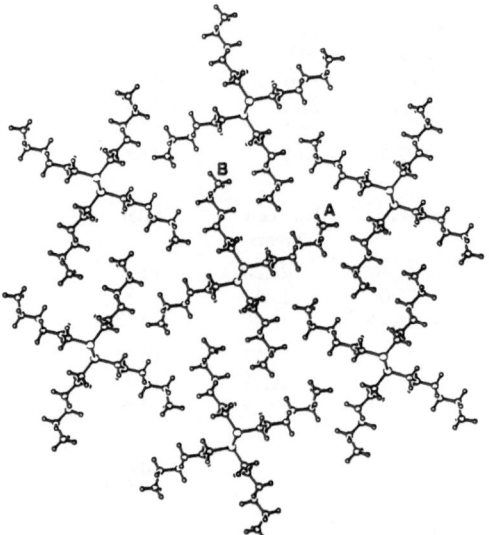

Fig. 8. Packing in poly(di-*n*-hexylsilane) [25]

a conformationally and orientationally disordered structure takes place. These transitions are consistent with the energetics found by calculation.

Silanes with longer alkyl groups have interesting packing [25]. The side groups are relatively extended, but not planar (Fig. 8). Calculations have not yet been made on the crystals, but isolated chain calculations show much the same situation as with the dimethyl case [25]. In the di-*n*-hexyl polymer, the preferred chain conformation is found to be somewhat distorted from all trans, a 7/3 helix. However, as in the dimethyl case, the experimentally observed structure is a planar skeleton. Again, the crystal packing efficiency appears to be able to distort the chain conformation.

7.4 Aliphatic Polyesters

These polymers, of formula, $-[O-(CH_2)_x-O-CO-(CH_2)_{y-2}-CO]_n-$, were some of the first to be studied by X-ray diffraction. They showed an interesting structural effect. The c axis repeat distance was found to be that expected for a planar zig-zag conformation in some examples but in most there was a distinct shortening. The shortening was later ascribed to the presence of gauche bonds in the conformation. In fact, the gauche bonds must occur as pairs, in most cases as two gauche bonds of opposite sense separated by a trans bond, i.e., a G^+TG^- 'kink'. Calculations have been made on a number of them, including the x, y = 2 − 4, 2 − 6, 2 − 8, 4 − 6, 6 − 6, 6 − 10 examples [8]. Attempts were made to find competitive structures incorporating all planar zig-zag conformations. Comparison of the kink containing structure with the most competitive all planar zig-zag conformation found is made for the 2−4 polymer in Figs. 9 and 10 and in Tables 1 and 2. Results for the 2−6 and 2−8 examples are also shown in the Tables. It turns out that the energy penalty associated with gauche bonds in the '2' diol is small or non-existent. Further, the kink structures pack better than the planar conformation. The ester group significantly disrupts the efficiency of the planar conformation packing.

The crystal structure of poly(pivalolactone) has been investigated by methods similar to the above, where the internal structure is allowed to be optimized along with the crystal packing [26].

7.5 Isotactic Polypropylene

Isotactic polypropylene (iPP) was the first stereo regular polymer to be synthesized. The determination of its crystal structure was instrumental in establishing that the synthesis of the polymer was indeed stereo specific. The TG helical conformation, the 3/1 helix, is the only conformation free of serious steric repulsions in the isotactic chain. Finding this as the observed helix in the crystal

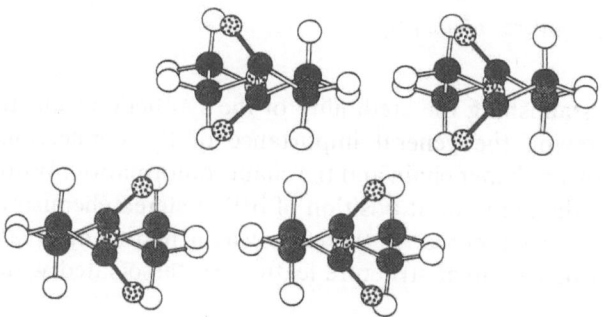

Fig. 9. Packing in actual kinked structure of 2,4 aliphatic polyester [8]

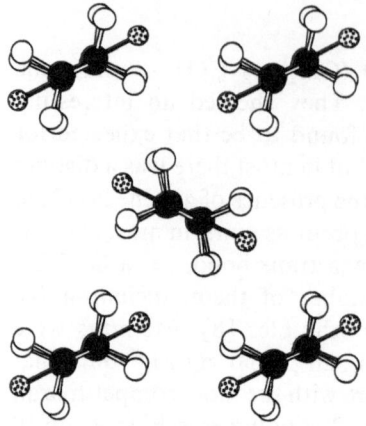

Fig. 10. Packing in a hypothetical planar conformation of 2,4 aliphatic polyester [8]

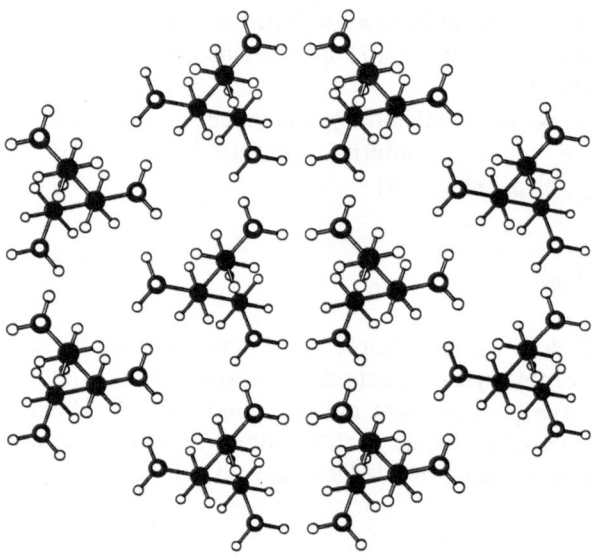

Fig. 11. Packing in isotactic polypropylene, α-form [27]

went a long way towards establishing the credibility of the synthetic result. It was also important in showing the general importance of the connection between the stereo content of a polymer chain and the chain conformation. With this seminal demonstration, the paramount position of iPP in stereo chemistry and conformational analysis was secured. It therefore was a matter of great astonishment when still another seminal structure feature was associated with the polymer.

This polymer has long been known to crystallize in several polymorphs. The best known is the α-form. Its packing, from a simulation [27], is shown in

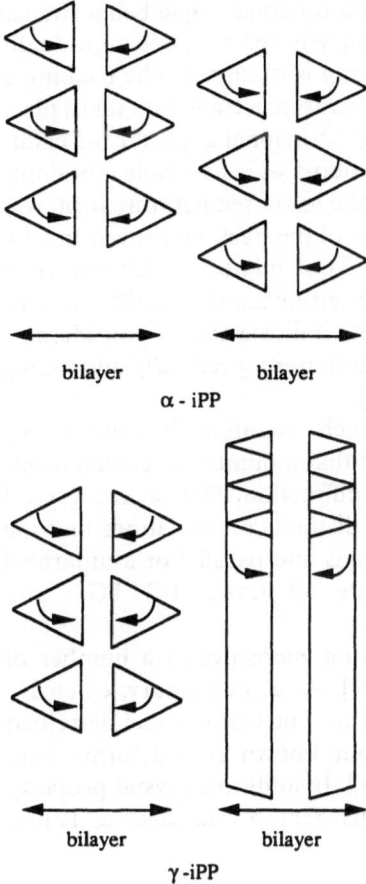

bilayer   bilayer

α - iPP

bilayer   bilayer

γ -iPP

Fig. 12. Comparison of packing in α- and γ-forms of iso-tactic polypropylene. The upper panel is an idealization of the pattern in Fig. 11, each molecule is represented by a triangle with the helix direction indicated. Every two rows can be considered a bilayer. In the γ-form, lower panel, the bilayers alternate chain direction. The prisms represent chain directions nearly orthogonal (± 80°) to the triangles [30]

Fig. 11. Recently it has been found that the γ-form has a unique structural feature [28, 29]. Although the chains are in the familiar 3/1 helix, they do not pack as parallel arrays. There are two sets of chains, with non-parallel axes, that coexist in the lattice. The basic plan of the each form is shown in Fig. 12.

Energy calculations have been used to advantage in rationalizing the structural possibility of this phenomenon [30]. The results are shown in Table 2. Comparative calculations were made for several appropriate space groups. It may be seen that the energies associated with the γ form are directly competitive with those of the α form.

7.6 Fluorocarbon Polymers

In polytetrafluoroethylene, the substituent fluorine atoms are small enough that steric crowding is not severe. However the repulsions between adjacent fluorine atoms are appreciable and the result is the chain torsional angle being twisted slightly from 180°. Below 19 °C the torsional angle is 163.5°. Thus a gradually twisting helix (13/6) rather than the planar zig-zag is observed. The packing is pseudo or metrically hexagonal. Above this temperature a transition takes place to a trigonal structure with a 15/7 helix and 165.8° torsional angle for the chain. Although very high quality X-ray diffraction patterns were available, obtaining a satisfactory representation of the structure of the lower temperature form was difficult. However application of energy analysis of the packing proved invaluable in proposing a satisfactory structure [31–33]. A mixture of left and right handed helices along with a distribution of chain setting angles were found to be necessary. The calculations and the experiments indicate that a considerable degree of disorder is present. A Monte Carlo simulation, specifically addressing the disordering issue, has been carried out [34].

Poly(vinylidine fluoride) has attracted much attention because of its piezoelectric properties. It is also complicated in that a number of crystal forms exist and have been carefully studied by X-ray diffraction. The various crystal forms arise because the chain can exist in three distinct conformations and the chain alignment can be up or down along the axis and parallel or antiparallel about the axis. The three conformations are all trans, TG^+TG^- and $T_3G^+T_3G^-$.

A very comprehensive analysis of the packing energetics of a number of crystal forms of PVDF has been carried out [9]. As a preliminary, extensive force field parameter development was undertaken. The CSM model described above in Sect. 4 was utilized. Structures of four known crystal forms were investigated and five new stable forms were found. In addition, crystal property evaluation along the lines described above under Sect. 5 was effected. Tables 1 and 2 summarize the results.

Acknowledgements. The author would like to express his gratitude to the Polymers Program of the National Science Foundation (USA) for support of his work in this area and to the Utah Supercomputing Institute for providing computational facilities. He would like to thank B. Farmer and S. Patnaik for providing Figs. 7 and 8. He also thanks Y. Jin for his help.

8 References

1. Two good desciptions of rigid chain packing are: (a) Tripathy SK, Hopfinger AJ, Taylor PL (1981) J Phys Chem 85: 1371; (b) McCullough RL (1974) J Macromol Sci Phys 9: 97
2. Sorensen RL, Liau WB, Boyd RH (1988) Macromol 21: 194
3. Boyd RH, Kesner L (1980) J Chem Phys 72: 2179
4. (a) Boyd RH, Kesner, L (1981) J Polym Sci, Polym Phys Ed 19: 375; (b) Boyd RH, Kesner L (1981) J Polym Sci, Polym Phys Ed 19: 393
5. Boyd RH, Kesner L (1987) Macromol 20: 1802
6. Sorensen RL, Liau WB, Kesner L, Boyd RH (1988) Macromol 21: 200
7. Smith GD, Boyd RH (1990) Macromol 23: 1527
8. Liau WB, Boyd RH (1990) Macromol 23: 1531

9. Karasawa N, Goddard WA (1992) Macromol 25: 7268
10. Taylor PL, Xu BC, Oliveira FA, Doerr TP (1992) Macromol 25: 1694
11. Born M, Huang K (1954) Dynamical theory of crystal lattices. Oxford U. Press, Oxford
12. Boyd RH, Pant PVK (1991) Macromol 24: 4073
13. Jungnitz S, Jakeways R, Ward IM (1986) Polymer 27: 1651
14. Gaur U, Wunderlich B (1981) J Phys Chem Ref Data 10: 1001
15. Theodorou D, Suter UW (1985) J Chem Phys 82: 995 ·
16. Brown D, Clarke JHR Clarke (1991) Macromol 24: 2075
17. Broadhurst MG, Mopsik FI (1991) J Chem Phys 54: 4239
18. Swan PR (1962) J Polym Sci 56: 403
19. Tadokoro H (1979) Structure of crystalline polymers. Wiley-Interscience, New York
20. Yemni T, McCullough RL (1973) J Polym Sci, Polym Phys Ed 11: 1385
21. Farmer BL, Eby RK (1976) Am Chem Soc Polym Prep 17(2): 131
22. Kobayashi M, Morishita H, Shimomura M, Iguchi M (1987) Macromol 20: 2453
23. Patnaik S, Farmer BL (1992) Polymer 33: 5121
24. Lovinger A, Davis DD, Schilling FC, Padden FJ, Bovey, FA, Ziegler JM (1991) Macromol 24: 132
25. Patnaik S, Gresco AJ, Farmer BL (1992) Polymer 33: 5115
26. Ferro DR, Brueckner S, Meille SV, Ragazzi M (1990) Macromol 23: 1676
27. Lee S (1992) B.S. Thesis, Dept Mat Sci Eng, U Utah
28. Brueckner S, Meille SV (1989) Nature 340: 455
29. Meille SV, Brueckner S, Porzio W (1990) Macromol 23: 4114
30. Ferro DR, Brueckner S, Meille SV, Ragazzi M (1992) Macromol 25: 5231
31. Weeks JJ, Clark ES, Eby RK (1981) Polymer 22: 1480
32. Farmer BL, Eby RK (1981) Polymer 22: 1487
33. Weeks JJ, Eby RK, Clark ES (1981) Polymer 22: 1469
34. Yamamoto T (1985) J Polym Sci, Polym Phys Ed 23: 771

Received: June 1993

Atomistic Dynamics of Macromolecular Crystals

B.G. Sumpter, D.W. Noid, G.L. Liang, and B. Wunderlich
Chemical and Analytical Sciences Division, Oak Ridge National Laboratory,
Oak Ridge, TN 37831-6197, USA, and
Department of Chemistry, The University of Tennessee, Knoxville,
TN 37996-1600, USA

In this article we review recent computational results on the dynamics of macromolecular crystals. From these studies it has been demonstrated that conformational defects can be created at temperatures as much as 100 K below the melting point of crystalline polyethylene and the concentration of the defects continues to increase (exponentially) with temperature, ultimately leading to a disordered crystal along the polymer chains (CONDIS crystal). Although the rate of formation of conformational defects is relatively high, approximately 1×10^{10} s^{-1} at 350 K, these defects do not by themselves lead to any macroscopic motion that could give rise, for example, to lamellar thickening. The mechanism appears to involve coupling of the large-amplitude torsional motions with the transverse and longitudinal vibrations of the crystal lattice, which can subsequently lead to the formation of short-range twists in the chains (twist defects). Defects like the twist can, under the correct conditions, move coherently toward the end of the crystal, thereby causing a chain diffusion process that leads to lamellar thickening or deformation processes. For smaller systems such as paraffins, disorder occurs by a collective twisting (so-called rotator phase) of the chains, which is not strongly influenced by conformational defects. The hexagonal or pseudo-hexagonal structure of the asymmetric motifs is caused by a dynamic multidomain arrangement of the twisting chains. Overall, a more accurate description of the thermodynamic, spectroscopic, and kinetic behavior is possible and gives a new understanding of the deformation, relaxation, annealing, and motion in macromolecular crystals.

Advances in Polymer Science, Vol. 116
© Springer-Verlag Berlin Heidelberg 1994

1 Introduction

An understanding of both the microscopic structure *and* dynamics of matter is essential for a full description of its macroscopic properties. In this regard, recent studies of the internal dynamics of paraffin- and polyethylene-like crystals by simulation with supercomputers have established a link between microscopic motion and macroscopic effect [1–9]. It was shown that conformational, rotational, and diffusional motion can start considerably below the melting or disordering transition temperatures. These types of motion underlie much of the observed macroscopic properties of polymers, and thus it is essential to develop an understanding of how such motion occurs on the microscopic level (mechanisms) and the rates of changes it introduces (kinetics).

The macroscopic structure of matter can be assessed, for example, by optical microscopy and can then be linked to its microscopic origin through X-ray, neutron, or electron diffraction experiments and the various forms of electron and atomic-force microscopy. A factor of 10^3–10^4 separates the atomic, nanometer scale from the macroscopic, micrometer scale. Macroscopic dynamic techniques ultimately linked to molecular motion are, for example, dynamic mechanical and dielectric analyses and calorimetry. In order to have direct access to the details of the underlying microscopic motion, one must, however, use computational methods. A realistic microscopic description of motion has recently become possible through accurate molecular dynamics simulations and will be described in this review. It will be shown that the basic large-amplitude molecular motion exists on a picosecond time scale (1 ps = 10^{-12} s), a factor at least 10^9 away from direct macroscopic observation. This large-amplitude motion is to be distinguished from the vibrational motion that, by itself, does not lead to mass transport, but is the cause of the largest part of the heat content of a solid (typical vibration frequencies extend to 10^{14} Hz).

In the research to be reviewed in this article, the atomistic details of the internal dynamics of crystalline polymers are examined using classical mechanics methods (molecular dynamics). Semiclassical molecular dynamics (where an initial energy is placed in a specific vibrational state) and ab initio quantum mechanical methods are used to establish the accuracies of the multidimensional potential energy surface describing the dynamics of the system. These detailed investigations have included the study of the structural and dynamic properties of crystal defects [1–9], the diffusion and relaxation processes [4, 5] as well as the vibrational spectroscopy [2]. Overall, a reasonable description has been derived of how defects are formed and destroyed and how they cause relaxation processes and exist in polymer and paraffin crystals, causing disorder, lamellar deformation and thickening. In addition, the effects of external perturbations such as laser interactions, stress, molecular collisions, and an atomic force microscope probe, have been studied. In the following section, Sect. 2, a brief description of the computer simulation techniques used to carry out the computations is given. In Sect. 3 the major results are presented, and the article is closed by a set of conclusions in Sect. 4. An overview of the literature on the defect state of polymers and its connection to the new results from the molecular dynamics simulations is collected in Sect. 6, the Appendix. The literature for the discussion of the simulations is collected in Sect. 5, the references for the literature overview are presented in Sect. 6.3.

2 Computational Methods and Procedures

Large-amplitude anharmonic motion is inherent in, and underlies, dynamic disorder in molecular crystals [1, 2]. The motion of atoms under the influence of their anharmonic potential is, in addition, affected by dynamic processes such as kinetic and resonant coupling and dynamic and phase space barriers. All of these processes have been shown to have pronounced effects on the kinetic and structural properties of molecular systems. While there are a number of simulation techniques that can be used to help understand structure and statistical properties, such a molecular mechanics and Monte Carlo studies, most of these methods do not permit the study of dynamical processes. Since the internal dynamics clearly plays a key role in the overall behavior of a given system, it is essential to use a method that provides an adequate representation of such motion. The molecular dynamics method is a powerful technique for studying the structure and dynamics of many-body systems that satisfy these requirements.

2.1 Molecular Dynamics

The molecular dynamics (MD) method is well-known and for the interested reader several recent reviews are available on the procedure as well as its merits [10]. In short, it involves the numerical solution of Hamilton's equations of motion (or any other formulation of the classical equations of motion), starting from some chosen initial positions and velocities of the atoms or particles of the system. In our MD simulations the integrations of the equations of motion are carried out in Cartesian coordinates, thus giving an exact definition of the kinetic energy. Integrations are performed using an up to 12th order predictor/corrector routine with variable time steps. This method (ODE) allows the integrations to be performed to a high accuracy. For example, conservation of all constants of motion (total energy, total angular and linear momenta) can be obtained to as many as four digits without the need for any ad hoc scaling of the momenta.

Fundamental to the molecular dynamics method is the necessity of describing the potential energy surface. The total molecular Hamiltonian is composed of the kinetic and potential energy for the system. We have used the following

Hamiltonian:

$$H = \sum_{n=1}^{m} \left[\sum_{i=100(n-1)+1}^{100m} \frac{p_{x_i}^2 + p_{y_i}^2 + p_{z_i}^2}{2M} + \sum_{i=100(n-1)+1}^{100m-1} V_{bond}(r_{i,i+1}) \right.$$

$$+ \sum_{i=100(n-1)+1}^{100m-2} V_{bend}(\theta_{i,i+1,i+2})$$

$$\left. + \sum_{i=100(n-1)+1}^{i=100m-3} V_{introt}(\tau_{i,i+1,i+2,i+3}) \right]$$

$$+ \sum_{i=1}^{100m} \left[\sum_{j} V_{nonbonded}(R_{i,j}) + \sum_{k=1}^{100l} V_{nonbonded}(R_{i,k}) \right]. \qquad (1)$$

Equation (1) refers to a crystal of m dynamic chains of 100 CH_2-groups each. The second brackets represent the nonbonded interactions among all dynamic molecules i with j, where the latter are the dynamic atoms not considered in Eqs. (2, 4, and 5) for the given i. The last term, summing over the interactions of the atoms i with k, is only needed if the crystal is enclosed with l chains of rigid (nondynamic) atoms to keep the volume constant. The potential functions for the adjacent bonded atoms (i and $i + 1$) are:

$$V_{bond}(r) = D(1 - e^{-\gamma(r-r_e)})^2 \qquad (2)$$

where r_e represents the equilibrium bond length. The interaction between any two nonbonded atoms separated by a sequence of three atoms or more is represented by:

$$V_{nonbonded}(R) = 4\varepsilon \left[\left(\frac{\sigma}{R} \right)^{12} - \left(\frac{\sigma}{R} \right)^6 \right] \qquad (3a)$$

or:

$$V_{nonbonded}(R) = A\exp(-bR) - \frac{C}{R^6} \qquad (3b)$$

where R represents the appropriate distance between i and j or k. For the bending of a three-atom, bonded sequence the potential is:

$$V_{bend}(\theta) = \tfrac{1}{2}K_\theta (\cos\theta_0 - \cos\theta)^2 . \qquad (4)$$

The torsional coordinate τ between four successive atoms is linked to the internal rotational potential:

$$V_{introt}(\tau) = \text{constant} + \alpha\cos\tau + \beta\cos^3\tau . \qquad (5)$$

The four-body potential term [Eq. (5)] has two parameters (α and β) that can be fitted to give a desired barrier height and rotational isomer energy difference (for example, in the paraffins and polyethylene models discussed here, a *cis* barrier of

16.7 KJ/mol and a *gauche-trans* energy difference of 2.5 kJ/mol). The constants D, and γ [see Eq. (2)] represent the usual Morse oscillator parameters. The nonbonded terms ε and σ [Eq. (3a)] represent the Lennard–Jones parameters, b and C [Eq. 3b)] are related to the overlap and dispersion of the atoms i and j, and A is a parameter related to the position and well-depth of the interaction. The bending force constant is K_θ, and θ_0 [Eq. (4)] indicates the equilibrium value of the angle formed by the three atoms of interest. The above potential energy functions [Eqs. (2–5)] have been demonstrated to yield good spectro-scopic, thermodynamic, and kinetic data, as well as to provide the atomistic details of temperature-dependent phase transitions for crystalline polymers.

For the majority of our calculations, we have chosen to collapse the CH_2 groups into a single particle of mass 14 amu. The potential parameters for the collapsed atom models are given in Table 1 and those for the models with explicit hydrogens, for which Eqs. (1–5) must be appropriately modified, are given in Table 2.

For all simulations, the initial conditions involved imparting a randomly chosen momentum, subject of a zero center-of-mass velocity constraint. Thus, the total energy is initially purely kinetic. Due to the large number of atoms in the system, and thus the density of vibrational states, thermal equilibrium, as described by the Boltzmann distribution, is rapidly achieved. An effective temperature can then be determined from the average kinetic energy $\langle KE \rangle$ by:

$$\langle KE \rangle = \frac{3}{2}NkT = \sum_{i=1}^{100m} \frac{p_{x_i}^2 + p_{y_i}^2 + p_{z_i}^2}{2M_i} \tag{6}$$

Table 1. Potential energy parameters for the collapsed atom model [Eqs. (1–5)]

Two-body bonded constants

  D = 334.72 kJ/mol
  r_e = 0.153 nm
  γ = 19.9 nm^{-1}

Three-body constants

  K_θ = 130.122 kJ/mol
  $\cos \theta_0$ = 0.4335 nm

Two-body nonbonded constants

  LJ 6–12 Potential
    ε = 0.4937 kJ/mol
    σ = 0.4335 nm
  Exp-6 Potential
    A = 2.2238 MJ/mol
    b = 623.85 MJ/mol
    C = 36.0 nm^{-1}

  Four-body bonded constants

    α = $-$ 18.4096 kJ/mol
    β = 26.78 kJ/mol

Table 2. Potential energy parameters for the models with explicit hydrogen atoms[a, b]

Intramolecular bonds	Angle bends
$D_{C-C} = 80.0$ kcal/mol	$1/2K_{\theta\,CCC} = 72.3$ kcal/rad^2 mol
$D_{C-H} = 106.7$ kcal/mol	$1/2K_{\theta\,HCC} = 41.85$ kcal/rad^2 mol
$\gamma_{CC} = 1.94$ Å^{-1}	$1/2K_{\theta\,HCH} = 38.9$ kcal/rad^2 mol
$\gamma_{CH} = 1.75$ Å^{-1}	$\theta^e_{CCC} = 111°$
$r^e_{CC} = 1.53$ Å	$\theta^e_{HCC} = 108.9°$
$r^e_{CH} = 1.09$ Å	$\theta^e_{HCH} = 109.3°$

Nonbonded Interactions

Term	C(Å^6)	A(kcal/mol)	b(Å^{-1})
H–H	35.967	2,556.53	3.74
H–C	289.75	14,313.4	3.67
C–C	534.61	149,104.5	3.60

[a] The parameters were adapted from Barnes J, Fanconi B (1978) J Phys Ref Data 7: 1309; Sorensen RA, Liam WB, Boyd RH (1988) Macromolecules 21: 194; see also more recent results by Mayo SL, Olafson BD, Goddard WA III (1990) J Phys Chem 94: 8897; Karasawa N, Dasgupta S, Goddard WA III (1991) J Phys Chem 95: 2260
[b] To convert to SI-units: 1 cal = 4.184 J, 1 Å = 0.1 nm

where k is Boltzmann's constant, T is the temperature, p the Cartesian momentum, and M the mass corresponding to atom i. In general, analyses of the simulations were started after allowing the kinetic energy to redistribute according to the dynamics of the system for about 5 picoseconds, thus giving both phase-averaged momenta and coordinates.

2.2 Analyses

Analyses of the data generated by the molecular dynamics method is an important part of the overall simulation. We have developed a number of very useful tools for the examination of the results such as to extract the maximum information. These include high resolution spectral estimators and the application of neural/fuzzy systems. A more detailed description of the methods and examples of application can be found in a recent review article [11].

2.2.1 Spectral Estimators

The spectral estimators provide a means to compute vibrational spectra from molecular dynamics results (from the time series of a correlation function) to any desired resolution [2]. This eliminates much of the negative aspects of the nonparametric methods, such as the standard fast Fourier transform (FFT) that needs a very long time series for good resolution (the resolution is $\approx 1/t$, where t is the total time of the series). Furthermore, due to the dynamic nature of polymers, chain folding and large amplitude motions define a time series that is very chaotic, even on a picosecond time scale, and this leads to a FFT that yields a spectrum that can be very noisy and extremely difficult to interpret. Thus, in order to compute vibrational spectra from a molecular dynamics generated time series, it is necessary to transform *very short* time slices. This would mean that the resolution would be poor if the FFT were used, but by employing spectral estimators (parametric methods), any desired resolution can be obtained.

2.2.2 Neural Networks

Neural networks and adaptive fuzzy systems are model free estimators [12]. This quality makes it possible to find relationships between a series of examples and the desired results [13]. For example, one might be interested in predicting the long-time behavior of a polymer chain, say to the millisecond. By providing a number of examples of the dynamics on a short time scale, say picoseconds, a neural network is able to map out the important relationships and thus make interpolations and extrapolations to times that were not given as examples. Furthermore, in the same manner, the size discrepancy between most simulations and their experimental counterparts, can be closed by using the predictive power of neural/fuzzy systems.

3 Results and Discussion

The importance of understanding the microscopic motion of matter has been expressed in the introduction of this article. The ramifications of dynamic disorder on the processes of polymer physics and chemistry are, indeed, very broad. In this regard, many defect types have in the past been proposed for polymer crystals (see the Appendix, Sect. 6). Often they were based on more or less extensive molecular mechanics calculations. Most of these defects were thought to explain some piece of the information needed for developing an understanding of the deformation of polymer crystals, but there were always some missing parts.

In order for a model to be a comprehensive representation of the defect polymer crystal, it should satisfy a number of rather rigid requirements:

1) There should be some mechanism and/or driving force for the defect creation.

2) The energies for both the creation and persistence of the defects have to be low.

3) The defects should have general applicability and not be structure-specific.

4) The defects should explain the observed relaxation behavior, both in terms of mechanism and of experimental observation.

5) Both the mechanism and the time dependence of relaxations, lamellar thickening, annealing, and plastic deformation need to be accounted for.

6) The models must yield quantitative accord with all experimental results (i.e., vibrational spectroscopy, calorimetry, X-ray diffraction, mechanical and dielectric relaxation measurements, NMR relaxation time determinations, and neutron scattering).

Most of the early proposed models satisfied only one or several of the above requirements, but not all. Since deformation and relaxation involves the dynamics of the polymer chains, it is by using the molecular dynamics method, quantum dynamics, or variants thereof that one can begin to derive a clear and complete description of what is happening inside a polymer crystal. In this regard, we have performed a number of detailed, systematic dynamics computations on paraffin- and polyethylene-like crystals.

The atomistic details of the anharmonic vibrational motion have been primarily examined by using accurate molecular dynamics calculations. In some cases, analyses were enhanced by using spectral estimators and neural networks [1, 2]. From these results we have been able to formulate a better understanding of the defect crystal.

3.1 Simulation of Paraffin Crystals

The simulations of paraffins made use of models of crystals consisting of $192-C_{50}H_{100}$-chains (pentacontylene) with as many as 9 600 (united atom approximation) to 28 800 atoms (using explicit hydrogens). The simulation times ranged up to 100 picoseconds. The initial crystal structures for the simulations, orthorhombic (ORTH) and monoclinic (MONO), represent perfect crystals constructed with unit cell parameters derived from X-ray diffraction at room temperature. Figure 1 shows a projection of these initial structures along the chain direction. A randomized setting-angle of the planar zig-zag chains in the orthorhombic structure was also used to simulate an initial state, perhaps representing the "rotator" hexagonal (HEXA) phase [9]. The size of each crystal was initially about $6.0 \times 6.0 \times 6.3 = 227$ nm^3. This is sufficiently large to serve also as a model for polymer crystals which are typically between 2 and 20 nm thick [14]. The simulations were conducted in a temperature range from 55 to 410 K for 10 to 100 ps with the nonbonding potential represented by Eq. (3) being cutoff at 1.0 nm to conserve computation time.

In simulations in which the crystals were allowed to fluctuate without external pressure, the systems quickly reached a steady-state, enabling us to compare dynamic and structural behavior with experimental results. After about 4 ps the radial distribution functions were not dependent on time and became indistinguishable among different initial structures, which means that all three

Projection of the Initial Crystals along the Chain-axis

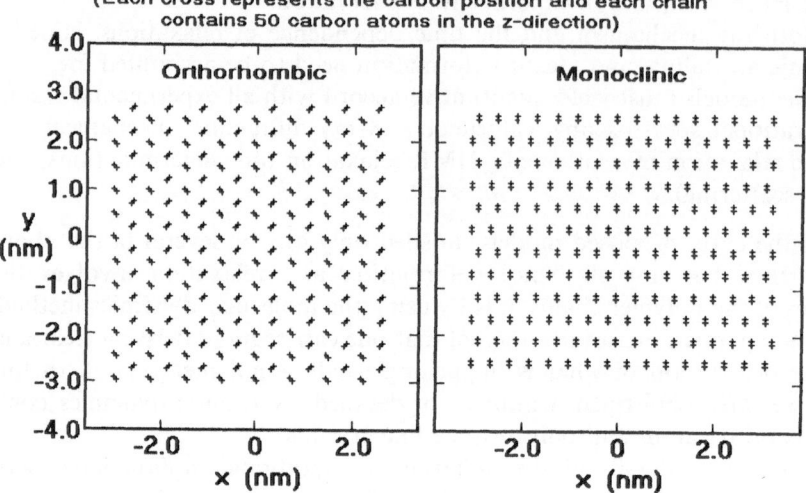

Fig. 1. Crystal structures at the beginning of the simulation, as derived from room temperature X-ray data

starting structures (ORTH, MONO, and HEXA) had been transformed to the same "steady-state", as can be seen from the radial distribution curves of Fig. 2.

In contrast to experiment, however, the resulting structure is a representation of the hexagonal "rotator" phase at all temperatures. The cause of this difference may have several reasons: (1) the use of the simplified united atom model; (2) the small size of the modeled crystal; (3) the initial conditions for the simulations.

The united atom approximation changes the three-center C_{2h}-symmetry of the crystal motif, the CH_2-group, to a single-center sphere [see Eq. (3)]. It was established earlier that the basic reason for the orthorhombic or monoclinic structure of paraffins is the packing of the protrusions of the CH_2-groups into corresponding hollows of neighboring chains. This leads to the unique paraffinic packing, discussed, for example, in Sect. 2.3.6 of [14]. Initial simulations with explicit hydrogen atoms did not, however, avoid the hexagonal phase. This can be an indication that the hydrogen potential energy surface is not well enough represented (needs the long-ranged electrostatic interactions) to predict crystal structures (densities were also not in full agreement with experiment). Naturally, one of the other two reasons may, in addition, be contributing. Enclosing the crystal in a rigid shell of fixed, orthorhombically placed CH_2-chains retains (with considerable disorder) the orthorhombic structure and is, perhaps, an indication that much larger crystals favor the equilibrium structure. Attaching a mobile layer on a rigid surface also keeps the orthorhombic structure for a number of layers, before changing to hexagonal. There exists some morphological evidence for a crystal-structure change with crystal size in the predominant close-to

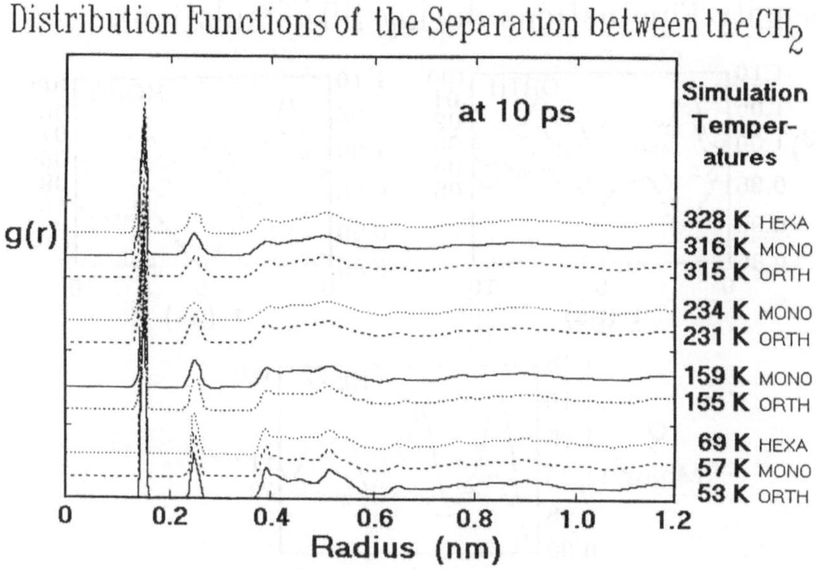

Distribution Functions of the Separation between the CH_2

Fig. 2. Radial distribution functions at different temperatures and initial structures

hexagonal twinning symmetry found on fast crystallization – see Chap. 3 of [14]. It will be shown below that the rotation of the setting angle is coupled to the fundamental breathing mode of the crystal. The start of the simulations (initial conditions), however, may excite this mode more strongly than expected in thermal equilibrium. This may thus contribute to the more predominant presence of the hexagonal phase in the short simulations.

Work is going on to try to resolve this problem of the wide existence range of the hexagonal phase on molecular dynamics simulation. In the meantime, these simulations offer a unique opportunity to study the behavior of the hexagonal phase of paraffins and polyethylene. Note, that due to the choice of Eqs (2, 4, and 5) one expects close agreement with experiment in all effects that involve the vibrational spectrum: stretching, bending, and torsional vibrations, as well as conformational, large-amplitude motion. Less precision is expected from effects involving the cut-off range of 1.0 nm in Eq. (3) and the united atom approximation. This was borne out, as will be shown below, by the agreement with experiment in vibrational spectra, speed of sound, *gauche*-conformation concentration, heat capacity, melting temperature, and density.

In the simulations with the united atom model (CH_2 represented by an appropriate atom of mass 14), the crystal expands or shrinks, depending on the starting temperature and then shows a breathing-mode vibration. The density plots in Fig. 3 show such oscillations about the average which is close to the experimental density of polyethylene. At about the same temperature, different initial structures have slightly different density fluctuations in direction and amplitude, but the same frequency ($\approx 3 \times 10^{11}$ Hz). Assuming that one sees the

Density Fluctuations during MD Simulations

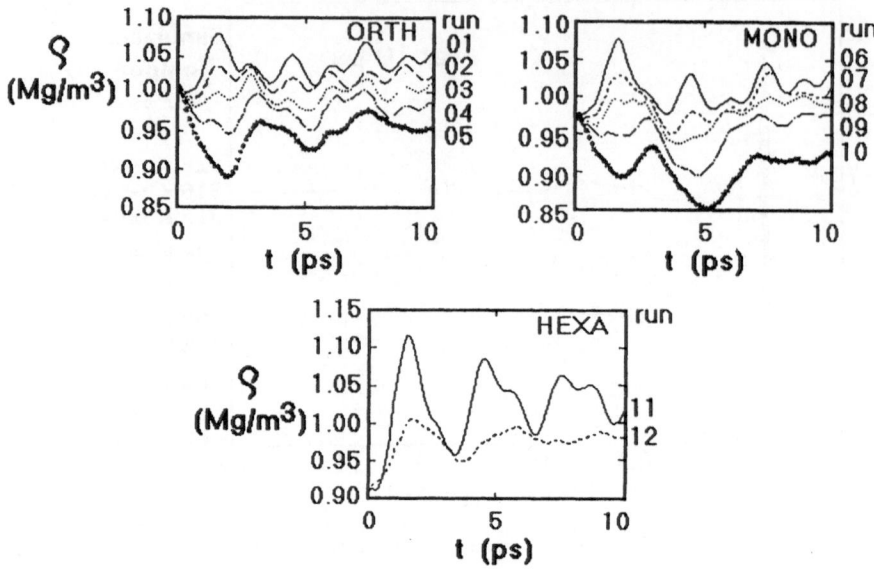

Fig. 3. Fluctuation of density as functions of time and initial structures. (Temperatures: run 01, 53 K; run 02, 155 K; run 03, 231 K; run 04, 315 K, run 05, 410 K; run 06, 57 K; run 07, 159 K; run 08, 234 K; run 09, 234 K; run 10, 403 K; run 11, 69 K; run 12, 328 K)

first fundamental vibration of the crystal with a wavelength of about 12 nm (twice the crystal size), one can estimate the speed of the wave to be about 4 km/s which is typical for speed of sound measurements in polymers (1.4–5.9 km/s) [15].

As temperature increases, the crystal breaks up into domains of different orientation of the CH_2-chains zig-zag planes (setting angle), as is shown in Fig. 4. At 57 K (Fig. 4a) the crystal structure already deviates from its initial, perfect monoclinic structure that has all zig-zag planes parallel to the yz plane. Not only are the projections of the $(CH_2-)_{50}$ chains broadened by vibrations, one can also see a break-up into at least two domains, separated by boundaries of two to three layers of chains. At 234 K the crystal becomes more disordered and develops easily recognizable domains (Fig. 4c). The disorder increases with temperature and becomes significant at 316 K where, in addition, a "fuzzy" surface structure is evident (Fig. 4d). This occurs at a temperature that is still about 50 K below the experimental melting point. The scale of disordering is not uniform throughout the entire crystal, but seems to vary from one domain to another. It will be shown, below, that these nanometer-size domains are dynamic and change orientation and boundaries on a picosecond time scale. A magnified side-view of the crystal reveals a decreasing density as one approaches the crystal surfaces.

Disorder in an Initially Monoclinic Crystal at Different Temperatures

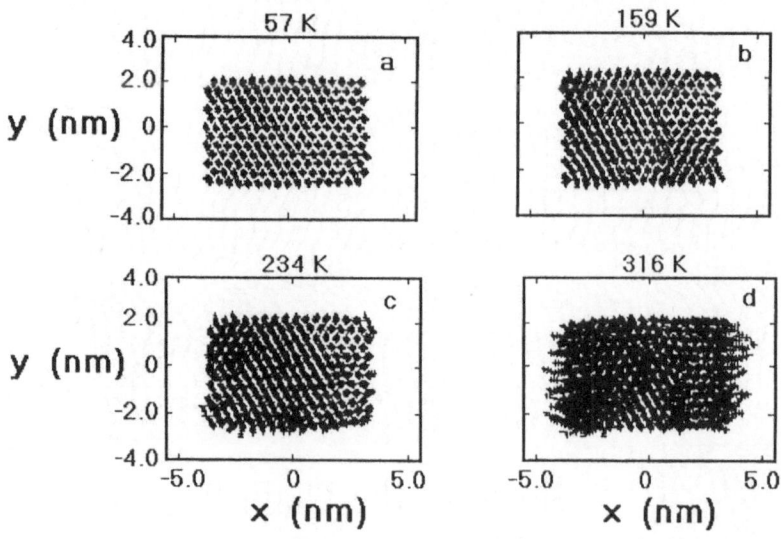

Fig. 4. Structures of MONO in the *x-y* projection at different temperatures after 10 ps simulations

From the present simulations, it is found that *gauche* defects in the otherwise all-*trans* chains become noticeable at about 165 K. The rate of formation of these *gauche* conformations was found to be about 10^{10} s^{-1} at room temperature [5]. There is an exponential increase of the *gauche* concentration with temperature, which is consistent with experimental results gained by IR analysis of paraffins. The *gauche* concentration for monoclinic C_{50} by experiment is, for example, 0.21% at 323 K and 1.7% at 365 K [16], and compares to about 1.5 and 4.0%, respectively, in the simulation of the hexagonal phase [9b]. The low concentration is the result of lifetime in the picosecond scale. Obviously, *gauche* bonds of such low concentration are insufficient to cause the domain structure, but are sufficient to explain the heat capacity increase below the melting temperature. For $C_{50}H_{102}$ the heat capacity starts to show a first increase beyond the vibrational heat capacity at 260 K. Close to melting, the heat capacity has reached a value about 20% larger than expected from vibrational motion alone [17].

Although the majority of bonds are at any one time close to *trans* "conformers", each of them is found to deviate frequently more than expected from normal vibrations. The zig-zag chains often obtain a twist by sequential deviations in the same rotation direction. The formation of such twists is demonstrated in a spatial plot in Fig. 5. A series of time snapshots of the center paraffin chain of the simulated crystal is drawn at 316 K. As can be seen, the chain undergoes rapid

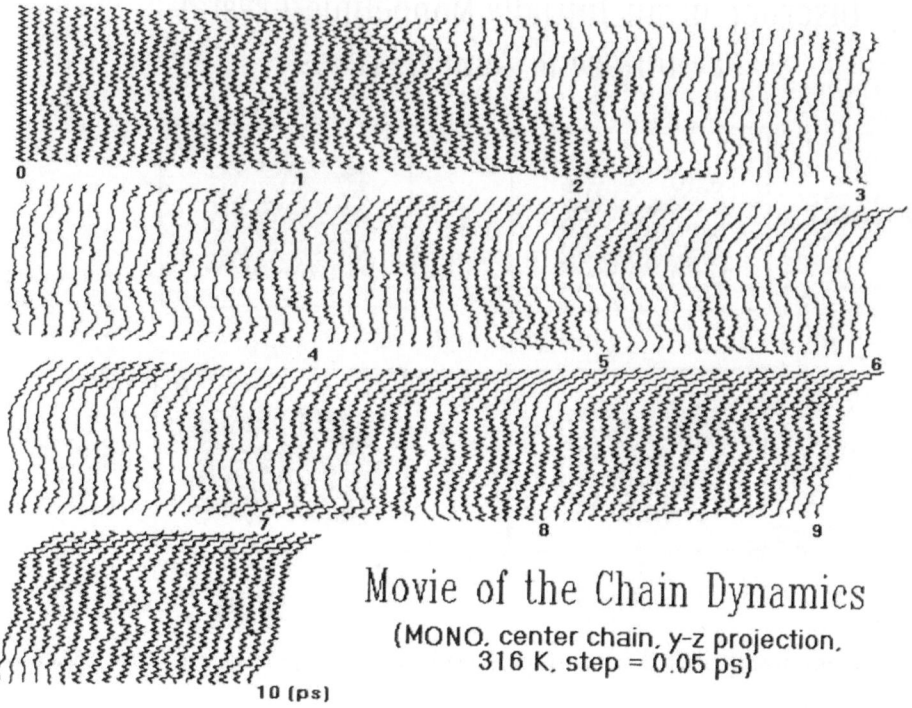

Fig. 5. The time evolution of the dynamics of the center chain for an initially MONO

twisting motion that results in a change of the orientation of the zig-zag plane of the molecule. After about 1.5 ps the chain is rotating via segmental motion, starting at the top of the crystal. Within about 0.5 ps the chain has turned by about 90°. Similarly, a new side-on view of the zig-zag develops between 7 and 8 ps, indicating another overall 90° rotation, turning roughly back to the initial orientation. That the rotation was not a full 180° turn can be deduced from the direction of the terminal group of the zig-zag chain in Fig. 5. As temperature increases, the frequency and magnitude of this "rotational" motion increases. In addition, the accordion mode of vibration causes diffusion of the whole chain in the crystallographic c-direction, and is followed by tilting of the chain-heads, as has also been seen experimentally [16].

The multi-domain structure in the hexagonal paraffin crystals can be further characterized by an analysis of the distribution of its time-averaged lattice parameters. Spatial plots, such as are shown in Fig. 5, reveal that, due to the constant twist motion of the molecular chains, constant setting angles exist only for a short time and over a nanometer-size domain. Therefore, the repeating-unit symmetry was averaged in the simulated crystals by finding the primitive unit cell, centered at the midpoints of the C–C-bonds. The results are illustrated in Fig. 6. They yield, on average, a hexagonal structure ($a \approx b$, $\alpha \approx \beta \approx 90°$,

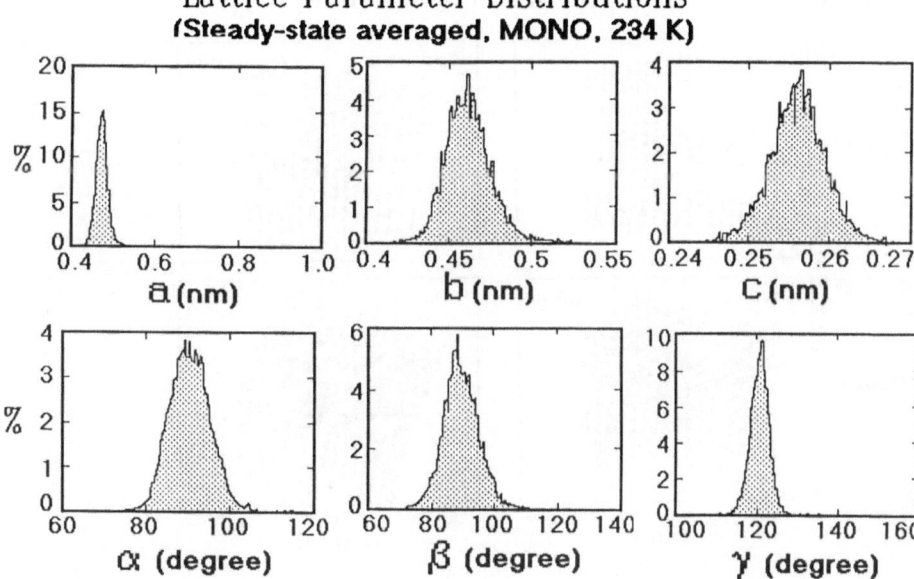

Fig. 6. Lattice parameter distribution (primitive unit cell, disregarding the instantaneous setting angle, example simulation at 234 K, averaged over the time range from 5 to 10 ps)

$\gamma \approx 120°$). A typical X-ray beam diameter is in the micrometer range, i.e., about 10^6 times larger than the cross-section of the simulated crystals, and the experimental time is at least 10^9 times longer than the lifetime of any domain seen in the simulations. The crystal motif, the CH_2-group, thus appears to be commensurate with a highly symmetric lattice because of the long-time and positional average seen by X-ray diffraction.

In the simulation at higher temperature, the beginning of melting of the crystal can be observed. Figure 7 shows the results of a simulation designed to test the crystal behavior about the melting temperature. It began at a temperature approximately 50 K above the experimental melting point of pentacontane (T_m = 365 K). The system quickly responded by expanding to a steady-state volume. The simulation was then run for 100 ps. The spatial coordinates were plotted every picosecond and animated in order to study the structure change as a function of time. A complete movie on VHS video is available for inspection. In contrast to the lower temperature simulations, the temperature did not remain constant in this case. The integration continually loses energy because of fusion and loss of cohesive contact beyond the 1.0 nm van der Waals interaction, so that the temperature is gradually decreased. In Fig. 7, selected time-temperature sequences are shown in the $x-y$ projections of the originally ORTH crystal. Progressive disordering, followed by renewed ordering, is visible. At the lower left, a close-to hexagonal arrangement is seen in Fig. 7c (42 ps, 383 K) and may be considered to be a recrystallization site. The simulation was repeated with

Change of Disorder in the Orthorhombic Crystal

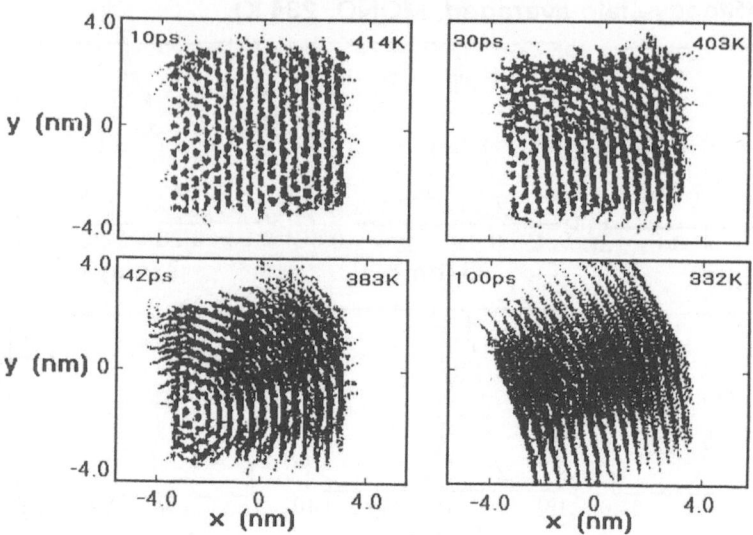

Fig. 7. The time evolution of the x-y projection of an ORTH crystal in a melting simulation

different sets of random numbers for initial momentum distribution, but always seems to give similar results [9c].

In general, this simulation yields information on what appears to be a melting process of an instantaneously superheated paraffin. It involves the development of large-scale disorder throughout the entire crystal, including rotational reorientation and translation of the CH_2-chains. The expected conformational disordering starting at the crystal surfaces seems to be a much slower melting process. Although the entire crystal symmetry has been destroyed quickly, the inner chains are unable to form random coils. The disordering and reordering of the inner chains (recrystallization) are found to be fast, competing processes. Throughout the melting simulation the maximum *gauche* concentration was found to be only slightly larger than that of stable crystals at that temperature and ultimately, it dropped to the typical crystal value at 332 K (at 100 ps) [9c].

The simulations of paraffin crystals was also carried out with a full-atom model in which hydrogens are explicitly included, making the size of a crystal about 30,000 atoms [18]. As expected from the proper rotational potentials used in the united atom model, the *gauche* concentration and lattice parameter distribution are not significantly different when the H-atoms are included explicitly. The observed crystal structure was, surprisingly, as mentioned above, also not close to experimental equilibrium paraffin structure, an indication that further work is needed to find the subtle adjustment needed in the intermolecular force field that governs the stable crystal structure, or develop simulations of larger crystal structures. Overall, this series of simulations has provided a detailed look at the dynamics of a small crystal.

3.2 Simulation of Polyethylene-like Crystals

The next set of simulations to be discussed were actually performed earlier to test the validity of the crystal models and the potential energy surface [5b]. It started with modeling of chains of usually 100 CH_2-groups with increasing numbers of additional chains, establishing an increasingly realistic crystal environment. Figure 8 displays the c-axis projection of such an arrangement of a total of 19 mobile chains, surrounded by 18 chains that were fixed to retain a constant volume. In general, the analyses were focused on the center chain since its environment should be the most representative of a chain in a polyethylene crystal.

The immediate observation, even for the case of a single chain surrounded by fixed chains only, was that more than 100 K below the expected melting temperature a large number of conformational defects could be seen [1c]. Earlier it was assumed, based on molecular mechanics calculations, that such defects should be rare or impossible. Rotating a single bond from *trans* to *gauche* should completely remove the ability to fit into the array of parallel chains of the crystal. Figure 9 illustrates how such disorder is, however, spread over larger areas of the crystal with torsional and bending vibrations being involved in the existence of the conformational defects. Figure 9 represents results from simulations at 80 and 430 K, the latter a temperature above actual melting. Due to the presence of the static ring of 12 additional chains that maintain a constant volume, melting is

Arrangement of a Polyethylene Crystal

(projection along the crystallographic c-acis)

19 mobile
chains
surrounded
by 18
rigid chains

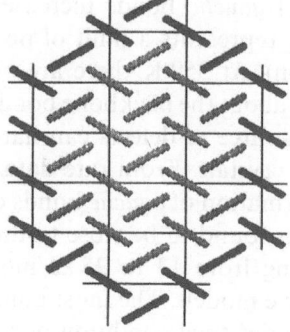

Fig. 8. Top view of the chains of a polyethylene-like crystal. The lamellar thickness is 100 CH_2-groups and the crystal consists of 19 inner, mobile chains, surrounded by a ring of 18 rigid chains

Simulation of a Polyethylene Crystal

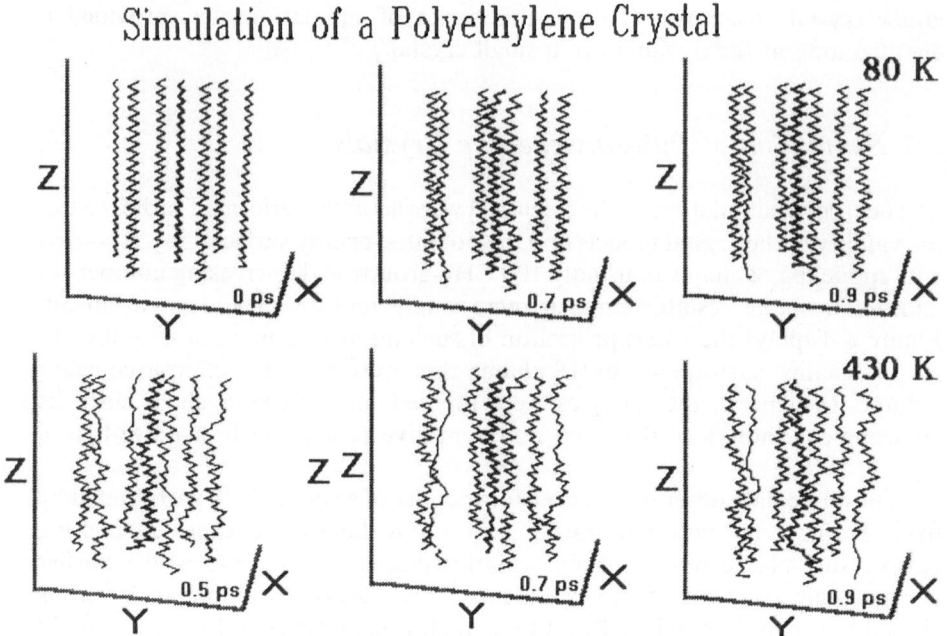

Fig. 9. Conformational disorder in a seven-chain simulation at the given temperatures and times. At low temperature, sceletal vibrations are obvious; at high temperature, the motion is more chaotic and shows occasional conformational defects

avoided. The short-time simulation can in this way show several defects. The creation and quenching of many rotational isomers could thus be studied. A rotational isomer was counted as soon as the rotational angle τ exceeded $\pm 90°$ from the *trans* conformation. Their rate of formation as a function of the size of the model crystal and temperature is displayed in Fig. 10. As the number of surrounding chains increases, the restriction due to the rigid crystal decreases, and the rate of formation of *gauche* bonds increases by 0.5 to 2.0 orders of magnitude. The upper curve represents a limit of no enclosure of rigid chains (constant pressure simulation). At 350 K there are of the order of 10^{12} large, internal rotations per second about the backbone bonds, or about 10^{10} per bond. Such an enormously active source of defects can naturally drive many defect-linked processes in polymer crystals. From rate data, as shown in Fig. 10, the activation energies for the formation of *gauche* bonds could be derived using the transition state theory. It was found to be close to the potential energy barrier introduced in Eq. (5), varying from 13 to 25 kJ/mol, with 16.7 kJ/mol most common for the less restrictive models. The most immediate question might be: how is it possible to have, under such condition of continual defect formation, any crystal and be in accord with the experiments that never gave information on the frequent creation of *gauche* defects? The answer lies in the relatively low

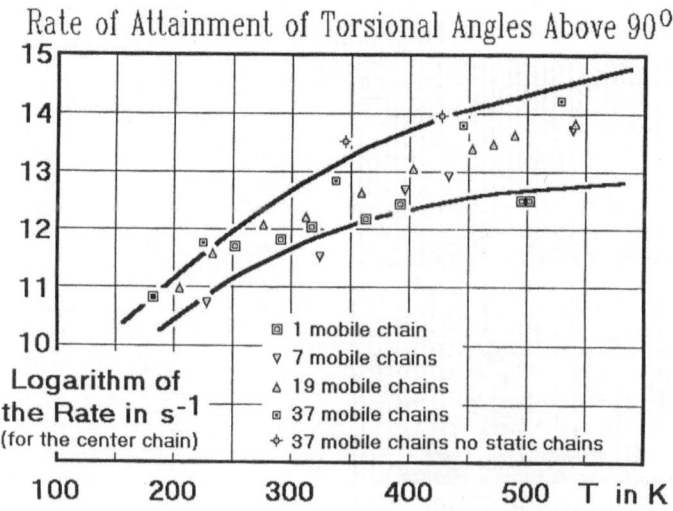

Fig. 10. Rate of formation of conformational defects for crystal models of different size

steady state concentration of *gauche* conformations. This also means that the lifetime of a *gauche* defect has to be extremely short. So short, in fact, that experiments that average over long time periods do not observe the deviations.

The population of rotated bonds for the constant-volume 19 mobile chains surrounded by 18 static chains, with each chain consisting of 100 CH_2 groups, is shown in Fig. 11 as a function of temperature. Adding the shallow peak about the *gauche* angle at 322 K, for example, leads to a total concentration of only about 0.5%. Note that the small deformation angles, indicative of the vibration amplitudes, are truncated heavily (with the actual percentages written at the top of each standard interval). Note also that the crystal environment has changed the *gauche* maximum from $\cos \tau = 0.5$ to $\cos \tau \approx 0.4$ (rotation from 180° to 66° or 294° instead of 60° or 300°). The lifetimes of the defects must be very short, in the picosecond range, because of the fast rate of formation documented in Fig. 10. Somewhat longer lifetimes are observed for less restrictive intermolecular forces, i.e., for simulations with larger numbers of mobile chains (see Fig. 10). The increased potential energy due to the conformational defects was closely duplicated by the observed increase in heat capacity in the same temperature range [1c].

Overall, these molecular dynamics crystal simulations showed that random, uncorrelated conformational disorder was governed by three processes: (1) the intramolecular dynamics leading to local isomeric transitions; (2) the number of intermolecular collisions; and (3) the restrictiveness of the crystal environment [5b]. These initial conformational defects do not correspond to a potential energy minimum and thus cannot easily be predicted by molecular mechanics calculations. They are the result of the dynamic interaction of skeletal vibrations

Occurrence of Angles τ for 19 Mobile Chains

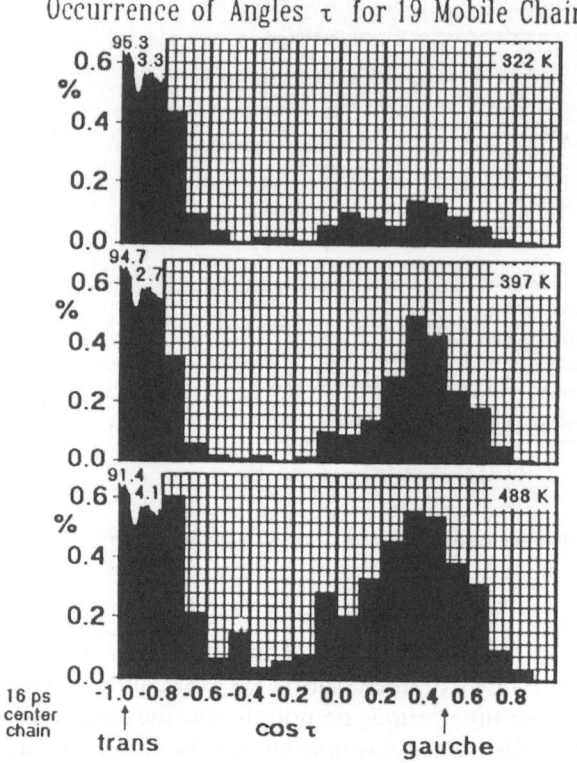

Fig. 11. Population of conformational defects at different temperatures after 16 ps of simulation

(such as longitudinal, torsional and transverse vibrations) and internal rotations [Eq. (5)].

One may examine the dynamics depicted, for example, in Figs. 9–11 for possible point defects and a microscopic mechanism for structure-sensitive macroscopic processes. The crystals that were treated up to now are models of equilibrium crystals (extended-chain crystals) and the simulations usually did not exceed 30 picoseconds. No strongly correlated motion could be observed. Figure 12 shows a typical plot of the lifetime of *gauche* defects in the central mobile chain of a very restrictive crystal of 7 dynamic chains, surrounded by 12 static chains (constant volume simulation). The *gauche*-defects are created and destroyed very rapidly. The high simulation temperature was chosen in order to display many such defects. Besides the random appearance of the conformational defects, there also exists no correlation with defects of neighboring chains. The defects are largely isolated and show no diffusion to different locations. The lack of appearance of any stable progression of these defects may be a signature of chaotic motion in these many-body systems.

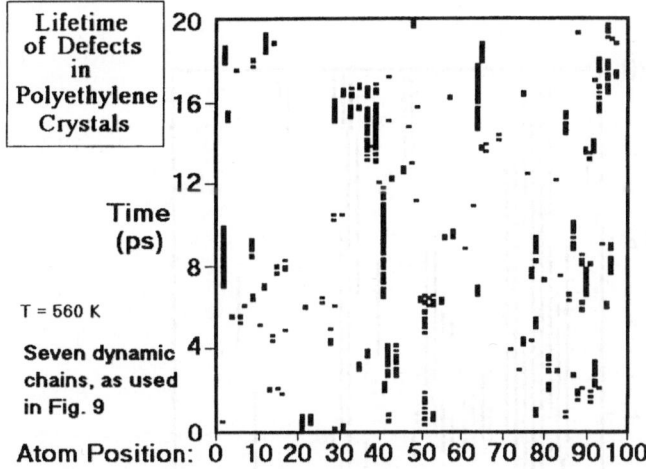

Fig. 12. Simulation of the lifetime of conformational defects with seven dynamic chains at high temperature (constant volume simulation)

Increasing the number of dynamic chains leads to the defect situation depicted in Fig. 13. The lifetime of the defects is increased and predominant sequences appear that can be characterized as $g^{\pm}tg^{\mp}$. Such sequences of conformational defects are called kink defects (see Appendix) and cause a shortening of the chain by $c/2$ in addition to some displacement in the a and b crystallographic directions. From the measured average chain-length at 440 K there is a shortening of about 1 nm or an equivalent of four 2g1 defects, in approximate agreement with Fig. 13. When geometric defects (by 180° rotation of an initial two-bond sequence) or kinetic defects (by addition of a kinetic energy pulse at a given position, sufficient to cause one or more defects) are introduced at the beginning of the simulations, no lasting effect was introduced. The initially created defects disappear as quickly as the randomly generated thermal defects. Even at temperatures as low as 135 K, where there is no creation of thermal defects observed, any artificially induced point defect does not freeze in the structure, but shows also only lifetimes of less than one picosecond. *Defects of such types within single chains are thus unstable* [5a, c].

Mechanistic details for the formation of a 2g1 kink can be derived from Fig. 14. The first 1.1 ps of a simulation at 320 K are displayed for the bottom portion of a chain in a crystal. One can easily recognize the process that leads in the end to the defect formation, a transverse vibration initiated at time 0.1 ps, a compression of the chain at 0.7 to 1.1 ps, and a torsional vibration moving into the field of view at time 0.5 ps. Their unique collision between 0.7 and 1.0 ps causes the 180° turn of the chain end at 0.8 ps and the indicated localized 2g1 kink at 1.1 ps. The kink caused the remaining end of the chain to be in register 1/2 unit cell length removed upwards. In this particular event, registered in Fig. 14, the lifetime of the 2g1 defect was about 2 ps [5c].

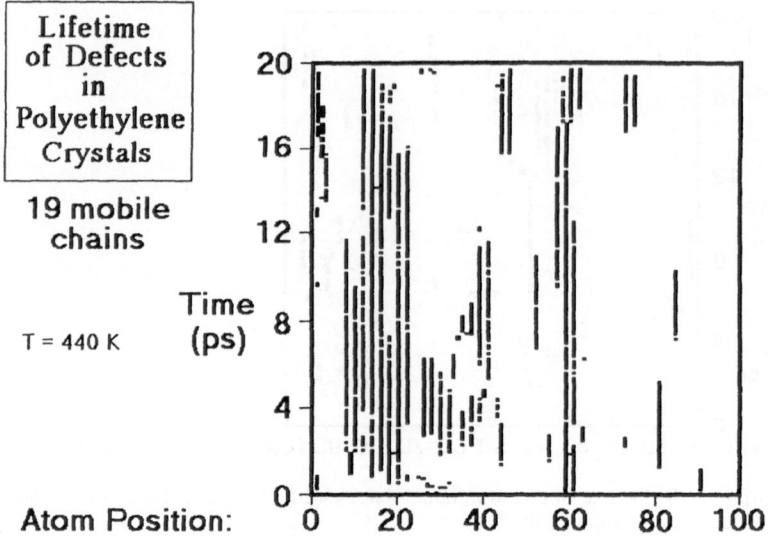

Fig. 13. Simulation of the lifetime of conformational defects with 19 dynamic chains as shown in Fig. 8

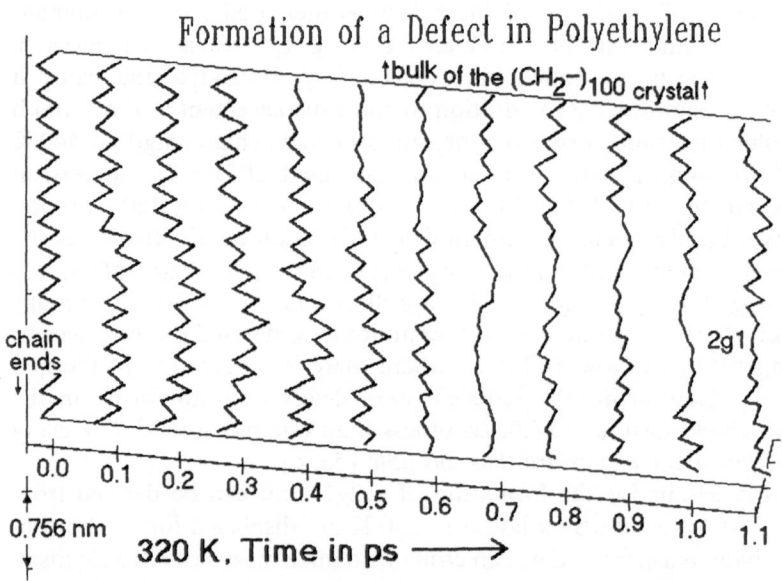

Fig. 14. Formation of a defect in polyethylene (bottom portion of a chain from a seven dynamic chain, constant volume simulation)

The discovery and study of a high-yield thermal source of conformational defects was followed by an effort to link these defects to chain diffusion and crystal deformation. An initial simulation was designed to study the fate of a twist defect once it was produced in a crystal [4a]. This is a defect, thought to be of major importance for chain-transport in polymer crystals (see the Appendix). Since the energy of its formation is estimated by molecular mechanics calculations to be rather high, one could not expect to observe the formation of such a specific defect during the rather short-time molecular dynamics simulations. A solution to such problems is to start the simulation with a defect already present. Figure 15 shows a sequence of perspective drawings of the lower half of a simulated seven dynamic chain crystals into which such a soft twist defect (180° twist over 12 CH_2-groups) has been introduced (the defect is located in the center chain). The atom which is marked with a circle at the bottom of the crystal shows that, below the defect, the chain is not in crystallographic register, thus providing a potential energy for motion. The lifetime of the defect after initiation of the simulation is less than 2 ps and the travel of the defect goes to the nearest crystal surface in the simulation shown at 320 K. Even at 50 K the defect was not frozen, but was quenched by motion to the end of the crystal, also within about 2 ps. Once any of these types of larger defects have been formed by overcoming the energy of formation, their lifetime is thus proven to be very short. However, it appears that if such a defect is created with a free energy gradient for motion toward only one crystal edge, the rapid transport could lead to chain diffusion as is needed for lamellar thickening, annealing, or deformation. In the thermally generated defects shown in Figs. 12 and 13 no gradient existed and consequently no coherent travel was observed.

The next step of the analysis was the introduction of an external force to one chain end. Adding a constant mechanical pull on the center chain, produces the

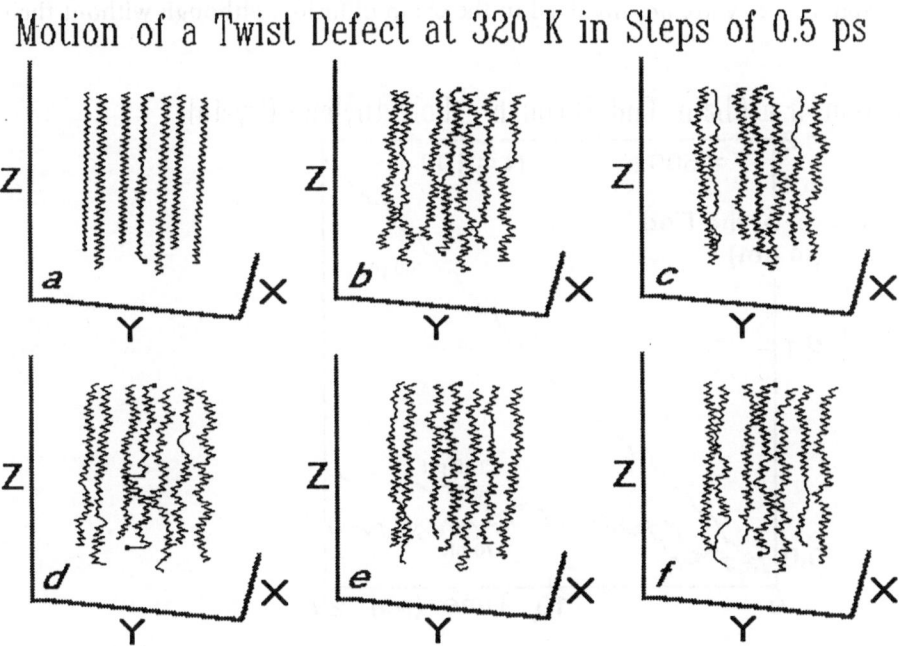

Fig. 15. Simulation of a twist defect in a seven dynamic chain, constant volume simulation

mass transport needed for crystal deformation [5a]. Figure 16 shows an example of a pull with 70, 140, and 280 pN at 500 K on the central chain of a crystal consisting of seven dynamic chains with 100 CH_2-groups each. These forces are of the order of magnitude needed for large-amplitude changes of the torsional coordinate [Eq. (5)], but should not significantly affect the bending and stretching coordinates [Eqs. (2) and (4), respectively]. For comparison, typical forces for cold drawing of polyethylene are smaller still by an order of magnitude. This is perhaps not surprising since there is no stress concentration in the model, and the chosen crystal has a very restrictive intermolecular potential. Increasing the size of the simulated crystal, thereby reducing the restrictiveness of the intermolecular potential, causes a significant decrease in the activation energy for the motion of a polyethylene chain by one lattice step. For a model with 37 dynamics chains at constant pressure, for example a 1 pN force resulted in the complete removal of a 100-CH_2-group chain in 0.25 nanoseconds. In Fig. 16, a hopping mechanism is observed at about 5 and 13 ps for 140 pN and 5 ps at 70 pN. At the highest force a more constant velocity is attained. This illustrates that the sliding of the chain out of the crystal is superimposed on the longitudinal vibration.

Adding a free energy gradient on the thermal motion of the crystal permits a diffusion of the chains in the direction of the gradient [5c]. To study the mechanism in more detail, the defects were noted in Fig. 17, simultaneously with the positions of the upper and lower chain ends in Fig. 18, and the end-to-end distances in Fig. 19. The illustrated crystal has a chain length of 50 CH_2-groups. The immediate observation is that the *gauche* defects do not move through the crystal, i.e., they are *not* involved in the chain diffusion, although without their

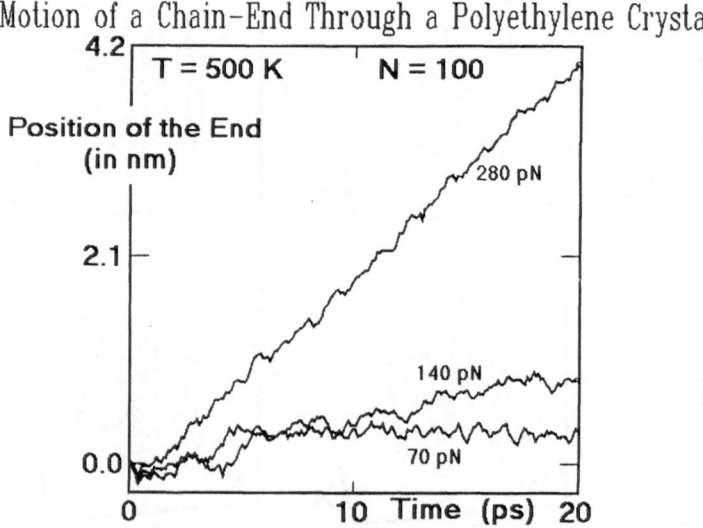

Fig. 16. Motion of a chain on application of an external force

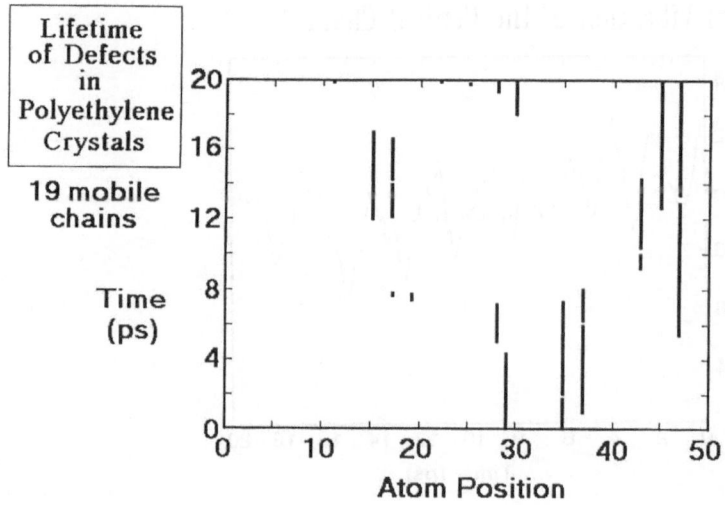

Fig. 17. Lifetime of defects in the center chain, exposed to a free energy gradient in the *c*-axis direction

Motion of the Central Chain Relative to six Neighbors

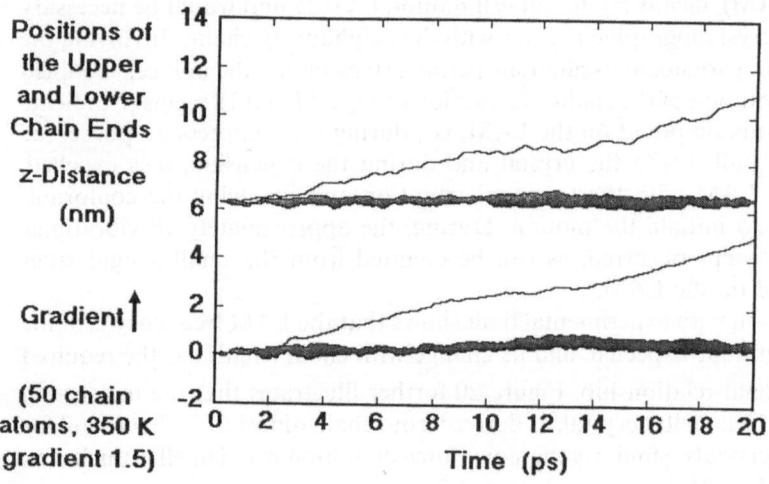

presence the transport appears to stop. The actual chain transport must thus be based on gradual twists or accordion-like chain vibrations, initiated in the gradient direction at the *gauche* defects. From Fig. 18 it can be seen that the chain moved about 5 nm, or more than half way out of the crystal. The six nearest

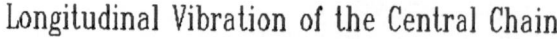

Longitudinal Vibration of the Central Chain

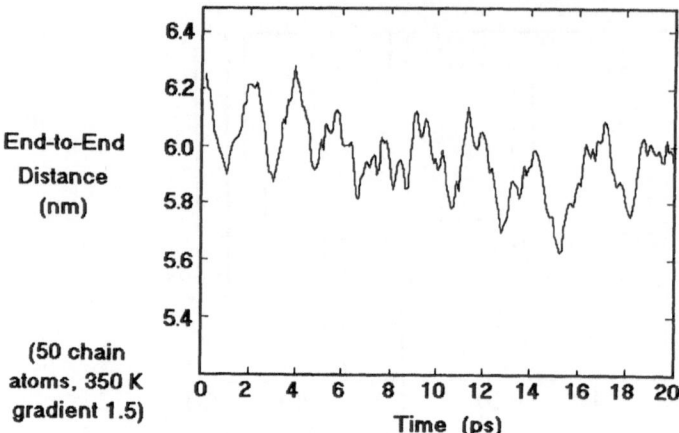

Fig. 19. Longitudinal vibrations of the center chains displayed in Fig. 17

neighbor positions are marked by the shaded horizontals and have, outside the longitudinal vibration, no translational motion. A move in the chain direction, making use of the amplitude of the longitudinal vibration (longitudinal acoustic vibration, LAM), means a full unit cell motion (≈ 0.25 nm) would be necessary to maintain crystallographic register with the neighboring chains. Involving the lower frequency torsional oscillations permits steps of half the unit cell, coupled with a 180° rotation of the chains. Inspection of Figs. 17 and 18 seems to indicate steps of c/2 superimposed on the LAM, i.e., during the compression phase, the chain end is pulled into the crystal and during the expansion, it is expelled, coupling the LAM with the torsional vibration and involving the conformational defect to initiate the motion. During, the approximately 10 vibrations, about 40 c/2-steps occurred, as can be counted from the small irregularities superimposed on the LAM.

Comparison with experimental data shows that the LAM frequency is of the order of magnitude expected, and its change with chain length has the required $1/\sqrt{n}$ functional relationship. Figure 20 further illustrates that the increase in thickness of the lamellar crystal, as derived from chain diffusion, is that found for annealing of crystals. Similarly the slow-down of motion with lamellar thickness, as shown in Fig. 21, is what is expected from experiment.

Relaxation phenomena in polymer crystals were simulated by the application of a static electric field on a given chain that was decorated with a dipole moment [4b]. The electric field was applied in a direction that was perpendicular to the direction of the dipole moment, and thus could result in a twisting of the polymer chain if the field is strong enough to overcome the activation energy for the formation of such a defect. The results from these simulations showed that twists of the chain can be observed under the influence of a strong electric field. For

Temperature Dependence of Chain Transport

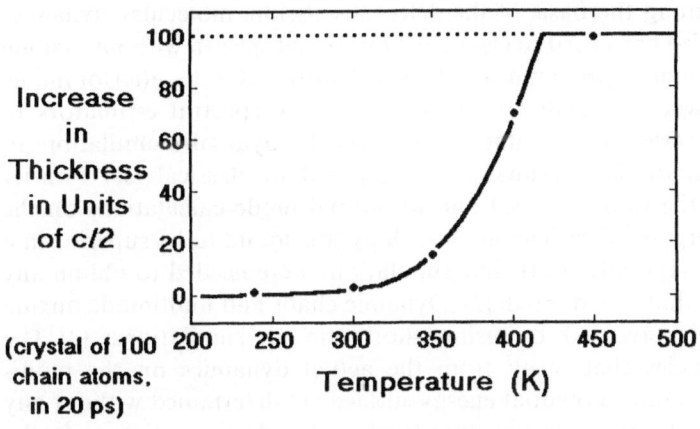

Fig. 20. Chain transport as a function of temperature

Lamellar Thickness Dependence of Chain Transport

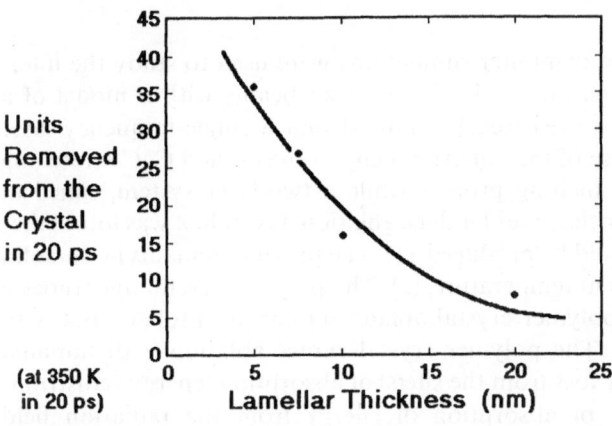

Fig. 21. Lamellar thickness effect on the diffusion of chains as in Fig. 20, at 350 K after 20 ps simulation

intermediate fields, other than full alignment reaches stability for reasonable lengths of time to cause macroscopically observable effects. If a small external force is applied toward one of the crystal edges, a net diffusion of the chain can be observed, and the temperature and chain length dependencies that are observed in dielectric relaxation experiments, can be qualitatively reproduced similar to the results in Figs. 20–21.

3.3 Other Molecular Dynamics Studies of Relevance

In addition to forming the basis of the defect discussion, molecular dynamics simulations have also been used to calculate vibrational spectra without making use of the normal mode approximation [2]. The densities of states $g(\omega)$ for model polymer crystals were calculated by using a class of spectral estimators to transform the time series data obtained from molecular dynamics simulations to the frequency domain. The results were compared to classical fast Fourier transformation of the same data set and to normal mode calculations on the same potential energy surface. The methodology was found to be superior in a number of aspects: (1) Only short time simulations were needed to obtain any spectral resolution that was desired; (2) Dynamic chaos and multimode mixing of signals could be avoided by using short enough simulations; (3) The vibrational frequencies that result from the actual dynamics on the multi-dimensional, anharmonic, potential energy surface are determined without any assumptions on the basic mode structure (such as harmonic motion as in the normal mode description); (4) The vibrational spectrum could be studied as a function of external variables such as pressure, temperature, stress, etc.

3.3.1 Laser Interactions

Molecular dynamics-based computer simulations were used to study the inter-action of one and two high-powered infrared laser beams with a model of a crystalline polymer surface [19] Results showed that a single-frequency laser beam system could, because of the anharmonicity of the excited C–C-stretching vibrations, only induce a melting process while a two-laser system, tuned at somewhat different frequencies, could induce ablation. Overall, it was found that four different situations could be produced: (1) The polymer remains in the solid state, but with an increase in temperature; (2) The polymer crystal undergoes a melting process; (3) The polymer crystal ablates (bonds are broken), but with significant melting; or (4) The polymer crystal shows ablation with minimal melting. The rate of energy loss from the site(s) of absorption (energy redistribu-tion) relative to the rate of absorption of energy from the radiation field determines whether the polymer crystal melts or ablates. Low absorption rates lead to melting or no change, and high rates lead to ablation. A sufficiently large rate of energy absorption was found to be accessible only through the use of two lasers with properly chosen frequency difference.

3.3.2 Collisional Dynamics

The interactions of polymer surfaces with atoms of high velocity on collisions has been examined in order to try and elucidate relevant information to surface degradation [20]. The collisional dynamics were examined for various impact velocities and clearly revealed that there is a high efficiency of collisional energy and momentum transfer from the collider to the surface. The atomistic details of the internal dynamics showed that the absorbed collisional energy first pulses down the impacted chain on a picosecond time scale, followed by redistribution of the energy within the chain and, finally, by intermolecular energy redistribution to the surrounding chains. The overall ultra-fast energy relaxation process does not exhibit significant surface degradation. After a sufficiently large amount of energy has accumulated in the surface, however, degradation in the form of chemical bond-breaking can occur. In addition, in many cases, the colliding molecule (nitrogen) penetrates the polymer surface and becomes trapped between the chains, followed by slow diffusion over long periods of time.

3.3.3 Melting

Initial progress was also made to describe the melting and crystallization kinetics [6]. Most of the prior simulations dealt with some dimensionality restrictions or started at or near the transition state. From these studies, it was observed that at very high superheating, a melting kinetics without reversibility occurs. In contrast, if the equilibrium melting temperature was approached, the melting-crystallization interplay could be simulated. Initial stages of a melting process have also been observed for the full dimensionality process for the paraffin crystal and was briefly discussed above (Fig. 7). In summary, the overall disorder that is preliminary to melting tends to increase substantially as the temperature increases beyond that of the melting temperature. The disorder primarily arises from conformational defects and begins, to a minor degree, as much as 100 K below the melting point. At higher temperatures, translational and rotational disorder occur in addition. For the model of a paraffin crystal that has an experimental melting temperature of about 365 K, the molecular chains start to show some translation diffusion along the chain axes at about 315 K. In addition, the surface chains start to coil (see Figs. 4 and 5). At about 400 K, the crystals show notable conformational, rotational, and translational disorder that starts at the surface and works its way inward.

3.3.4 AFM Probing

The atomistic structure and dynamics of the interaction of an atomic force microscopic probe (AFM) with a crystalline polyethylene surface was examined by using the molecular dynamics method coupled with ab initio quantum calculations [21]. A set of force parameters and guidelines have been derived from the extensive computer simulations, and these results were used to help explain some of the AFM images. In general, AFM experiments can be performed in a nondestructive mode with a reasonable resolution, provided that the forces of interaction between a typical-size tip and sample are kept within the range of 10^{-11}–10^{-8} N. This broad load-force range marks the upper and lower limits for the conditions necessary to achieve optimum resolution and minimal surface/tip deformation.

The AFM probing of polymeric surfaces can, besides imaging the surface, also produce a number of anomalies. Surface contaminants, such as those caused by adsorbed polar molecules, were found to cause significant perturbations on the images produced. From the calculations, it appears that when the AFM tip encounters a polar defect, it is initially attracted (phenomenon) and becomes "trapped" for a short period of time. This type of stick-slip phenomenon leads to an enhanced frictional energy dissipation which, in turn, causes an increase in the surface temperature of both the AFM tip and the polymer surface. The increase in temperature can subsequently induce rotational defects in a polymer chain and ultimately cause deformations on a long time-scale.

In addition to the surface perturbation caused by frictional dissipation, the presence of adsorbed polar molecules may also cause a substantial change in the AFM image. In this case, the adsorbed polar molecules act like attractors and lead to an image which appears to have large and sharp features imbedded on the surface topology. Such structural distortions were, indeed, observed in AFM images obtained from experimental studies of crystalline polyethylene [22]. Although the size of the observed distortions in the experimental observations may also be due to much larger structural defects (such as sharp protrusions on the surface or tip), rather than being caused by adsorbed polar molecules, it has been clearly demonstrated that more precise interpretation of AFM images of polymeric surfaces is possible by simulation of the proposed features by molecular dynamics calculations. In all of the studied cases reasonable correlation with experiment was possible.

4 Conclusions

This review of extensive molecular dynamics simulations of systems with relatively large numbers of atoms illustrates how it is now possible to begin the study of cooperative molecular motion involving interactions between skeletal vibrations and conformational isomers. Supercomputers have accomplished the extreme slow-motion display necessary to analyze simultaneous, cooperative changes. One can extrapolate that one more order of magnitude in atoms, and the extension of the analyzed time into the nanosecond range, can resolve many of the basic problems of the defect solid state of linear, flexible macromolecules.

Many connections have been established between microscopic MD simulations and macroscopic experimental properties of polymers, such as seen in conformational disorder and heat capacities [1], molecular motion and the vibrational spectrum [2], stress-induced frequency shifts and conformational changes [3], twist motion and the dielectric α-relaxation [4], molecular diffusion and the lamellar thickening [5], and the melting process in polyethylene crystals [6]. The results have also provided a much better understanding of the correlation of rotational isomers and defect generation and motion in polyethylene crystals [7]. As expected from the chosen potentials, Eqs. (2–5), density, speed of sound, vibrational frequencies, *gauche*-defect concentrations, and even melting temperatures are close to the established experimental data. The importance of the understanding of motion in the picosecond time scale for the link between macroscopic property and microscopic cause is, for the first time, securely documented.

Acknowledgements. This work was supported by the Division of Material Sciences, Office of Basic Energy Sciences, U.S. Department of Energy, under Contract No. DE-AC05-84OR21400 with Martin Marietta Energy Systems, Inc., and by the Polymer Program of the National Science Foundation, present Grant No. DMR-92-00520. Computations were performed on the UTCC computers at the University of Tennessee, Knoxville, the Cray X-MP at Oak Ridge National Laboratory, the Cray Y-MP at the National Center for Supercomputing Applications, and the Cray Y-MP at the Pittsburgh Supercomputer Center. We thank Dr. A. Xenopoulos for many useful discussions and for providing input on the current state of the literature on defect polymer crystals.

5 References

Note that the references of our laboratory have been cited with title to permit easy and appropriate selection of more detailed information. The references for the literature review are collected in Sect. 6.3

1. (a) Wunderlich B, Möller M, Grebowicz J, Baur H (1988) Conformational motion and disorder in low and high molecular mass crystals. Springer, Berlin Heidelberg New York (Adv. Polym. Sci. 87); (b) Noid DW, Sumpter BG, Wunderlich B (1990) Molecular dynamics simulation of the condis state of polyethylene. Macromolecules 23: 664; (c) Noid DW, Sumpter BG, Varma-Nair M, Wunderlich B (1989) Molecular dynamics results for a polyethylene-like crystal. Makromol Chem Rapid Commun 10: 377; (d) Sumpter BG, Noid DW, Wunderlich B (1990) Theoretical studies of the effects of anharmonicity on polymer dynamics: temperature dependence of heat capacity. Polymer 31: 1254; (e) Noid DW, Pfeffer GA (1989) Dispersion curves from short time molecular dynamics simulation: stressed polyethylene results. J Polymer Sci Part B: Polymer Phys Ed 27: 2321
2. (a) Noid DW, Sumpter BG, Wunderlich B (1990) Molecular dynamics calculation of the density of states for poly(ethylene): collective versus local modes. Anal Chem Acta 135: 143; (b) Roy R, Sumpter BG, Noid DW, Wunderlich B (1990) Estimation of dispersion relations from short-duration molecular dynamics simulations. J Phys Chem 94: 5720; (c) Roy R, Sumpter BG, Pfeffer GA, Gray SK, Noid DW (1991) Novel methods for spectal analysis. Phys Rep 205: 109
3. (a) Pfeffer GA, Noid DW (1990) Stress-induced infrared frequency shifts in polyethylene, Macromolecules 23: 2573; (b) Pfeffer GA, Sumpter BG, Noid DW (1992) Conformational changes in polyethylene model under tension and compression. Polym Engr & Sci 32: 1278
4. (a) Noid DW, Sumpter BG, Wunderlich B (1991) Molecular dynamics simulation of twist motion in polyethylene. Macromolecules 24: 4148; (b) Noid DW, Sumpter BG, Liang GL, Wunderlich B (1992) Molecular dynamics simulations of electric field induced motion in crystalline polyethylene, in Plastics Shaping the Future. Proceeding of the SPE 50th Annual Technical Conference & Exhibition, p 1982.
5. (a) Noid DW, Sumpter BG and Wunderlich B (1990) Molecular dynamics studies of the lamellar thickening process for polyethylene. Polym Commun 31: 304; (b) Sumpter BG, Noid DW, Wunderlich B (1990) Computer experiments on the internal dynamics of crystalline polyethylene: mechanistic details of conformational disorder. J Chem Phys 93: 6876; (c) Sumpter BG,

Noid DW, Wunderlich B (1992) Computational experiments on the motion and generation of defects in polymer crystals. Macromolecules 25: 7247

6. (a) Sumpter BG, Noid DW, Wunderlich B (1990) Molecular dynamics study of the rate of melting of a crystalline polyethylene molecule: effect of chain folding. Macromolecules 23: 4671; (b) Noid DW, Pfeffer GA, Cheng SZD, Wunderlich B (1988) Computer simulation of the melting process in linear macromolecules. Macromolecules 21: 3482

7. (a) Xenopoulos A, Noid DW, Sumpter BG, Wunderlich B (1990) The correlation of rotational isomers in polyethylene-like crystals. Makromol. Chem. 191: 2261; (b) Wunderlich B, Xenopoulos A, Noid DW, Sumpter BG (1991) Defect generation and motion in polyethylene-like crystals, analyzed by simulation with supercomputers. Mat Res Soc Proc 209: 147

8. (a) Noid DW, Sumpter BG, Wunderlich B, Pfeffer GA (1990) molecular dynamics simulations of polymers: methods for optimal Fortran programming. J Comp Chem 11: 236; (b) Noid DW, Sumpter BG, Cox RL (1991) Computational strategies for molecular dynamics simulations of polymer processes: numerical integration schemes. J Comp Polym Sci 1: 161

9. (a) Liang GL, Noid DW, Sumpter BG, Wunderlich B (1993) Dynamics of a paraffin crystal. Makromol Chem Theory Simulation 2: 245 (b) (1993) Molecular dynamics simulation of the hexagonal structure of crystals with long methylene sequences. J. Poly. Sci. B., Poly. Phys. 31: 1909 and (c) (1993) Atomistic details of disordering processes in superheated polymethylene crystals. Acta Polym. 44: 219

10. Klein ML (1985) Ann Rev Phys Chem 36: 525; Hoover WG (1983) Ann Rev Phys Chem 34: 103

11. Sumpter BG, Noid DW, Wunderlich B (1993) Computer simulation and modeling of polymeric crystals. Trends Polym Sci 1: 160

12. Hertz J, Krogh A, Polymer RG (1991) Introduction to the theory of neural computation, Addison-Wesley, Redwood City, CA; Zurada JM (1992) Introduction to artificial neural systems, West, New York

13. (a) Sumpter BG, Getino C, Noid DW (1992) A neural network approach to the study of internal energy flow in molecular systems. J Chem Phys 97: 293; (b) Noid, Darsey JA (1991) Neural net simulation of polymer dynamics. J Comp Polymer Sci 1: 157

14. Wunderlich B (1973, 1976, and 1980) Macromolecular physics, Vols. I–III; Academic Press: New York

15. Brandrup J, Immergut EH (1989) Polymer Handbook, J. Wiley: New York

16. Kim Y, Strauss HL, Snyder RG (1989) J Phys Chem 93: 7520

17. Jin Y, Wunderlich B (1991) The heat capacity of paraffins and polyethylene. J Phys Chem 95: 9000

18. Liang GL, Noid DW, Sumpter BG, Wunderlich B, (1993) Dynamics and structure of polymethylene crystals with explicit hydrogen atoms. J. Comp. Poly. Sci. 3: 101

19. Sumpter BG, Voth GA, Noid DW, Wunderlich B (1991) Infrared laser-induced chaos and conformational disorder in a model polymer crystal: melting vs. ablation. J Chem Phys 93: 6081

20. Gelb A, Sumpter BG, Noid DW (1990) Computer simulation of molecular collisions with a polymer surface. J Phys Chem 94: 809; (1990) Molecular dynamics calculations of energy transfer to polymer surfaces. Chem Phys Lett 169: 103

21. Sumpter BG, Getino C, Noid DW, Wunderlich B (1993) Computer simulations of atomic force microscopy: crystalline polymers and the effects of surface contaminants. Makromol Chem Theory Simul 2: 55

22. Annis BK, Reffner JA, Wunderlich B (1992) Atomic force microscopy of extended-chain crystals of polyethylene. J Polymer Sci Part B: Polymer Physics 31: 93; Annis BK, Noid DW, Sumpter BG, Reffner JA, Wunderlich B (1992) Application of Atomic Force Microscopy (AFM) to a block copolymer and an extended chain polyethylene. Macromol Chem Rapid Commun 13: 169

6 Appendix: Review of the Literature and Comparison with the MD Simulations

6.1 Overview of the Defect Types

Information about crystals, gained by X-ray diffraction experiments, have led to the derivation of hundreds of ideal polymer crystal structures [1]. While for nonpolymeric crystals structure-insensitive properties like density, heat content, modulus, etc, could be approximated by such ideal crystal structures, this was not possible for polymer crystals. An added level of gross imperfection needed to be recognized, that of the "crystallinity". A two-phase model has been commonly used over the last 50 to 60 years. One of the two phases being represented by rather perfect crystals, the other by a fully amorphous phase. Structure-insensitive properties vary approximately linearly with the weight or volume fraction of crystallinity. Typically, polymers may show crystallinities from 5 to 95%. Since such two-phase structures in a one-component system cannot, according to the phase rule, be in equilibrium, there must be constraints hindering further crystallization of the amorphous phase, i.e., there must be a connection between the two phases (microphases). Because of this link between crystal and amorphous phase, the latter has also been called a three-dimensional "amorphous defect" [2]. Only at larger distance from the crystal may one reach the well-defined amorphous phase [3]. Typical for an amorphous phase connected to a crystal is its increased glass-transition temperature when compared to noncrystalline samples. The less mobile, restrained amorphous polymer layer has also been named the "rigid amorphous fraction" [4] and may often be separated from the "mobile amorphous fraction", detectable by various experimental techniques. The structure of the molecules outside of the crystal proper may thus be further subdivided. Their structure and mobility is still not well understood, but may in the future be amenable to MD simulation of the type presented in the bulk of this review.

Besides this three-dimensional defect that is characteristic for polymeric materials, there are also those of two dimensions, represented by surfaces. External surfaces are always the prime defect of a single crystal. For polymer crystals they are particularly important because of the usually small crystal size. The highly regular surfaces of solution-grown crystals could be analyzed in detail by optical and electron microscopy, and more recently also by atomic force microscopy. These investigations gave much information about chain folding, tie molecules, fold-sector boundaries, stacking faults, twin boundaries, kink bands, etc. [5]. Although more difficult, much could also be learned about surfaces created by melt crystallization. The research involved the study of fracture surfaces, microtomed sections, etched samples, and decorated interfaces [5].

Decreasing the dimensionality of the defects leads one to dislocations (one-dimensional defects). The classical edge and screw dislocations of crystals of small molecules could also be documented in crystals of macromolecules. Particularly obvious are growth spirals with a Burgers vector of the size of the lamellar thickness, i.e., 5–50 nm, which are linked to the crystal growth mechanism [6]. Such large screw dislocations were observed first for paraffins [7]. Dislocations with a Burgers vector of the unit-cell dimension were seen somewhat later by electron microscopy with the help of moiré patterns and were also linked to the accommodation of chain ends in the crystals [8]. A link to deformation processes as was found in crystals of small molecules was, however, not possible. Most dislocations in polymer crystals seem to be formed by terminated growth planes and are not free to glide. The glide planes for motion of dislocations are limited to planes parallel to the chain direction because of the strong bond-anisotropy and find further interference with the fold geometry and the presence of tie molecules [9]. The most important deformation process of polymer crystals is that occurring on drawing, cold-extrusion, and rolling. These processes involve, however, a much more chaotic deformation process, leading perhaps to a "pseudomelting" [10], as is illustrated in Figure A-1. Based on the discussion of the MD simulations in this paper, the defects involved in these processes seem to involve the coupled skeletal vibrational and conformational motion and fall into the class of point defects.

The zero-dimensional or point defects were the major object of this review. They are much different for polymer crystals when compared to small-molecule

Attempt at an Illustration of the Fiber Formation on a Microscopic Scale as given by Peterlin

(The affine deformation region is very limited, but one can recover deformation due to the tie molecules)

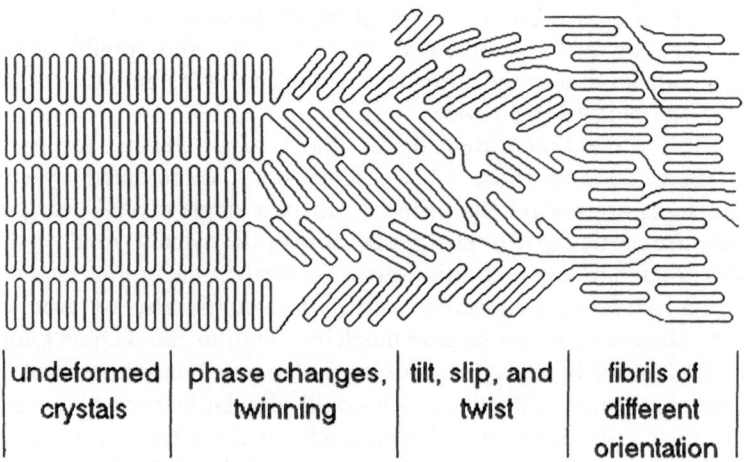

| undeformed crystals | phase changes, twinning | tilt, slip, and twist | fibrils of different orientation |

Fig. A-1. Deformation of a polymer crystal as proposed by Peterlin

crystals and provide the ultimate element in the understanding of structure-sensitive properties, such as diffusion, annealing, strength, mass transport, and chemical reactivity. In small-molecular crystals, isolated vacant lattice sites, interstitial atoms, and foreign motifs on interstitial and substitutional positions form the basic point defects. In macromolecular crystals differences arise from the connectiveness of the permanent covalent bonds of the backbone chain. Vacancies can only arise together with other defects such as kinks, chain ends, or wrong substituents in the chain (copolymers). Interstitial atoms must be linked with chain disorder, copolymerization, or monomeric impurities. Similarly, substitutional positions must be produced by prior copolymerization or by cocrystallization with small molecules (rare) [2]. Chain disorder, as described by kinks, folds, jogs, torsions (twists), etc., are thus the basic point defects in macromolecular crystals [11].

The macromolecular crystals of interest to the present review are largely polyethylene-like, so that electronic defects, such as electron holes, interstitial electrons, and exitons, need not be considered.

A starting point for defect evaluation can be the rotational isomer model [12]. It makes use of the three-fold minimum in rotational energy [see also Eq. (5)]. To produce a defect in an all-*trans* sequence of a polymer crystal such as polyethylene, one can, for example, represent a defect conformation as a sequence of *gauche* (g^+, g^-; $\tau = 300°$ and $60°$, respectively) and *trans* (t; $\tau = 180°$) conformations. A simple defect in polyethylene is a "2g1 kink", represented by the sequence of $g^{\pm}tg^{\mp}$. The symbol 2g indicates the number of *gauche* conformations, and the second numeral, the amount of shortening of the chain in multiples of $c/2$ (c is the unit cell dimension along the chain axis). A 2g1 kink displaces the chain in addition by about 0.2 nm in the directions at right angles to the zig-zag plane. The latter displacements were proposed to be removed by distortion of the chain. When the displacement are bigger, as in 2g2, they reach a neighboring zig-zag plane and are called jogs. The intermolecular energy of a 2g1 kink was estimated from the increased volume multiplied with the cohesive energy density. Overall it was estimated that an equilibrium concentration of 1 kink per 250 carbon atoms should be possible at 400 K, in close agreement with the simulations reviewed in the text [13].

Defects that do not involve rotational isomers, but gradual twists, compressions or expansions of the chain have also been proposed earlier. Successive small rotations about the backbone bonds and bond angle deformations are involved in such defects. A 360° rotation about the chain axis leads to proper register with the neighboring chains in the crystal above and below the defect. Defects of this type are called disclinations.

Finally, a rotation of 180° can be combined with a translational defect of $c/2$ to achieve crystalline register above and below the defect. Such combination of disclination and dislocation is called a dispiration. The nomenclature of chain dislocation, disclination and dispiration was developed by Reneker and Mazur [14].

In the following part, the literature of the point defects of interest to the present review is summarized and briefly discussed in the light of the reviewed MD simulations.

6.2 Overview of the Literature

6.2.1 Experimental Investigations of Gauche Defects

Vibrational spectroscopy can be used to observe directly the *gauche* conformations involved in defects of polymer crystals. The methods used involve associating specific absorption bands active in the IR or Raman spectra with *gauche* conformations or even *gauche-trans* sequences. Selective deuteration is often used to identify the position of the defect.

Zerbi and coworkers have studied, among others, the transition of *n*-nonadecane ($C_{19}H_{40}$) to the rotator phase [15]. It involves the onset of a roto-translational (screw) motion, where chain ends are squeezed out of the crystal, pushing the lamellae apart. Twisted chain ends with gauche defects at the position 2–3 from the chain end are formed, while no appreciable concentration of defects exists in the center of the chains. The same research group also calculated [16] the IR spectrum corresponding to a polymethylene chain with a twist defect called twiston. Some agreement was found between the calculated spectrum and the one measured for $C_{19}H_{40}$ in the rotator phase. This was taken as a sign that a soft twist, i.e., a defect without *gauche* conformations, might be a realistic model.

Maroncelli et al. have similarly studied conformational disorder in the high-temperature phases of *n*-alkanes [17]. The transition to these phases involves a step-wise increase in the concentration of *gauche* bonds. The defects found consisted almost entirely of 2g1 kinks. Their concentration is high at the chain ends and decreases exponentially going toward the middle of the chain, but does not reach zero, as suggested in [15]. Based on lattice-model calculations, it was deduced that the dominant factor for the *gauche*-bond distribution is the longitudinal displacement of the chains. More recent work on longer *n*-alkanes that do not exhibit a high temperature phase ($C_{50}H_{102}$ and $C_{60}H_{122}$) [18] demonstrates that the onset of disorder is not associated with the transition, but occurs gradually in all paraffins. Both the concentration and the distribution of the defects is similar to that found in the shorter alkanes, indicating that the mechanism of the formation of *gauche* defects is essentially the same. As was pointed out in the present review, the energetics of the existence of *gauche* defects is also in agreement with the increase in heat capacity beyond that expected from vibrational energy alone. Figures 9–15 are in accord with the observed *gauche*-concentrations and Fig. 5 documents the tendency of translational chain diffusion as well as the bending and twisting of chain ends.

6.2.2 Calculations of Point Defects, Energies, and Geometries

An early point defect was proposed by Reneker in 1962 [19]. This defect is described by a linear segment of a polymer chain that is compressed at both ends. A compression can cause a buckling of the chain at some localized position, with the uncompressed region of the chain remaining in crystallographic register. Such a defect is an interstitial-like dispiration [14]. This is sometimes called the Reneker defect and has an energy of 49.2 kJ/mol. The buckle is accommodated in the crystal by a small distortion of the neighboring chains. In polyethylene, this simple point dislocation results from twisting the PE molecule by 180° about the chain axis and incorporating one additional methylene unit in the compressed portion. The formation of such a defect was thought to be able to begin at the end of a chain and then advance along the chain, and perhaps, remain stationary somewhere within the crystal. If it ultimately travels to a fold, this would result in a length-increase of the segments joined at the fold by half a repeat distance (c/2).

The energy required to form a Reneker defect was estimated based on the number of bonds that have to be rotated from their equilibrium position and on the strain introduced in the lattice for the accommodation of the defect. It was estimated that at 370 K the equilibrium concentration should be about one defect per 400 carbon atoms. More detailed calculations and generalization of this defect concept has recently been given [14], expanding the number of defects to five types, two dispirations, two chain dislocations and one disclination. The crystallographic defects are described for an isolated chain of crystalline polyethylene. In addition to the interstitial-like dispiration described above (Reneker defect), there could be a similar vacancy-like dispiration by removing one CH_2-group from the sequence of chain units at an energy of about 88.4 kJ/mol. The dislocations consist of mismatches in translational symmetry and show much higher defect energies (98.9 and 196.2 kJ/mol), while the disclinations have chain twists at a constant number of CH_2-groups (72 kJ/mol). Two chain twist-boundaries and a partial dislocation boundary were also described (about 32 kJ/mol). The latter are partial defects, in that they cannot exist alone, but can combine to form a defect. It was suggested that the propagation of these defects can result both in chain rotation, needed for relaxations, and translation, involved in lamellar thickening.

The molecular dynamics simulations of Sect. 3 indicate that the activation energies (estimated from transition-state theory) for the production of defects that seem to be similar in structure and function is much lower ≈ 10–20 kJ/mol) than the overall defect energies calculated (≈ 50–200 kJ/mol). This discrepancy points to the limitation of molecular mechanics (lowest energy configuration) calculations for the description of cooperative, dynamic processes. The proposed geometries correspond in many cases, however approximately, to the observed, more irregular defects.

6.2.3 Kink Theories

The kink model of Pechhold [13] constitutes an integrated approach to the description of condensed matter with a detailed microscopic model. Besides describing the crystal defects in terms of kinks, jogs, and folds (see Sect. 6.1), it made it possible to describe also liquid-like states using the so-called meander model. In this meander (or bundle) model, disorder is achieved in small crystal-like blocks by the introduction of kink isomers. Some bonds are thus always away from their equilibrium positions, but the chains are essentially kept parallel over the dimension of the block, so that a largely extended low energy macroconformation remains. Melting into a meander is thought to occur when the defects in the crystal reach a certain critical concentration. The short-range order that is usually observed in the melt is therefore explained by the fact that the chains are still parallel and the properties of the liquid are modeled by the correlated motion of the basic blocks of the parallel segments.

During annealing the kinks in a crystal are thought to be thermally activated to move, resulting in its thickening. Moreover, they can reorientate, for example by changing from left-handed (g^+tg^-) to right-handed (g^-tg^+) and this hopping might give rise to relaxation processes. It has been proposed that the α-relaxation is caused by rearrangement of clusters of kinks (kink blocks) consisting of 2g2 and 2g3 kinks, while the γ-relaxation arises from reorientation of 2g1 kinks.

As shown in Sect. 3, kinks exist in the predicted concentrations (Figs. 12 and 13), but are relatively immobile in the simulations, and thus relaxation and annealing are more heavily linked to the skeletal vibrations of the molecular chains than the motion of conformational defects, although the presence of the kink defects seems to play an important role in the activation of chain diffusion (see Figs. 16–18).

6.2.4 Twist Theories

Most of the more recent defect models involve chain perturbations that are gradual in nature and extend over many bonds. These so-called "twist" ideas are based on work performed in the past on dielectric relaxations of small molecules [20].

The first such model put forth was the "Utah" twist [21], named after the institution where it was developed. It was documented through detailed molecular mechanics calculations on an array of $C_{22}H_{46}$ molecules of orthorhombic crystal structure and compared to the previously suggested models. The main reservations against the previous models, primarily from relaxation studies, were summarized as follows. Point defect mechanisms assume an equilibrium or induced number of stable defects that are thermally activated to give rise to various motions. It was shown by NMR however, that all the crystalline chains participate and not just a small number of defective segments.

Moreover, the intensity of a relaxation due to such stable defects would be small and should strongly increase with temperature, which is not observed experimentally. Finally, the strong crystal thickness dependence of the activation free energy of the α-relaxation is also difficult to explain by localized defects. The additional fact that there is almost a single relaxation time suggested to the authors some two-site reorientation of a single chain. Therefore a screw rotation of 180° about the axis with an advancement of $c/2$ along the axis was implied. This defines the initial and final states without specifying the exact mechanism for going from one state to the other. A rigid chain motion with all CH_2-units in a crystal segment turning simultaneously would have a $\log(f_{max})$ dependence on crystal thickness at constant temperature, decreasing linearly with thickness. The observed behavior is, however, an initial drop-off with a subsequent plateau, showing that the relaxation frequency is independent of thickness for very thick crystals. Most of the experimental results are explained quantitatively by the gradual twist model. Some comparisons were made with the Reneker defect, since both contain relatively localized 180° twists, the latter being buckled and not smooth. The main difference is that the soft twist in only an intermediate state and there are no equilibrium concentrations, as are assumed by Reneker. If the twist is also considered to be an intermediate state, then the two become energetically competitive in the case of thick crystals, while for thin crystals the soft twist is favored. It was suggested that defect hopping is not associated with the relaxation mechanism, but perhaps with the thickening mechanism.

Soon after its introduction, the twist defect was interpreted as a "soliton" [22, 23]. The Hamiltonian for the system was obtained from the sine-Gordon model for a linear chain of harmonically coupled particles, and the solutions of the equation of motion yielded propagating solitons. Both of these models were initially applied only to polyethylene, but were later extended to isotactic and syndiotactic polypropylene and isotactic polystyrene [24]. Both the existence of solitons, as well as the activation energy of the resulting α-relaxation, were predicted for isotactic polypropylene and polyethylene, while the absence of the relaxation is also predicted for the other two cases. Three criteria for the existence of solitons were proposed: (1) the energy barriers in the interstem potential should be low, (2) the elongational stiffness of the polymer stem should be high, and (3) the number of bonds per repeat unit should be small.

A soft twist with a helicity reversal was also proposed by Taylor and coworkers for the case of polytetrafluoroethylene (PTFE) [25, 26]. The "ambidextron", as it was termed, came naturally as a result of the equations of motion for the system. The difference to the other soft twists, already discussed, is a chain-length dependence. Free motion of the perturbation is only possible when there is a high ratio of free ends to stems in the crystal lamella, as is the case for PTFE.

Zerbi and Longhi [16] have introduced the term "twiston" to describe the soft twists discussed. The twiston is the mobile defect determined numerically for a realistic model of the polymer, as opposed to the soliton, which is the ideal solitary wave obtained analytically from an idealized model of a polymethylene chain.

Overal[1], the soft twist models have gained much popularity since they seem to explain more experimental data than the other models. Nonetheless, Mansfield and Boyd's statement against equilibrium concentrations of point defects might be stronger than needed. As is shown in Figs. 10 and 11, due to the large rate of formation and short lifetime, defects seem to be available everywhere. There is experimental evidence for the existence of *gauche* bonds in the crystals (infrared spectroscopy, X-ray diffraction, and NMR) and it was shown that their concentration does increase with temperature. The predominance of the soft twist as a model for defects does not necessarily preclude other defects, their relative importance and detailed function must, however, be evaluated case by case. The chain conformations revealed by Figs. 5, 9, 14, and 15 show many "soft twists", and prove their involvement in the crystal dynamics and the importance of the concept. Their characterization defies, at least at present, the link with the much idealized models discussed in this section.

The interpretation of the soft twist as a soliton might be instructive, but is, again, very much idealized. In principle, solitons will propagate freely only in the absence of external friction, i.e., interchain and soliton–soliton interactions. This is most likely not going to be the case in polymer crystals.

6.2.5 Paracrystallinity

The concept of paracrystallinity, developed by Hosemann [27] is another integrated approach to disorder in the solid state. A paracrystal is thought to have short-range, but no long-range order. The structure is quasiperiodic, in the sense that the perturbation increases with distance and finally long-range order is lost. Paracrystallinity broadens the X-ray reflections in a characteristic manner. While paracrystallinity represents a geometrical description of a distorted lattice, it does not specify its actual origin. It has been suggested that statistically distributed dislocations produce the observed paracrystallinity in polymers [28]. These ideas were obtained from line-width analysis of WAXD data and provide the needed link with actual physical defects. A relevant question to be resolved is whether the simulated defect crystals of this review yield, as a time average, a paracrystalline lattice.

6.2.6 Other Defect Concepts

Attempts to understand the experimentally observed rate of crystal thickening on annealing were summarized in [6]. Diffusion of defects from chain ends would produce, in contrast to experiment, a very strong molecular weight dependence and can thus be excluded from the discussion.

An improvement in the agreement with experiment of predictions of crystal thickening rates on annealing as a function of long-period and temperature was suggested by Peterlin [29]. He proposed collective mass transport by simultaneous jumps of the whole, straight fold-section between the fold planes, as opposed to the Reneker-type succession of individual, defect jumps. The underlying reason favoring such a collective phenomenon should be the anisotropy of forces in the polymer crystal, with the bonding forces along the chain being much larger than the lateral forces between adjacent chains. The activation energy for such a jump is lE_m/d_0, where E_m represents the barrier for a single repeating unit jump, and l and d_0 are the lamellar thickness and repeating unit distance, respectively. The experimental data were fitted satisfactorily by giving E_m the proper temperature dependence. The resulting decrease of E_m as one approaches the melting temperature seems reasonable.

Parallel to the thickening of crystals by chain extension, it must also be resolved how vacancies created by the chain diffusion are to be treated. Formation of macroscopic holes, diffusion of amorphous chains into an extending crystal, and mass transport between stacked crystals have been reported (see, for example [6]).

The simulations shown in Figs. 17–21 may allow one to interpret the jumps required by Peterlin as the abrupt end-to-end distance-changes superimposed on the LAM (see Fig. 19). Inclusion of a single fold in the simulated crystal revealed that transport through the fold without motion of the fold is possible. The lateral diffusion needed to fill the space left by the chain diffusion could be followed in some of our simulations and involves cooperative adjustments of all chains surrounding the created vacancy. It may thus be possible to progress in understanding this important problem of annealing of polymer lamellar crystals with future, specially designed MD simulations involving multiple chains and additional folds.

Structural effects caused by defects associated with a random arrangement of chain ends were discussed by Predecki and Statton. They described dislocations specifically caused by chain ends [30]. For an orthogonal unit cell with chains represented by straight lines, these were coupled-screw (unique to long-chain molecules), coupled-edge, screw and edge dislocations. Figure A-2 shows such arrangement of chain ends and dislocations. The defects were proposed to rearrange, given enough segmental mobility, and to cluster into line defects or grain boundaries. The present simulations did not include sufficient chain ends to comment on the importance of this model.

Some more specific ideas were proposed based on work by Hoffman et al. on polychlorotrifluoroethylene (PC3FE) [31] and Sinnot on single crystals of polyethylene [32]. The intensity of the low temperature γ-relaxation was seen to decrease and the maximum temperature of the loss peak to increase with annealing temperatures above 370 K, where the long period (measured by small-angle X-ray diffraction) and the mobile fraction (measured by broad-line NMR measurements) markedly increase. The explanation of these and other observations were that part of the γ-relaxation, normally assumed to originate only from the amorphous fraction, is caused by the crystal. In the case of n-paraffins the crystalline γ-relaxation involves the reorientation of a "defect" chain, probably accompanied by a small, lengthwise translation. For polyethylene, an additional

Defects in Nylon 6,6

**Disruptions
of a crystal
caused by
chain ends**

**15,000 MW
10^{20} ends
per cm^3,
2.2 nm
between
chain ends,
three layers
are shown**

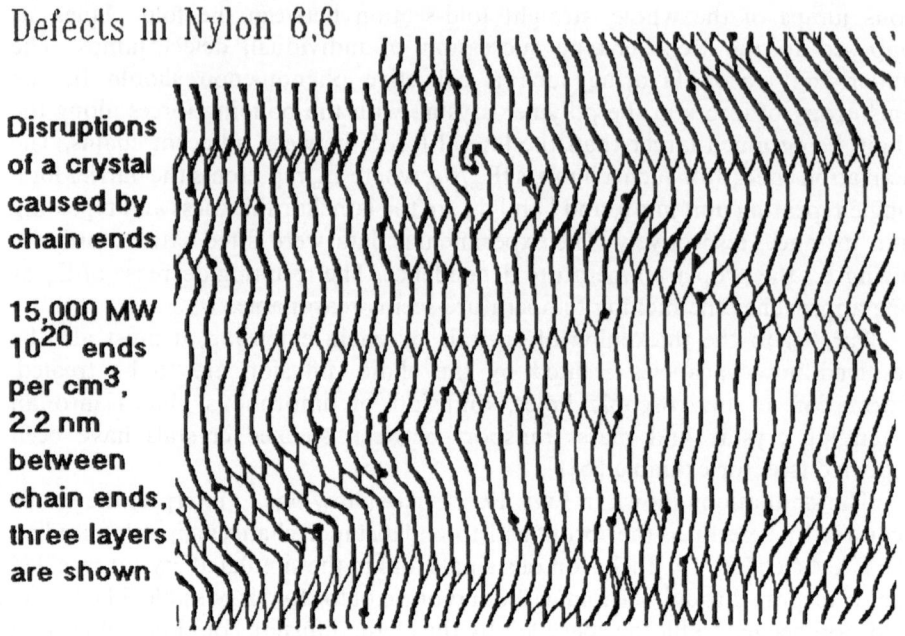

Fig. A-2. Chain ends in a polymer crystal as proposed by Predecki and Statton

reorientation of the chain folds was assumed, which would be independent of the fold period. As the annealing temperature increases, the number of folds decreases, resulting in the observed decrease of relaxation intensity. The increase of the relaxation temperature with annealing temperature is due to the need to reorient the now longer linear chain segments.

It has to be noted that crystal thickening itself cannot be discussed in terms of sub-glass-transition relaxations and has customarily been associated with the α-relaxation. The reason is that at the temperature of the γ-transition there is no large-scale motion, needed for thickening to occur (see Fig. 9 at 80 K). It seems, however, that for PC3FE (primarily used for justification) there is some crystalline contribution to the γ-relaxation.

Another model for annealing, involving suction of folds into the crystal, was proposed by Dreyfuss and Keller, based on diffraction data on solution-crystallized nylon 6,6 [33]. Such a mechanism does not involve a coordinated displacement of the whole fold segment as discussed in Sect. 6.2.6. The mechanism was proposed following the observation that the polyamides studied increased their fold lengths with difficulty and, when they did, it was by a factor of two and four. Considering three consecutive folds of a molecule, the middle one at the top of the lamella, the other two at the bottom it was proposed that the middle fold can recede, extending the other two. Ultimately the middle one is being consumed, leading to a single fold of double length. Such annealing would be favored energetically, particularly if complementary processes would occur at the same time in the crystal layers above and below at matching positions. Intermediate stops at double the fold length on annealing were also observed by electron microscopy of isolated solution grown lamellae of polyethylene (for a discussion see [6], Sect. 7.2.1).

The MD simulations discussed in Sect. 2 are capable of realistically representing models such as the travel of folds through the crystal. Simulations as in Figs. 17–19 were also extended to the study of two chains connected by a single fold within the otherwise standard crystal. Forcing one chain towards the fold produced an accumulation of the diffusing chain segment outside of the crystal, and forcing one chain away from the fold pulled the chain through the fold without changing the fold location. The crucial quantitative study of the force needed to dislodge the fold by pulling on both stems of a fold has not yet been performed. Simultaneous motion of the two segments in the same direction without an external force is, however, expected to be rare. Although this model appears as an elegant solution to the experimental observation of integral thickening, it seems unlikely in the light of the present simulations, at least in its idealized form.

6.2.7 Dynamics Studies

The isomerization kinetics of linear alkanes has enjoyed a number of important contributions [34–55]. These studies employed the techniques of molecular dynamics, Monte Carlo, Brownian/stochastic dynamics or equivalent approaches and provide a large amount of useful data. We can discuss only a few specific cases among the large number of results.

The n-alkanes have been studied not only as a model for polyethylene, that would have required extremely CPU-extensive calculations, but also because much is known experimentally about their high-temperature rotator phase. Molecular dynamics simulations were performed on infinite alkane chains (using periodic boundary conditions) starting from an orthorhombic structure [34]. At low temperatures a stable structure exists and the calculated thermal expansion was in agreement with experimental results. Above 375 K, one observes in these simulations liquid-like diffusion along the chain and rotational diffusion about the chain axis, giving a clear four-site orientational distribution. These drastic changes occurred without a spontaneous transition to a hexagonal, rotator phase. The rotator phase was studied in simulations performed on the bilayer phases of tricosane ($C_{23}H_{48}$) [35, 36], and its structure was found to be similar to the high-temperature orthorhombic phase of the previous simulation. Many intramolecular conformational defects were observed, predominantly at the chain ends, but their effect on the structure was only modest. The dominant factor affecting the change at higher temperatures is judged to be the longitudinal chain motion, giving a structure akin to liquid crystals.

The rotator phase of n-alkanes was also studied by Monte Carlo simulations [37]. A four-site orientational distribution was found for the orthorhombic rotator phase, but a six-site distribution was seen in the case of the hexagonal rotator phase. In both cases the structure is composed of many ordered domains, wherein the chains are parallel. Considerable rotational oscillation about each of the six preferred orientations, as well as active translational motion, exist in the hexagonal phase, in contrast to the orthorhombic phase where only jumps among the permitted orientations could be seen.

The dynamics and structure of C_{10}-n-alkanes in the solid state have also been examined by using the molecular dynamics method [38]. In an investigation of 31 molecules of a C_{10}-n-alkane at temperatures ranging from 150–275 K, the influence of chemical defects, such as methyl groups or chlorine atoms, on the dynamics of the system was also considered. Conformational defects (gauche defects) were observed for temperatures larger than 210 K and these defects had lifetimes of the order of 3.5 ps at 275 K. Analyses showed that the defects that had a lifetime greater than 0.5 ps at 275 K consisted of $g^{\pm}$ (40%) and $g^{\pm}tg^{\mp}$ kinks (22%) of all cases (other types of defects were also detected, but were less predominant). The setting angles of the zigzag plane was found to undergo a rotating mobility, increasing with temperature. However, no correlation between rotational mobility of the molecules, their average orientation, and the position in the structural cell was observed. The so-called chemical defects caused significant dynamic changes: the average torsional amplitude was less and only trans-conformations appeared at the bonds nearest to the defect.

The above results can be seen to agree fairly well with those of this review. In particular, the formation of conformational disorder (involving gauche defects) in a crystal and the longitudinal mode-assisted "phase" change. For the small chain length systems, such as the n-alkanes, the fact that isomeric transitions are frequently observed, yet do not play a clear role in facilitating the transition to a high temperature phase, was also discussed in Sect. 3.1 (Figs. 2–7). Furthermore, the translational and rotational disordering at the higher temperatures and the formation of domain structures was found in the simulations of the pentacontylene in Sect. 3.1, providing further evidence for the many-site distribution discussed for the orthorhombic rotator phase of the n-alkanes.

In addition to these studies on n-alkanes, a number of research groups have been active in perusing detailed information on the atomistic structure and dynamics of polymeric materials [55–76]. Overall these studies agree in their more limited scope with the results presented in Sect. 3. We briefly discuss a small number of these studies and make comparisons of some of the results to those of this review.

The work of Mattice et al. [57–60] has clearly demonstrated that the intramolecular correlation between isomeric transitions decay rapidly as a function bond separation, with large correlations occurring for the i + 2 types (next nearest neighbors). Similar results have also been found by many other research groups for the case of smaller systems [42, 48, 52]. Overall, there is good agreement with the results discussed for Figs. 10–14 of this review, where 2g1 defects were shown to dominate the defect generation in the relatively large crystals examined. The rate for isomeric transition, however, varies somewhat for each simulation. The trans-gauche transition rate was reported in a range of

$1.3 \times 10^8 – 1.0 \times 10^{10}\ s^{-1}$ at temperatures between 300 and 400 K and models consisting of backbone chains with 4–200 carbon atoms. This range is in good accord with that discussed in Sect. 3.2 and shown in Fig. 10.

Rigby and Roe have also examined the dynamics of chain molecules [71–76]. They have systematically studied amorphous polymer molecules above and below the glass transition temperature. A threshold temperature for the onset of isomeric transitions was observed at 130 K and the activation energy determined for the transitions was about the same as that used to describe the potential to a *trans-gauche* isomerization. Rates for isomerization were found to vary in the range of $6 \times 10^8\ s^{-1} – 7 \times 10^{10}$, depending on the length of the chains and on the density. For higher densities and longer chain lengths, the threshold temperature, activation energy for isomerization, and the rates were in very good agreement with those of this review, derived for crystals (Sect. 3.1–2).

6.3 References to the Appendix

1. Tadokoro H (1979) Structure of crystalline polymers. Wiley, New York; Miller RL (1989) Crystallographic data for various polymers. In: Brandrup J, Immergut EH (eds) Polymer Handbook. Wiley, New York; or [2]
2. Wunderlich B (1973) Macromolecular physics, crystal structure, morphology, defects. Academic, New York
3. Wunderlich B, Poland D (1963) J Polymer Sci Pt A, 1: 357
4. Cheng SZD, Wu ZQ, Wunderlich B (1987) Macromolecules 20: 2801
5. Geil PH (1963) Polymer single crystals. Wiley-Interscience, New York; Woodward AE (1988) Atlas of polymer morphology. München; or Ref. [B]
6. Wunderlich B (1976) Macromolecular physics, nucleation, crystallization, annealing. Academic Press, New York
7. Dawson IM, Vand V (1951) Proc Roy Soc A206: 555
8. Agar AW, Frank FC, Keller A (1959) Phil Mag 4: 32; Predecki P, Statton WO (1966) J Appl Phys 37: 4053
9. Keith HD, Passaglia E (1964) J Res Natl Bur Stand 68A: 513
10. Peterlin A (1971) J Mat Sci 6: 490
11. Reneker DH (1962) J Polym Sci 59: S39; Schatzki TF (1962) J Polymer Sci 57: 496; Pechhold W, Blasenbrey S, Woerner S (1963) Kolloid Z Z Polymere 189: 14; Illers KH (1964) Rheol Acta 3: 202; McMahon PE, McCullough RL, Schlegel AA (1967) J Appl Phys 38: 4123
12. Volkenstein MV (1963) Configurational statistics of polymeric chains. Wiley-Interscience, New York; Birshtein TM, Ptitsyn OB (1966) Conformations of macromolecules. Wiley-Interscience, New York; Flory PJ (1969) Statistical mechanics of chain molecules. Wiley-Interscience, New York
13. Pechhold W, Blasenbrey S, Woerner S (1963) Kolloid Z Z Polymere 189: 14; Pechhold W, Blasenbrey S (1970) Kolloid Z Z Polymere 241: 955; Pechhold W, Liska E, Grossmann HP, Hägele PC (1976) Pure Appl Chem 127
14. Reneker DH, Mazur J (1988) 29: Polymer 3
15. Zerbi G, Magni R, Gussoni M, Moritz KH, Bigotto A, Dirlikov S (1981) J Chem Phys 75: 3175
16. Zerbi G, Longhi G (1988) Polymer 29: 1827
17. Maroncelli M, Strauss HL, Snyder RG (1985) J Chem Phys 82: 2811
18. Kim Y, Strauss HL, Snyder RG (1989) J Phys Chem 93: 7520
19. Reneker DH (1962) J Polym Sci 59: S59
20. Fröhlich H (1958) Theory of dielectrics, Second edition, Oxford University Press, Oxford
21. Mansfield M, Boyd RH (1978) J Polym Sci, Polym Phys Ed 16: 1227
22. Mansfield ML (1980) Chem Phys Lett 69: 383
23. Skinner JL, Wolynes PG (1980) J Chem Phys 73: 4022
24. Syi J-L, Mansfield ML (1988) Polymer 29: 1827
25. Wright NF (1988) Bull Am Phys Soc 33: 248
26. Heinonen, Taylor PL (1989) Polymer 30: 585

27. Hosemann R (1950) Z Phys 128: 1465; Hosemann R, Bagchi SN (1962) Direct analysis of diffraction by matter. North Holland, Amsterdam
28. Wilke W (1983) Coll Polym Sci 261: 656
29. Peterlin A (1963) J Polym Sci, Polym Lett Ed 1: 279
30. Predecki P, Statton WO (1966) J Appl Phys 37: 4053
31. Hoffman JD, Williams G, Passaglia E (1966) J Polym Sci, Polym Symp 14: 173
32. Sinnot KM (1966) J Appl Phys 37: 3385
33. Dreyfuss P, Keller A (1970) J Polym Sci, Polym Lett Ed 8: 253
34. Ryckaert J-P, Klein ML (1986) J Chem Phys 85: 1613
35. Ryckaert J-P, Klein ML, McDonald IR (1987) Phys Rev Lett 58: 698
36. Ryckaert J-P, McDonald IR, Klein ML (1989) Mol Phys 67: 957
37. Yamamoto T (1985) J Chem Phys 82: 3790; (1988) 89: 2356
38. Mazo MA, Oleynik EF, Balabaev NK, Lunevskaya LV, Grivtsov AG (1984) Polymer Bulletin 12: 303
39. Edberg R, Morriss GP, Evans DJ (1987) J Chem Phys 86: 4555
40. van Gunsteren WF, Berendsen HJC, Rullmann JAC (1981) Mol Phys 44: 69
41. Helfand E, Wasserman ZR, Weber TA (1979) J Chem Phys 70: 2016
42. Helfand E, Wasserman ZR, Weber TA (1980) Macromolecules 13: 526
43. Weber TA (1978) J Chem Phys 69: 2347
44. Weber TA (1978) J Chem Phys 69: 4277
45. Fixman M, Kovac J (1974) J Chem Phys 61: 4939
46. Fixman M (1978) J Chem Phys 69: 1527
47. Ryckaert JP, Bellemans A (1978) Faraday Discuss Chem Soc 66: 95
48. Levy RM, Karplus M, McCammon JA (1979) Chem Phys Lett 65: 4
49. Evans GT, Knauss DC (1979) J Chem Phys 71: 2255
50. Brooks BR, Bruccoleri RE, Olafson BD, States DJ, Swaminathan S, Karplus M (1983) J Comput Chem 4: 187
51. Wielopolski PA, Smith ER (1986) J Chem Phys 84: 6933
52. Montgomery JA, Holmgren SL Jr., Chandler D (1980) J Chem Phys 73: 3688
53. Edberg R, Evans DJ, Morriss GP (1986) J Chem Phys 84: 6933
54. Karaborni S, O'Connel JP (1990) J Chem Phys 92: 6190
55. See for example papers in, Computer Simulations of Polymers, Ed by Roe RJ, Prentice Hall, Englewood Cliffs, NJ (1990), and references therein.
56. Brown D, Clarke JHR (1986) J Chem Phys 84: 2858; (1991) Macromolecules 24: 2075; Brown D. Clarke JHR (1991) Comput Phys Commun 62: 360
57. Yongjian Z, Mattice WL (1992) J Chem Phys 96: 3279
58. Dodge R, Mattice WL (1991) Macromolecules 24: 2709
59. Zúniga I, Gahar I, Doge R, Mattice WL (1991) J Chem Phys 95: 5348
60. Cho D, Neuburger NA, Mattice WL (1992) Macromolecules 25: 322; Neuburger N, Bahar I, Mattice WL (1992) Macromolecules 24: 2447
61. Plathe FM (1992) J Chem Phys 96: 3200
62. Takahashi N (1992) Rep Prog Polym Phys Japan 35: 187
63. Takahashi N (1992) Rep Prog Polym Phys Japan 35: 183
64. Takahashi N, CAMSE92 (in press)
65. Kremer K, Grest GS (1990) J Chem Phys 92: 5057; Kremer K, Grest GS (1990) J Phys., Condens Matter 2: 295
66. Depner M, Schürmann BL, Auriemma F (1991) Mol Phys 74: 715
67. Jung J, Schürmann BL (1989) Macromolecules 22: 477
68. Geyler S, Pakula T (1988) Makromol Chem Rapid Commun 9: 617; Pakula T (1988) Macromolecules 21: 1665; Pakula T, Geyler S (1987) Macromolecules 20: 2909
69. Tiller AR (1992) Macromolecules 25: 4605
70. Hahn J, Pertsin AJ, Grossman HP (1992) Macromolecules 25: 6510
71. Rigby R, Roe R-J (1987) J Chem Phys 87: 7285
72. Rigby R, Roe R-J (1988) J Chem Phys 89: 5280
73. Rigby R, Roe R-J (1989) Macromolecules 22: 2259
74. Takeuchi H, Roe R-J (1991) J Chem Phys 94: 7446
75. Takeuchi H, Roe R-J (1991) J Chem Phys 94: 7458
76. Blonski B, Brostor W (1991) J Chem Phys 95: 2890

Received June 1993

Brownian Dynamics Simulations of Local Polymer Dynamics

M.D. Ediger and D.B. Adolf[1]
Department of Chemistry, University of Wisconsin, Madison, WI 53706, USA

This article describes stochastic computer simulations of the local segmental dynamics of synthetic polymers. Particular attention is given to local dynamics in solution. Related work involving experimental methods, analytical theory, and molecular dynamics simulations will also be discussed. An introduction to the concepts involved in stochastic simulations will be presented. Methods of characterizing local segmental dynamics will also be described. The main portion of the article describes various features of the simulated dynamics. The approximations inherent in stochastic approaches and their influence on the observed dynamics will be discussed. Issues of cooperativity in conformational transitions will be highlighted.

---

[1] Current address: University of Leeds, UK

Advances in Polymer Science, Vol. 116
© Springer-Verlag Berlin Heidelberg 1994

1 Introduction

The local segmental dynamics of polymer chains have an important influence on the macroscopic properties of polymeric systems. For bulk polymers this influence is readily apparent. The glass transition temperature T_g, the temperature dependence of the bulk viscosity, and the toughness of polymer glasses are all directly related to local polymer dynamics. The local dynamics of polymer chains in solution also have an effect on macroscopic properties. For example, differences between the high frequency limiting solution viscosity and the solvent viscosity have been shown to be related to the time scales of local polymer motions [1, 2]. In this case, fast local motions can indirectly influence the intrinsic viscosity and polymer motions on all length scales at finite polymer concentrations [3]. In addition, dilute polymer solutions provide a useful prototype for investigations of local dynamics in other environments. The effects of chain–chain interactions are minimized in dilute solution. As a result, attention can be focused on the internal motions of a single chain. Studies of dilute solution dynamics should provide guidance in unravelling the more complex dynamics of bulk polymers. Several reviews of local polymer dynamics have recently been presented [4–8].

Conformational transitions in synthetic polymers in dilute solution typically occur on the time scale of 100 ps to 10 ns. This time scale depends upon the chemical structure of the chain as well as the temperature and the solvent viscosity. Key questions remain to be answered about the manner in which chemical structure affects local polymer dynamics. A comparison with the dynamics of a smaller molecule illustrates why this is a challenging problem. Butane is the shortest alkane which can undergo a conformational transition. The reaction coordinate for butane is, to a good approximation, simply a rotation about the central carbon–carbon bond. Now compare this to a conformational transition in the middle of a polyethylene chain. Clearly the reaction coordinate must now be more complicated; a simple rotation about one bond would result in one end of the chain swinging in a very large arc. What is the reaction coordinate? Do two transitions occur in order to localize chain motion? In general, what degrees of freedom are cooperating and over what length scale does this occur? The answers to these questions likely depend upon the chemical structure of the chains. Only when these questions are addressed can we understand how chemical structure influences local dynamics.

Computer simulations provide an important tool for the exploration of local polymer dynamics and structure/property relationships in polymers. Since simulations provide the positions and velocities of all atoms at all times, any question about dynamics can be answered in principle. Many questions of interest, such as those in the previous paragraph, cannot be directly answered by experiment. The microscopic analysis of a simulation can reveal insights into the origin of important properties. However, such insights are trustworthy only to the extent that the simulation is realistic. The calculation of experimental observables from a simulation, and their comparison with experiment, provides an indication of how far a simulation should be trusted.

The primary topic of this review is Brownian dynamics (BD) simulations of local polymer dynamics. Most of the simulation work on local polymer dynamics has been done with the BD technique. We will review this work and also discuss work in closely related fields. The importance of various approximations made in the simulations will be addressed. Results from other types of simulations, and from analytical calculations, will be introduced as they are relevant. Experimental results will be discussed throughout the review. Indeed, the article is written from the perspective of a suspicious experimentalist.

This review will be significantly biased towards the local dynamics of synthetic polymers in solution. The specific interactions which are quite important in most biopolymers are difficult to handle in BD simulations (see Sect. 4.3). Many aspects of biopolymer dynamics are thus more suitably treated with molecular dynamics (MD) simulations. MD simulations are also more suitable for the study of local dynamics in bulk polymers; most of the computational advantages of BD simulations are lost in this case.

BD simulations are a powerful technique for examining static and dynamic properties of polymers on length scales large compared to the repeat unit [9]. This is a large field almost entirely ignored in this review. Such studies do not require atomistic models and thus do not fit well with the other contributions to this volume.

We close this section with a few comments about the organization of this review. Section 2 describes the BD technique and its relationship to other computer simulation methods. Section 3 introduces methods of characterizing local segmental dynamics. Most of the results on local dynamics obtained from BD simulations are discussed in Sect. 4. Section 5 highlights issues of cooperativity in local polymer dynamics. The questions posed in this introduction will be addressed there. Although the material in Sect. 5 logically could have been included in Sect. 4, it has been treated separately because of the central importance of cooperativity in local dynamics.

2 Stochastic Simulations

2.1 Fundamentals

A useful introduction to stochastic simulations is by comparison to the more familiar molecular dynamics (MD) simulations [10, 11]. Molecular dynamics simulations use Newton's equation of motion

$$m_i \vec{a}_i = \vec{F}_i \tag{1}$$

for a collection of particles (typically atoms) with masses m_i. $\vec{F}_i$ represents the sum of all external forces experienced by particle i and is determined from the gradient of the potential energy U. Given the position and momentum of a particle at time t_1, the position at time t_2 can be determined by numerically integrating Eq. (1). Performing this operation successively creates the trajectory of the particles as a function of time. In principle, U is a function of the coordinates of all the particles in the simulation. For a study of local polymer motions in dilute solution, this would likely include one polymer chain (or one section of a chain) and hundreds of solvent molecules.

The Langevin equation is the starting point for stochastic simulations [12, 13]:

$$m_i\vec{a}_i = \vec{F}_i - \zeta_i\vec{v}_i + \vec{N}_i(t). \tag{2}$$

The two new terms (relative to Eq. (1)) describe interactions between the particles of interest and a bath. In a stochastic simulation of local polymer dynamics in dilute solution, the polymer chain is the system of interest and the solvent is the bath. Bath particles are not represented explicitly. Rather, the bath damps the motion of the particles with friction terms ζ_i and supplies stochastic forces $\vec{N}_i$ which mimic the effect of collisions between solvent molecules and the polymer. Energy is not conserved in the system; stochastic forces exchange energy between the bath and the chain. In principle, $\vec{F}_i$ should now include a hydrodynamic interaction term [14]. In practice, this term is usually neglected in simulations of local dynamics and only the intramolecular potential energy of the polymer is used to determine the force. The stochastic forces in Eq. (2) are characterized by:

$$\langle\vec{N}_i(t)\vec{N}_k(t')\rangle = 2\zeta_i k_B T\delta_{ik}\delta(t - t')\mathbf{1}. \tag{3}$$

Here, k_B is Boltzmann's constant and $\langle\ \rangle$ represents an ensemble average. The average magnitude of the stochastic forces is fixed by the simulation temperature.

In this review, we will refer to simulations which use Eq. (2) as the equation of motion as Langevin dynamics (LD) simualtions. Often the high friction or diffusive limit of Eq. (2) is used in simulations:

$$\zeta_i\vec{v}_i = \vec{F}_i + \vec{N}_i(t). \tag{4}$$

In this case the inertial term on the left side of Eq. (2) is assumed to be negligible. We will refer to this type of simulation as a Brownian dynamics (BD) simulation. It has been shown that the neglect of the inertial term has an insignificant effect on conformational transition rates for alkane chains in solution [15].

Two properties of the "solvent" in a LD or BD simulation are specified by Eqs. (2)–(4). The constant coefficient of the velocity in Eqs. (2) and (4), and the temporal delta function in Eq. (3), together imply that the solvent is purely viscous (as opposed to viscoelastic). Solvent relaxation is taken to be infinitely fast compared to any relaxation process of the particles. The spatial delta function in Eq. (3) means that the neighboring particles always experience uncorrelated "collisions". Qualitatively, these approximations are valid at long times for large and slow objects moving in a solvent of small and fast objects. When the repeat unit of a flexible polymer is the same size or smaller than a solvent molecule, this approximation needs to be closely examined. On short time scales, one would expect both temporal and spatial correlation in solvent–polymer collisions. Attempts have been made to include some of these effects using the generalized Langevin equation. This equation incorporates temporal memory effects which replace the delta function correlation in Eq. (3). We will refer to these simulations as generalized Langevin equation (GLE) simulations [16, 17].

Algorithms for integration of stochastic differential equations such as Eqs, (2) and (4) have been discussed elsewhere [14, 16, 18–23].

2.2 Brownian Dynamics Simulations in the Spectrum of Approximations

BD, LD, and GLE simulations represent a sequence of simulation strategies of increasing sophistication and computational cost. Each of these approaches uses a bath to represent the solvent in studies of local polymer dynamics in solution. In contrast, an MD simulation would explicitly model the solvent. Such a simulation would be even more time consuming (because of the increased number of degrees of freedom) but is expected to yield more realistic results. A typical BD simulation [24] of the local dynamics of a polymer in dilute solution requires the equivalent of about 5 CPU hours on a supercomputer. The corresponding MD simulation might require 5 to 50 times as much computer time. Such MD simulations have not yet been performed for the long trajectories (several nanoseconds) needed to characterize conformational dynamics but would be feasible (using a lot of CPU time!) on current computers. Of course, even with an MD simulation there is no guarantee of reproducing experimental behavior. All types of simulations use imperfect potentials. Values of potential barrier heights are not well known and few experiments provide direct measurements of them. This is a significant consideration as it is expected that conformational dynamics are quite sensitive to potential barrier heights.

An orthogonal set of approximations involves the structural representation of the polymer. As an example, Fig. 1 shows four different ways in which the structure of polyisoprene might be represented in an MD or BD simulation. The top structure (a) is an atomic representation of the polyisoprene repeat unit. The second structure (b) results from collapsing all the hydrogens onto their parent carbons (the "united atom" approximation). The third structure (c) further collapses the three carbon centers in the fairly rigid double bond unit into a single pseudo-atom. Any of these three structures might be used in a simulation of local polymer dynamics. The fourth structure (d) shows a bead-

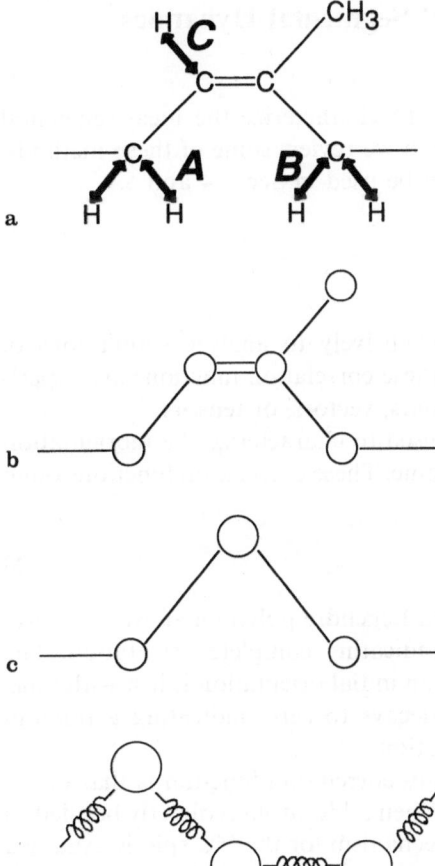

a

b

c

d

Fig. 1a–d. Various possible structural representations of polyisoprene for computer simulations. Motion of the C–H bond vectors labeled A, B, and C is detected in ^{13}C NMR experiments and will be discussed below

spring model for the polymer, with each bead representing several repeat units. At this level of representation, all differentiation on the basis of monomer structure has been lost, and local segmental dynamics cannot be studied. Such a representation is often used to study large distance scale properties of polymer chains [25].

The influence of the structural approximations shown in Fig. 1b, c on the simulation dynamics is not yet known [26]. The motivation for these approximations is two-fold. Fewer particles in the simulation leads to an increase in computational efficiency. In addition, the elimination of high frequency motions (C–H stretches) allows a larger time step to be used in the integration. It is likely that the combined effect of these factors would allow a simulation of structure c to run more than an order of magnitude faster than one of structure a.

3 Methods of Characterizing Local Segmental Dynamics

Many different methods have been used to characterize the local segmental dynamics of polymer chains. In this section we review some of these methods. The material discussed in this section will be used in Sects. 4 and 5.

3.1 Correlation Functions

Correlation functions have been used extensively to analyze simulations of polymer dynamics [24, 27, 28]. Some of these correlation functions are experimentally accessible. They can involve scalars, vectors, or tensors.

Vector autocorrelation functions are used to characterize the reorientation of some vector defined in the molecular frame. These correlation functions often have the general form:

$$CF(t) = \langle P_n(\hat{x}(0) \cdot \hat{x}(t)) \rangle. \tag{5}$$

Here, $\hat{x}(t)$ is a unit vector and P_n is the n-th Legendre polynomial. At time zero, the correlation function is equal to 1, indicating complete orientation. The correlation function decays as memory of an initial orientation is lost with time. At long times, the correlation function decays to zero, indicating a random orientation relative to the starting orientation.

One important example of a vector autocorrelation function is that measured by ^{13}C NMR experiments [29–31]. When a ^{13}C atom is directly bonded to at least one ^1H, the primary relaxation mechanism for the ^{13}C spin is usually a dipole–dipole interaction with the ^1H spin. If the ^1H spins are decoupled during the experiment, the T_1 relaxation time measured is sensitive to the decay of the correlation function shown in Eq. (5). In this case, n = 2 and $\hat{x}(t)$ is a unit vector in the direction of a particular C–H bond. This function can be written more explicitly as:

$$CF(t) = 3/2 \langle \cos^2 \theta (t) \rangle - 1/2. \tag{6}$$

Here $\theta(t)$ is the angle between the C–H bond vector at time zero and time t. This particular correlation function is referred to as a second order orientation autocorrelation function.

Often first order correlation functions (n = 1 in Eq. (5)) are also calculated from simulations. Under certain circumstances dielectric relaxation measures a simple P_1 correlation function of this form. An example is a dilute solution of a bromo-alkane in a non-polar solvent [32]. In this case, $\hat{x}(t)$ is a unit vector in the direction of the dipole moment vector (which is equivalent to the C–Br bond vector to a good approximation).

Helfand has characterized local segmental dynamics in polyethylene using orientation autocorrelation functions and a set of molecule fixed vectors [28].

These vectors are shown in Fig. 2. In the following sections, we will refer to both P_1 and P_2 correlation functions for these vectors. When we refer to structures other than polyethylene, the vectors in Fig. 2 will be defined relative to the backbone atoms only.

Tensor autocorrelation functions are measured in some experiments. A pertinent example is a coupled relaxation ^{13}C NMR experiment [16, 33–38]. In this case, Cartesian tensors fixed in the molecular frame can be used to express the results of the relaxation experiments. In a depolarized light scattering experiment, reorientation of the polarizability tensor determines the observed relaxation.

Correlation functions associated with local polymer motions almost always decay in a non-exponential manner. Often it is convenient to characterize the correlation function decay with a characteristic time. For a non-exponential decay, there are a number of possible choices for this characteristic time. Sometimes the time required for the function to decay to $1/e$ of its initial value is used. Often the time integral of the correlation function is used:

$$\tau_c = \int_0^\infty CF(t)\, dt . \tag{7}$$

In this review, we will refer to the integral of a correlation function as a correlation time τ_c. The $1/e$ relaxation time and τ_c are equal for an exponential function.

Under some circumstances, it is difficult to extract the shape of $CF(t)$ from experiments. This is the case for NMR measurements [31, 35]. However, in one limit (the extreme narrowing regime), the time integral of the correlation function (τ_c) can be measured unambiguously [31].

3.2 Conformational Transition Rates

Conformational transition rates can be obtained from computer simulations by simple counting, hazard analysis, or monitoring the decay of conformational correlation functions [39, 40]. The definition of a "transition" is somewhat arbitrary and different investigators have made different choices. One could

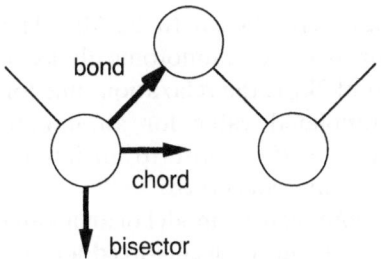

Fig. 2. Molecular fixed unit vectors used to characterize reorientation in flexible polymers. The out-of-plane vector (not shown) is orthogonal to both the bond and bisector vectors. Mathematical definitions of these vectors are given in [28]

choose a maximum in a torsional potential to be a simple dividing point between two conformational states. In this case, the transition rate would include a large number of barrier crossings which occur in small portions of the trajectory where the torsional angle is never more than a few degrees from the maximum. To avoid counting each of these multiple crossings, some investigators have defined a transition to occur the first time a potential *minimum* is encountered after a barrier crossing [24, 39]. Another reasonable choice (intermediate between the two above) would be to define the transition to occur when the torsional potential has dropped at least $k_B T$ below the barrier.

The ambiguity in the definition of a transition cannot be resolved by reference to experiment. There is no generally applicable experimental technique which measures conformational transition rates for polymers in solution. Almost all experiments measure the decay of some orientation correlation function. It is often assumed that the time between conformational transitions is proportional to the time required for decay of the correlation function (see Sect. 4.2). Statements about the time scales for conformational transitions such as the one made in the introduction to this article are not directly based on experiment.

3.3 Other Methods

Several authors have used the temporal and spatial proximity of conformational transitions to discuss mechanisms of conformational dynamics. We have recently introduced a related technique for evaluating cooperativity in local polymer dynamics. These approaches will be discussed in Sect. 5.

4 Local Polymer Dynamics as Viewed by Simulations

4.1 Molecular Weight Independence

One of the defining characteristics of *local* conformational dynamics is the molecular weight independence of these motions in the limit of high molecular weight. For long chains, the chain ends need not move in response to a conformational transition in the interior of the chain.

Fixman studied the behavior of alkane-like molecules as a function of chain length N [27]. In this study, the bond angles were chosen to be 90°. The relaxation times for the average bisector vector increased monotonically as N was increased from 6 to 25. For a barrier height of $2k_B T$, the relaxation time for N = 25 was only 20% faster than the extrapolated value for an infinite molecular weight chain. Thus, for this model, N = 25 is close to an infinite molecular weight chain insofar as local dynamics are concerned.

Xiang et al. used BD simulations to study a more realistic model of an alkane chain [16]. They calculated correlation times for various molecule-fixed tensors as chain length was increased from 5 to 27. Correlation times were calculated for both the center of the chain and the end. Surprisingly, these correlation times showed maxima between N = 15 and 20. These investigators also used the same molecular model to perform GLE simulations. Again, maxima in the relaxation times were observed for intermediate chain lengths.

NMR studies of local polymer dynamics have been performed on a number of polymers as a function of molecular weight [6]. These studies show a monotonic increase in relaxation time with molecular weight; above 30 to 100 monomer units, the local dynamics are independent of molecular weight. Experimental results for alkanes in p-xylene also indicate that the relaxation time for the central C–H vectors increases monotonically with molecular weight [41]. At N = 25, the relaxation time is about 33% faster than for high molecular weight chains. These experimental results for alkanes are in reasonable accord with Fixman's simulations. It is puzzling that the simulations with the more realistic molecular model give a different trend. It seems unlikely that the reason for this difference is the slightly different observables involved.

In BD simulations where the major interest is the local dynamics of the chains, special techniques are often used to minimize end effects. Periodic boundary conditions have been used to eliminate the effects of chain ends [28]. With this technique, the chain end-to-end distance is fixed to an average value and image chains are generated on each side of the simulated chain. A similar effect can be obtained by simulating a large ring [39]. These two approaches give the same transition rates for alkane-like chains of 200 backbone atoms [28, 39]. This result suggests that the high molecular weight limit for local dynamics has been reached in these simulations.

Results from MD simulations by Zuniga et al. support this conclusion [42]. They simulated linear and ring alkane chains in a vacuum. Transition rates were the same for linear and cyclic structures containing 100 carbons, and these rates also agreed with the result for a linear 50 carbon chain. Solvent friction plays the most important role in localizing conformational dynamics in solution; reasonable arguments indicate that the effect of inertia is negligible [15]. In a vacuum, only inertia acts to localize conformational dynamics. One would thus expect to reach transition rates characteristic of infinitely long chains at shorter chain lengths in solution than in a vacuum.

The molecular weight at which local motions become independent of molecular weight is a function of the chain structure and potentials [6]. Fixman's work shows that increasing the barrier for conformational transitions increases the chain length required to achieve asymptotic relaxation times [27].

4.2 Temperature Dependence

Another defining characteristic of local polymer motions in solution is a temperature dependence distinct from that of larger scale chain motions. Both global and local motions depend upon the viscosity of the solvent and hence the temperature dependence of the viscosity influences both types of motions. Local motions have an additional temperature dependence due to the potential barrier which is crossed in a conformational transition. Very large scale chain motions are not influenced by the torsional barrier height. As noted by de Gennes, this is a delicate point [43]. A rough explanation can be given in analogy to chemical kinetics. Transitions between conformational states occur very rapidly compared to large scale motions, establishing a pre-equilibrium; solvent friction in the large scale motion is the rate limiting step. Xiang et al. have compared the temperature dependence of local and larger scale chain motion in nonane (C_9H_{20}) [16]. They analyzed tensor correlation times and the correlation time for the end-to-end vector. Both BD and GLE studies showed that the local motions had a stronger temperature dependence, as expected.

The nature of this extra temperature dependence has been used as a key indication of the mechanism of local conformational dynamics. If two barriers were crossed simultaneously, as implied in the crankshaft mechanism [44] (see Fig. 7a, Sect. 5.1), the observed activation energy due to the local dynamics should be *twice* the barrier height. Experiments and simulations agree that this is not the case. An activation energy equal to a single barrier height has been almost universally observed. The most significant experiments on this point were performed by Morawetz and coworkers [45, 46]. They measured the rate of excimer formation in $R-Ph-CH_2-NR'-CH_2-Ph-R$. Here the R groups were either methyls or long polymer chains. These investigators found that the temperature dependence of excimer formation was the same in both experiments. Thus they concluded that the potential barrier for the conformational transition needed to form the excimer was the same in the polymer as in the small molecule.

Fixman analyzed simulations on alkane-like chains in terms of Kramers' theory [27]. This theory correctly accounted for the change in relaxation rates with barrier height, indicating an activation energy of one barrier height. Helfand showed the same conclusion more directly in simulations of polyethylene at multiple temperatures, holding the friction constant [39]. The calculated activation energy for the conformational transition rate was within 10% of a single barrier height. Helfand also calculated correlation functions for various vectors as a function of temperature [28]. We have analyzed the temperature dependence of the 1/e times for these correlation functions [47]. All four vectors have an activation energy of a single barrier height within statistical error.

In a simulation where all torsional barriers along the chain are identical, it is a simple matter to check whether the apparent activation energy is one barrier height. In more heterogeneous polymers, the results are not so easily interpreted. For example, Adolf and Ediger have simulated polyisoprene using three different torsional potentials in the chain backbone, each with at least two different barrier heights [24]. Figure 3 shows the temperature dependence of correlation times for three C–H vectors (defined in Fig. 1) and for the inverse of the three different conformational transition rates. A single temperature dependence was

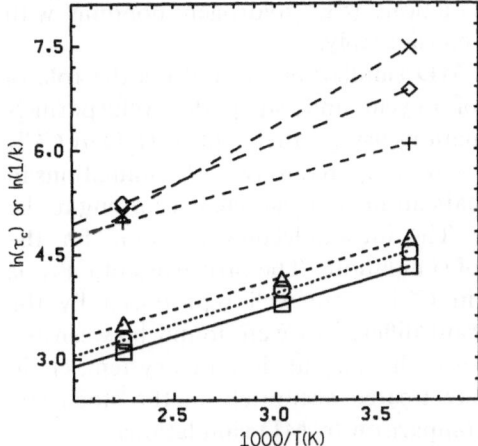

Fig. 3. Comparison of transition rates (k) for conformational transitions and P_2 correlation times (τ_c) for reorientation of C–H vectors in polyisoprene [24]. Correlation times are shown for vectors identified in Fig. 1: vector A, $\square$; vector B, $\bigcirc$; vector C, $\triangle$. Transition rates for the three carbon–carbon single bonds in polyisoprene are shown. They are identified by two letters which indicate the nearest C–H bond vectors on each side: AC, $+$; CB, $\diamond$; BA, $\times$. Both correlation times and inverse rates are given in units of ps. Estimated errors in both quantities are 10%

observed for correlation times for all C–H vectors. Surprisingly, from the same trajectories, three different activation energies are obtained for transition rates about the three different torsional potentials. (It is possible that the definition of a conformational transition influences this result – see Sect. 3.2.) This is of interest because experimentalists often assume that correlation times are proportional to the inverse of conformational transition rates. For the simulated polyisoprene system, this is not true. For this case, the calculated activation energies for the conformational transition rates cannot be easily compared to the multiple barrier heights. All of the observed activation energies for this system are somewhere between the highest and lowest barrier.

4.3 Specific Solvent/Polymer Interactions

The use of stochastic techniques in the study of local polymer motions entails certain approximations in the treatment of polymer/solvent interactions. The presence of a spatial delta function in Eq. (3) effectively implies that the solvent is being approximated as a continuum. Thus the solvent structure *per se* cannot have an effect on the simulated dynamics. Solvent properties are usually included in stochastic simulations only through the friction terms. In principle, the intramolecular potential should be modified by the solvent to produce a potential of mean force [48, 49]. In practice, this is usually not done. As discussed below, recent work has produced several examples which justify this practice in the absence of specific interactions between the polymer and solvent.

The presence of strong specific interactions (e.g., hydrogen bonding with aqueous solvent) complicates matters considerably.

Yun-Yu et al. used both LD and MD simulations to address the role of specific interactions in the dynamics of a cyclic undecapeptide, cyclosporin A (CPA) [49]. They performed MD simulations using either explicit H_2O or CCl_4 molecules as the solvent. These results were compared to two LD simulations in which the friction coefficients for the chain atoms were selected according to the characteristics of these two solvents. The intramolecular potential for the polymer was the same for both types of simulations. The properties of CPA as determined by the MD simulation in CCl_4 were well reproduced by the corresponding LD simulation. Significant differences were found between the MD and LD simulations when H_2O was the solvent. For this system, chain dynamics in the nonpolar solvent were treated accurately in the stochastic simulation, at least as judged by the comparison to MD simulations.

Two recent studies indicate the influence of strong specific interactions on the effective intramolecular potential. Depner et al. used both LD and MD to investigate the dynamics of polyethylene oxide (PEO) in water [50]. They found that the equilibrium distribution of torsional angles observed in the MD simulations differed from that predicted by the LD simulations. Widmalm and Pastor performed LD and MD simulations on one ethylene glycol molecule in water [51]. They found that the effective barrier for isomerization observed in the MD simulations was lower than that expected on the basis of intramolecular potentials alone.

Experimental work is consistent with the view that solvent specific corrections to the intramolecular potential are not required in the absence of strong polymer/solvent interactions. Glowinkowski et al. used ^{13}C NMR to study the local dynamics of polyisoprene in ten solvents as a function of temperature [31]. They measured correlation times due to the motion of different C–H vectors in the chain backbone. They found that these correlation times were determined by only the temperature and the solvent viscosity. Variables such as solvent shape, flexibility, and chemical functionality had no effect on the correlation times, except through the solvent viscosity. (Highly polar solvents were excluded from this study as they do not dissolve polyisoprene.) Similar conclusions have been reached in NMR studies of polybutadiene [52] and in time-resolved optical studies of polyisoprene [53] and polystyrene [54] with anthracene labels. NMR studies of PEO [55] have been interpreted as supporting this same conclusion [31] (except when the solvent was water [56]).

The application of stochastic simulation techniques to systems with strong specific interactions remains a challenging problem. It may be possible that stochastic simulations performed with a potential of mean force (perhaps obtained from short MD simulations) will reproduce the dynamics of simulations which explicitly model the solvent. This is by no means guaranteed. Another approach which has been utilized is the explicit modeling of a small shell of solvent molecules in a stochastic simulation [57]. Fortunately, many interesting problems concerning the dynamics of synthetic polymers do not involve strong specific interactions. These interactions are of course important in many problems involving biopolymers.

4.4 Viscous Solvent Approximation

As discussed in Sect. 2.1, the solvent is treated as purely viscous (as opposed to viscoelastic) in LD and BD simulations. This approximation will be valid when solvent relaxation times are much shorter than all relevant relaxation processes of the polymer chain. In BD simulations, an additional approximation is made by ignoring inertial effects. A consequence of these approximations is that all relaxation times in a BD simulation scale linearly with the solvent viscosity η.

Recent experimental evidence indicates that the view of the solvent as purely viscous is seriously in error where the local dynamics of flexible polymer chains are concerned. Glowinkowski et al. found that the motion of various C–H vectors in the polyisoprene backbone scaled as $\eta^{0.4}$, not the η^1 dependence expected for a purely viscous medium [31]. They argued that inertial effects could not be responsible for this discrepancy. Glowinkowski et al. also measured solvent rotation times in the same solutions used to observe the polymer dynamics. They observed similar time scales for polymer and solvent motion and attributed the sub-linear viscosity behavior to this observation. Thus conformational dynamics in polyisoprene occur on fast enough time scales that the solvent cannot be regarded as purely viscous. Studies of the isomerization of small molecules support this view [58]. Significantly sub-linear viscosity dependences have also been observed in NMR measurements on polybutadiene [2, 52] and PEO [31] and in optical measurements on labeled polyisoprene [53]. In contrast, the available data indicate that the local dynamics of polystyrene scale linearly with the solvent viscosity [2, 54]. The local dynamics of polystyrene are significantly slower than for the other polymers discussed in this paragraph. Slower polymer dynamics allow the separation of time scales between polymer and solvent motions to be re-established. This is consistent with a linear dependence of dynamics on solvent viscosity.

The viscoelastic nature of the solvent may limit the usefulness of the BD technique for the study of polymer chains with very fast dynamics (e.g., polybutadiene, polyisoprene, and PEO). At this time, it is not clear to what extent various features of the dynamics in such a BD simulation will be unrealistic as a result of the assumption of a viscous continuum. Careful comparisons between BD simulations and MD simulations will be required to address this question. Studies such as the one by Yun-Yu et al. [49] (see Sect. 4.3) could address this point if longer trajectories were run and conformational transition rates were compared. It seems likely that the BD technique will be adequate to simulate the behavior of polymer chains with relatively slow dynamics (e.g., polystyrene).

In principle, GLE or MD simulations can be used in cases where the solvent and polymer motions occur on comparable time scales. Grant and coworkers have used the GLE approach to model alkane chains [16]. Somewhat faster local dynamics were observed in the GLE simulations than in corresponding BD simulations. The GLE simulations reproduced experimental results for Cartesian tensor correlation times better than the BD work.

4.5 Effect of Repeat Unit Structure

Understanding the role of repeat unit structure in determining the local segmental dynamics of polymers is arguably the most important goal of simulations of local polymer motions. Even an understanding of this connection for polymers in solution would be a very significant step towards fundamentally understanding structure-property relationships.

The range of structures for which local conformational dynamics have been simulated is rather limited at this point. A number of workers have simulated polyethylene-like chains [27, 32, 35, 39, 42, 59, 60, 79]. For this structure, the effect of various torsion potentials have been explored [27, 39]. Recent work has added side-groups to this model [59]. Qualitatively, higher barriers and the addition of side-groups slow dynamics, as would be expected. Simplified versions of polyisoprene [24] and polybutadiene [61] (structures as shown in Fig. 1c) have also been studied. In this section we compare these simulations with each other and with experiments.

Tables 1 and 2 compare the dynamics of polyethylene, polyisoprene, and polybutadiene. The condition chosen for the comparison in Table 1 is a solvent with η of 1.6 cP at 273 K. The vectors referred to in the tables are defined in Figs. 1 and 2. As shown in Table 1, simulated correlation times for vectors sensed in NMR experiments are generally similar for the three polymers, varying by less than a factor of four. An initial reaction to these results might well be that the effect of polymer structure on the dynamics is disappointingly small. However, the structural differences between these three polymers are

Table 1. P_2 correlation times* for C–H vectors

	Simulation	Experiment
Polyisoprene		
τ_A (273 K)	81 ps[a]	92 ps[b]
τ_C/τ_A	1.41[a]	1.47[b]
τ_B/τ_A	1.18[a]	1.18[b]
Polybutadiene		
τ_A (273 K)	49 ps[c]	$\approx$ 31 ps[d]
τ_C/τ_A	1.44[c]	1.41[e]
Polyethylene		
τ (273 K)	$\approx$ 180 ps[f]	–

* ± 10%
[a] [24]; [b] [31]; [c] [61]; [d] [52]; [e] [66]; [f] [47]

Table 2. Correlation and 1/e relaxation times* (ps) for various correlation functions (425 K) [47]

	$\tau(P_2)$	$\tau(P_1)$	$\tau(P_1)^a$
Polyethylene			
Correlation Times			
Bond	39	542[b]	516
Bisector	32	48[b]	42
Chord	112	794[b]	748
Out-of-plane	32	43[b]	45
1/e Relaxation Times			
Bond	21	112	126
Bisector	12	37	30
Chord	35	280	290
Out-of-plane	11	35	32
Polyisoprene			
Correlation Times			
Bond	34	306	
Bisector	30	154	
Chord	53	528	
Out-of-plane	24	74	
1/e Relaxation Times			
Bond	25	100	
Bisector	20	66	
Chord	31	142	
Out-of-plane	17	58	

* $\pm 10\%$
[a] Calculated from [28]; [b] [24]

relatively small, and as indicated in Table 1, experiments also indicate that the differences in dynamics are not large. In contrast, experiments on polystyrene have found local dynamics more than an order of magnitude slower than for polyisoprene [2, 62]. It is not an accident that the first simulation studies of local dynamics have involved chains with quite rapid conformational dynamics; these systems require the least computational effort.

Table 1 indicates that the presence of a cis-double bond in the chain speeds up dynamics considerably (i.e., compare polybutadiene to polyethylene). Two competing effects are at work here. The double bond unit forms a "stiff link" in the chain which might be expected to slow dynamics. However, the potential barriers for rotation of the single bond next to the double bond is lowered by its presence, tending to speed up motions. These effects have been discussed by Skolnick and coworkers [63, 64].

Table 1 indicates that the ratios of correlation times for various C–H vectors in the backbone of polyisoprene and polybutadiene are well reproduced by the BD simulations. The fact that different backbone C–H vectors within the same repeat unit apparently have different dynamics might initially be regarded as surprising. Several proposals have been made to explain these ratios [65, 66]. As suggested by Monnerie and coworkers, it is likely that one important contribution is differing amounts of fast librational motions of C–H vectors at short times [66]. Figure 4 presents correlation functions decays obtained in a BD simulation for the three different C–H vectors in polyisoprene. The initial rapid drop in the correlation functions is likely due to the librational motions. The initial drop is largest for the C–H vector with the shortest correlation time. The differences in this initial drop partially account for the different correlation times of C–H vectors.

It is appropriate to include brief comments about the compilation of Table 1. The simulation correlation times for various vectors were calculated by directly integrating the appropriate P_2 correlation function. The corresponding experimental quantitites were obtained from NMR T_1 values [67]. Observed activation energies have been used to adjust some correlation times to 273 K. The comparison between simulated and actual *absolute* correlation times is problematic because of the different viscosity scaling (see Sect. 4.4). We have previously argued that a comparison at low viscosity is reasonable [24]. The experimental values correspond to a viscosity of 1.6 cP; this is a reasonable estimate of the viscosity implicit in the simulations. Experimental results for polyethylene near room temperature are not available; polyethylene crystallizes in typical organic solvents.

As this section has discussed the relationship between simulation and experiment, it is appropriate to mention work done on short alkane-like molecules (C_{27} and below) by Evans and coworkers [32], and Grant and coworkers [35]. This work has shown that BD simulations often calculate experimental observables within a factor of two. This is consistent with the results shown in Table 1. In addition, it has been shown that GLE simulations provide improved agreement with experiment [16]. The alkane chains used in

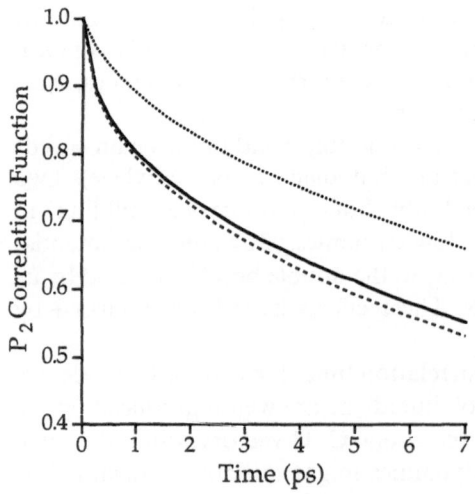

Fig. 4. Simulated P_2 autocorrelation functions for C–H bond vectors in polyisoprene (T = 425 K) [24]. The vectors are defined in Fig. 1: vector C, – – –; vector B, —— and vector A — — —. NMR experiments indicate that the correlation function decay is fastest for A and slowest for C, in good agreement with the displayed curves

these studies are too short to provide information about conformational motions of polymer chains in the high molecular weight limit.

4.6 Anisotropic Dynamics

The term "anisotropy" in the context of a discussion of dynamics can mean many different things. In this section, the term will refer to differences in the dynamics of vectors affixed to the chain backbone with various orientations relative to it. Helfand and coworkers analyzed simulations of polyethylene by calculating P_1 correlation functions for the vectors shown in Fig. 2 [28]. Correlation times and $1/e$ relaxation times calculated from the fits reported in [28] are shown in the right-most column in Table 2. The bisector and out-of-plane vectors lost orientation most rapidly and at about the same rate. As judged by the $1/e$ point of the correlation functions, the bond vector lost correlation four times slower while the chord vector was nine times slower. Although only values for 425 K are shown in Table 2, the same trends are observed at other temperatures. These results are consistent with the notion that motions perpendicular to the general direction of the backbone occur more rapidly than those along this backbone direction. The same trends were observed in NMR experiments and simulations (both BD and GLE) on nonane by Grant and coworkers [16, 35].

Some simulations have been done which investigate the effect of repeat unit structure on the anisotropy of the local dynamics. Darinskii et al. used BD simulations to study the effect of a side group (attached on alternate carbons) on the anisotropy of motion in a freely rotating alkane chain [59]. The ratio of P_1 relaxation times for chord and bisector vectors in the absence of side groups agree closely with the polyethylene results shown in Table 2. The addition of side groups decreased this ratio from nine to three.

Adolf and Ediger examined the behavior of various correlation functions in their BD simulations of polyisoprene [24] and polyethylene [47]. These results are also shown in Table 2. As expected, their polyethylene results matched Helfand's within statistical error [24, 68]. The results for polyisoprene were significantly different than those for polyethylene, however. While the same general trend in terms of $1/e$ times was observed in polyisoprene, it was found that the bond vector decayed only 1.4 times slower than the out-of-plane and bisector vectors. The chord vector decayed only 2.2 times slower than the two fast vectors. Given the results of Darinskii et al., the presence of a side group in polyisoprene may be responsible for these results (although the side group was only present in the simulation through an increased friction constant for the appropriate backbone atom). It seems likely that different torsional potentials for polyethylene and polyisoprene also play a role. Further work would be required to disentangle these factors.

Roe and coworkers have examined the anisotropy of motion for poly-ethylene chains in the melt using MD simulations [69]. Simulations with both a freely rotating chain and realistic torsional potentials were performed. They observed the same general trends as those discussed above. The chord vector decayed much more slowly than the bisector or out-of-plane vectors. They point out, however, that the ratio of these decay times in the MD simulations were more than 200, as opposed to the factor of 9 discussed above. Roe and coworkers ascribe this difference between the MD and BD simulations to the constraining effect of surrounding chains in the former. This explanation was supported by an additional MD simulation in which non-bonded interactions were eliminated; in this simulation the ratio of chord to bisector $1/e$ times fell to 1.7. It is not clear how to interpret this ratio relative to that observed in the BD simulations.

4.7 Short Time Dynamics

As discussed in Sect. 4.5, librational motions are believed to play an important role in the short time dynamics of polymer chains. Figure 4 shows some evidence of fast initial drops in the correlation functions for different C–H vectors in polyisoprene. These fast drops have been attributed to librational motions. Adolf and Ediger have performed a more direct analysis of these motions using the same polyisoprene simulation [24]. This is shown in Fig. 5. The evolution of the distribution of dihedral angles with time is examined for one of the single bonds adjacent to a double bond in polyisoprene. The ordinate presents the probability of a given change in angle during a certain time after an arbitrary starting time. Within one ps, the dihedral angles have already spread into a cone with a 50° FWHM. This feature of the dihedral angle distribution changes little as time evolves. With increasing time a new peak centered at 180° grows, indicating transitions to the other conformational state (the potential is two-fold). The two features shown in Fig. 5 indicate dynamics on two very different time scales. Fast motions within potential wells reach their equilibrium distribution before a significant number of conformational transitions occur. This is consistent with experimental inferences about librations [66, 67].

It is important to note that the approximations made in the derivation of Eqs. (2) and (4) are questionable for very short times. As a result, it is expected that short time dynamics from LD and BD simulations are likely to contain significant errors. For example, the equation of motion for BD simulations (Eq. (4)) neglects inertial terms. Because of this, all motions are damped by the same friction coefficient and must scale exactly with the viscosity. This is, of course, inconsistent with the experimental observation that high frequency motions such as bond vibrations are largely independent of the molecular environment. Although LD simulations retain the inertial term, spatial and temporal correlations in the stochastic forces at short times are neglected. Even GLE simulations, which contain some of these correlations, would not be expected to be quantitatively correct at short times.

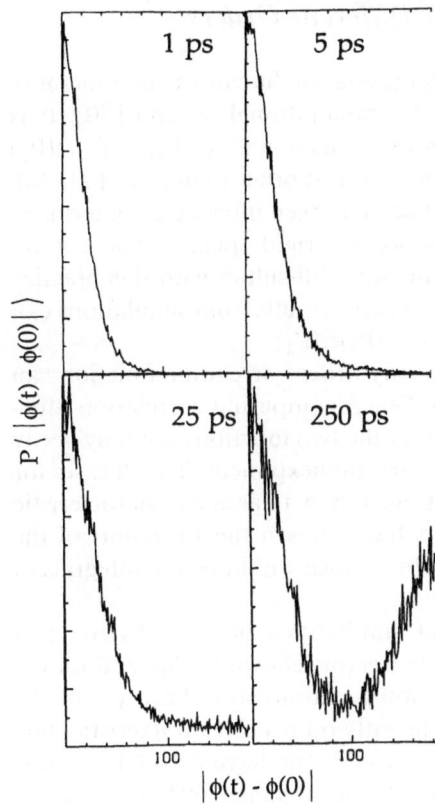

$P\langle\,|\phi(t) - \phi(0)|\,\rangle$

1 ps   5 ps

25 ps   250 ps

$|\phi(t) - \phi(0)|$

Fig. 5. Evolution of the distribution of dihedral angles for the single bond between vectors C and B in polyisoprene (see Fig. 1 for definition of vectors) [24]. Motions with a FWHM of 50° take place in less than one ps from an arbitrary starting time. Conformational transitions begin to build in at longer times

The importance of errors in the short time dynamics can be explained with reference to Figs. 4 and 5. The initial drop in the correlation functions shown in Fig. 4 would likely occur on a faster time scale in a more realistic simulation. These high frequency motions are too severely damped by the BD equation of motion. Likewise, it is expected that the separation of time scales shown in Fig. 5 is even more striking in reality. NMR experiments on polyisoprene support this conclusion. They indicate that motion on two time scales contribute to the correlation function decay, in qualitative agreement with the simulation results [67]. The separation of these two time scales, however, is larger in the experiments than in the simulations.

Errors in the short time dynamics of a stochastic simulation can be addressed by comparison to MD simulations, which are expected to be more realistic. This comparison has been performed for ethylene glycol in water by Widmalm and Pastor [51]. Xiang et al. compared GLE and BD simulations of nonane [16]. Each of these investigations indicates that the initial drop in the correlation function occurs more rapidly in the more sophisticated simulation. After this initial drop, correlation functions calculated from the various simulations decay at about the same rate.

4.8 Ratios of Correlations Times of Different Orders

The ratios of correlation times for P_1 and P_2 correlation functions are sometimes used as an indication of the mechanism of reorientational motion [70]. It is known that small step rotational diffusion of a sphere will lead to $\tau(P_1)/\tau(P_2)$ = 3. Large angle jumps of a sphere will lead to ratios other than three [71, 72]. Sometimes $\tau(P_1)/\tau(P_2)$ for flexible chain systems has been interpreted in terms of angular jump size by using models developed for rigid spheres. There is no rigorous justification for this procedure. Potential difficulties with this practice have been pointed out [32, 69]. As discussed below, results from simulations can provide some insight into the significance of $\tau(P_1)/\tau(P_2)$.

Since a complete trajectory is available, any order correlation function can be calculated from a computer simulation. When comparing correlation functions of different orders, it is usually found that the two functions not only decay on different time scales but have different shapes (nonexponential in all cases for flexible molecules). Thus an ambiguity exists as to how to assign a characteristic decay time to each function. Often authors have chosen the $1/e$ points of the functions for convenience. Another reasonable choice would be the integrals of the correlation functions.

Table 3 presents $\tau(P_1)/\tau(P_2)$ from BD simulations of polyethylene and polyisoprene. The ratios are presented for the vectors shown in Fig. 2. Both the $1/e$ points and the integrals of the correlation functions have been used. The results indicate that the ratio depends significantly on which characteristic time is chosen and varies from 1.4 to 14. Takeuchi and Roe have calculated ratios from $1/e$ times from their MD simulations of polyethylene [69]. Their ratios varied from 2 to almost 100.

What determines $\tau(P_1)/\tau(P_2)$ for a vector attached to a flexible polymer chain? Three important factors are present in this problem which are not present in the rotation of a sphere. First, intramolecular potentials in flexible polymers

Table 3. Ratios of characteristic decay times for P_1 and P_2 correlation functions[a]

	$1/e$ time	Integral
Polyethylene		
Out-of-plane	3.2	1.4
Bisector	3.5	1.5
Bond	5.9	14
Chord	8.6	7
Polyisoprene		
Out-of-plane	3.7	3.1
Bisector	3.6	5
Bond	4.1	9
Chord	4.9	10

[a] Calculated from Table 2

make possible rotational motion on two different time scales [66, 67]. On short
time scales it is likely that librational motion within the wells of the torsional
potentials cause a fast initial loss of orientation (see Sect. 4.7 and Fig. 4). On
longer time scales, conformational transitions facilitate complete loss of orientation. Since higher order correlation functions emphasize smaller angular displacements, this will naturally lead to $\tau(P_1)/\tau(P_2)$ greater than three. Second,
intermolecular potentials (e.g., surrounding chains) could cause a similar effect
as the one described for intramolecular potentials above [69]. Third, it is
unlikely that small sections of polymer chains can be approximated as spheres.
Ellipsoidal bodies which are completely rigid have $\tau(P_1)/\tau(P_2)$ unequal to three
even when diffusion takes place in a continuum by small steps [47]. These
factors indicate problems in interpreting $\tau(P_1)/\tau(P_2)$ solely in terms of jump
angles.

4.9 Constrained Bond Lengths and Angles

For reasons of computational efficiency, constrained bond lengths and bond
angles are sometimes used in stochastic and MD simulations. The reason for
this approximation is apparent if the time scales for various types of motions in
a polymer chain are compared. In a typical hydrocarbon chain, bond length
changes occur on time scales of about 30 fs, bond angle changes on time scales of
about 100 fs, and conformational transitions on time scales of 50 ps or longer.
The time step in a simulation must be chosen small enough to integrate
accurately the fastest motions. Usually a step size of about 1 fs is chosen if bond
length fluctuations are allowed. If conformational transitions are of interest,
a 1000 ps or longer trajectory may be required to generate adequate statistics.
Thus millions of time steps are required if no additional approximations are
made.

One might think that high frequency, low amplitude motions such as bond
length and bond angle fluctuations would have a negligible effect on conformational dynamics. Certainly there are significant computational advantages to
replacing these high frequency degrees of freedom with rigid constraints.
Simulations can be performed with time steps at least an order of magnitude
larger than in the unconstrained case [14, 59]. Interestingly, while it appears
that bond lengths can be constrained without perturbing the conformational
dynamics this is not the case for constraints on bond angles. The computation
savings from fixing only bond lengths is roughly a factor of four [23, 73].

The effects of constraints on *static* properties of chain molecules (e.g., the
equilibrium distribution of torsional angles about a particular bond) have been
carefully considered [14, 74, 75]. The introduction of fixed bond lengths or
angles requires the introduction of an additional potential term (the so-called
"Fixman potential") in the equation of motion in order to compensate rigorously for the effect of the constraints. It has usually been argued that the

compensating potential term is small and can be ignored [14, 73]. Pear and Weiner implemented this correction in a BD simulation of a freely rotating chain with fixed bond lengths and angles [60]. In the absence of the compensating potential, torsional angles were not evenly populated. The addition of the compensating potential produced torsional distributions which were reasonably uniform.

Even if the Fixman potential is used to correct the static properties of a chain with constrained bond lengths and angles, it does not follow that the *dynamics* of an unconstrained chain will necessarily be reproduced [76]. Work by Helfand and coworkers indicates that changing the bond angle force constant has a major effect on conformation transition rates [76]. Stiffer bond angle force constants cause lower transition rates. In contrast, changes in the bond length force constant have a minor effect on the transition rate. Thus constrained bond lengths (with the compensating potential) would be expected to yield the correct conformational dynamics, while constrained bond angles would not. Work by van Gunsteren and Karplus supports this conclusion [73]. They performed MD simulations of a protein for three cases: no constraints, bond length constraints only, and both length and angle constraints. The first two cases yielded reasonably similar dynamics while the third case showed significant deviations. Unfortunately, these simulations were not long enough to evaluate unequivocally the effect of the constraints on conformational transition rates.

The issue of constrained degrees of freedom in computer simulations is more than simply a technical point of optimizing algorithms. The results above contain physical insights about the nature of conformational transitions in chain molecules [76]. Fluctuations in bond angles facilitate conformational transitions. One might object that the time scales for the two motions are too different for this coupling to be efficient. Although the average time between conformational transitions is very long compared to the period of a bending motion, the critical time during a conformational transition is the time it takes to move across the top of the potential barrier (from $\approx k_B T$ below the barrier on one side to the same level on the other side). This critical barrier crossing time may be of the order of 1 ps. During this short time, a favorable fluctuation of a bond angle may significantly enhance the probability of a transition.

It is appropriate to mention in this section two other strategies for improving computational efficiency. Some investigators have used artifically loose potentials in order to be able to use a larger time step [24, 28, 39]. In a simulation of polyisoprene, Adolf and Ediger showed that using bond length and angle force constants three times smaller than realistic values introduced only a 20% decrease in absolute correlation times [24]. Fixman has introduced an implicit algorithm for BD simulations which allows large time steps and treats vibrational motion in an approximate manner [20]. In a simulation of an 11 carbon chain, the implicit algorithm was six times faster than the usual algorithm; no significant differences between the dynamics were reported for the two approaches.

4.10 Neglected Interactions

As mentioned in Sect. 1, hydrodynamic interactions (HI) are formally included as a part of the forces appearing in the equations of motion for BD or LD simulations. In practice, HI are not usually included in simulations where conformational dynamics are the major concern. As pointed out by Depner et al., the neglect of HI implies that translational and rotational motion of the entire chain will have the wrong dependence on chain molecular weight [50]. Neglect of HI is inherently inconsistent with the approximately non-draining behavior of polymers in dilute solution. Several authors have argued that this neglect should not have a significant effect on local polymer motions. Fixman tested this in a simulation of an 11 carbon chain. Correlation functions were calculated for the three outermost bisector vectors on the chain, both with and without HI. No difference (outside of statistical noise) was detected for the third bisector from the end. The outermost bisector correlation function decayed about 50% slower with HI. Thus it seems likely that the inclusion of HI would not significantly change conformational dynamics in the center of the chain.

Two other approximations have sometimes been made in stochastic simulations of local polymer dynamics. Excluded volume interactions are sometimes ignored for atoms separated by more than three carbons along the backbone [24, 39, 59], effectively simulating the motions of a "phantom chain." Independent torsional potentials have often been utilized for computational simplicity [24, 39, 59], even though it is well known that this approximation is unsuitable for calculating static properties of polymer chains. A careful evaluation of the errors associated with these approximations has not been made to our knowledge.

4.11 Relationship to Models

The local dynamics of a polymer chain with realistic intramolecular potentials can be viewed from two perspectives [8]. One perspective regards the addition of realistic potentials as a perturbation to a bead-spring (or Gaussian) model of a linear chain [20, 59, 77]. This approach concentrates on long length scale motion and compares the simulation results to normal mode calculations such as the one by Rouse. The second perspective regards conformational transitions as the essential feature of local polymer dynamics [32, 78, 79]. Since there are no conformational transitions in bead-spring models, models [80, 81] derived from a consideration of local chain structures and potentials are utilized. Both approaches have provided interesting results (and are sometimes both used in a single investigation). In general, investigators taking the second approach have used more realistic models. Investigators taking the first approach perform longer simulation runs in order to capture correctly large length scale and long time dynamics.

Fixman has compared relaxation rates for the normal modes of a Rouse chain to those obtained from BD simulations [27]. The results are shown in Fig. 6. The simulated chain had 90° bond angles and a three-fold torsional potential barrier ($E_b/k_BT = 0$, 2, or 4). As shown, the longest length scale motions (smallest k) correspond reasonably well to the Rouse model. Deviations for shorter length scales are more substantial as the barrier height increases. Higher barriers cause the dynamics to be dominated by local conformational transitions at large k. It is expected that more small k modes would follow the Rouse predictions if longer chains were simulated. Consistent with the discussion in Sect. 4.2, the small k modes are essentially independent of barrier height.

Qualitatively similar behavior was seen by Darinskii et al. [59]. They performed BD simulations of a freely rotating chain with tetrahedral bond angles. This basic model resulted in mode relaxation rates similar to those shown in Fig. 6 in the absence of a barrier. Darinskii et al. also performed simulations for the same model with side groups added on alternating carbons. For this case the results were more similar to those in Fig. 6 with a barrier of $4\,k_BT$. Thus alterations of the potential or the structure which tend to slow local motions have similar effects on mode relaxation rates.

A number of theoretical models have been derived specifically for local chain motions [82]. These models have been useful in stimulating thinking about the underlying mechanisms for local dynamics, and about how these mechanisms influence the shape of the correlation function. For example, it is now understood that non-exponential correlation functions can result from a variety of sources, including specific cooperative motions. Any model which resembles a one-dimensional diffusion equation will have a non-exponential decay [81].

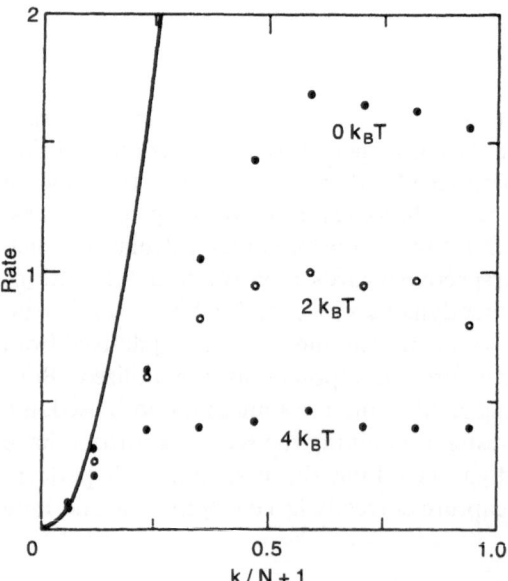

Fig. 6. The vector mode relaxation rates as a function of $k/(N + 1)$. Results shown are from BD simulations of a 16 carbon alkane chain with the indicated torsional barrier height. The line indicates the prediction of the Rouse model. (Reproduced from [27] with permission)

A model of local polymer dynamics can be useful in a predictive sense in at least two different ways. First, it could provide an absolute, accurate prediction of some dynamical property based on structure and potentials. None of the current models have been shown to have this capability, although some approaches may eventually attain this result [77, 83]. Most of the current models are based on quite idealized pictures of polymer chains. Realistic structures and potentials (which differ from chain to chain) are generally not included in the models. Almost all the models contain more than one free fitting parameter, further blurring the distinctions between them. It has been shown previously that two such models [4, 84] with very different underlying assumptions can have very similar shapes when the fitting parameters are freely varied. Thus a good fit to a simulated correlation function cannot be taken as evidence that the physical picture underlying the model is correct.

The second manner in which a model can provide some predictive power is if it can predict multiple observables. The Hall-Helfand model [80] for conformational dynamics contains two adjustable parameters. It has been shown, however, that if these parameters are adjusted to fit the conformational autocorrelation function from BD simulations, the same parameters can reasonably predict conformational cross correlation functions [28].

5 Cooperativity in Conformational Transitions

One of the key concepts for understanding local dynamics in polymers is cooperativity. It has long been recognized that individual conformational transitions cannot occur in the middle of a polymer chain if other degrees of freedom do not cooperate. A significant portion of the chain would have to swing rigidly in order for this to occur. Even in solution, the friction associated with this type of motion would be enormous. If this were the normal pathway for conformational transitions they could not occur on the nanosecond time scale and would not be independent of molecular weight. Somehow other degrees of freedom must cooperate with the rotating bond in order to localize the portion of the chain which is required to move.

Computer simulations provide unique insights into the mechanism of cooperativity in local polymer dynamics. Since the positions of all atoms are known at all times, any aspect of cooperativity can be examined if the right question can be formulated. Experiments are not nearly as flexible, and, as a result, many interesting questions about conformational dynamics cannot be directly addressed experimentally. Of course, the microscopic insights into the mechanisms of local dynamics which are extracted from a simulation should only be regarded as trustworthy to the extent that the simulation is realistic. This question of realism has not been adequately addressed in some of the literature. While one would hope to be able to extract general features from a simulation even in the absence of quantitative agreement with experiment, these insights must be regarded with some suspicion.

5.1 Cooperative Transition Pairs

One explanation of the cooperativity of conformational transitions postulates that conformational transitions occur in pairs. The two transitions involved occur nearby in space and time. In this manner, the ends of chains are not required to move. Figure 7 shows several proposals of this type. Figure 7a shows the crankshaft motion proposed by Schatzki [44]. Figure 7b, c shows two "crank-like" transitions proposed by Helfand [78, 85]. The crankshaft motion occurs about two parallel, collinear bonds. Thus the chain ends can remain precisely stationary. As discussed in Sect. 4.2, the crankshaft motion would result in an activation energy of two barrier heights. This is inconsistent with

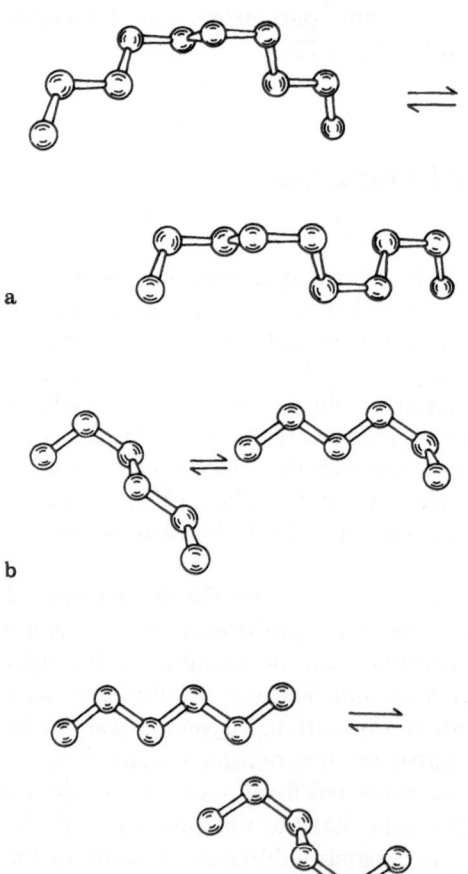

a

b

c

Fig. 7a–c. Schematic representations of possible mechanisms for conformational transitions. Sections of polyethylene chains are displayed with only carbon atoms shown explicitly. (Adapted from [85])

experiments and BD simulations. The crank-like motion occurs about parallel, non-collinear bonds. Because the two barriers are not necessarily crossed simultaneously, crank-like motions are consistent with the experimentally observed temperature dependence. In this type of motion, a translation of a portion of one chain end relative to the other is required.

Helfand and coworkers analyzed BD simulations of polyethylene in terms of cooperative transition pairs [39]. They found that, after a conformational transition occurred in polyethylene, a second transition occurred nearby along the chain within a very short time (3 ps at 425 K) in 29% of cases. The second transition always occurred about the second-neighbor torsion, relative to the initial transition. This second-neighbor coupling effect had been predicted previously by Helfand [85]. Almost all of the cooperative transitions were the motions shown in Fig. 7b, c. 71% of all the transitions occurred without a closely following transition. We will refer to these as "isolated transitions."

Zuniga et al. performed fully atomistic MD simulations of isolated polyethylene chains in a vacuum [42]. Chains of 50 and 100 carbon atoms were investigated. The results of their study were in qualitative agreement with Helfand's BD simulations of polyethylene. While cooperative transition pairs most often involved second-neighbors, overall most of the transitions were isolated.

Adolf and Ediger performed a similar analysis of cooperative transition pairs in a BD simulation of polyisoprene [24]. A total of 7% of the transitions occurred as cooperative pairs involving second-neighbors. An additional 13% of the transitions were cooperative with respect to first-neighbors. Most transitions occurred without an accompanying transition.

The above studies indicate that many conformational transitions occur as isolated transitions. For these motions, schematic representations like those shown in Fig. 7 are completely inadequate. Such representations assume that the rotational isomeric states (RIS) provide a reasonable basis set for understanding conformational dynamics. The observation that many transitions occur as isolated transitions cannot be explained within the RIS framework. In other words, many conformational transitions cannot be explained in terms of "cartoon pictures" like Fig. 7. More general methods for discussing cooperativity are examined in the next section.

We close this section with a caveat. The above discussion applies to random coil polymers. Polymers in ordered environments such as channels or crystals might be expected to have quite different dynamics [86–91]. In such cases, the RIS description of conformational transitions may be adequate.

5.2 More General Views of Cooperativity

All conformational transitions are cooperative in a certain sense. Even if only one conformational state changes in a given motion (an isolated transition), other degrees of freedom must adjust to localize the transition [78]. Which degrees of freedom adjust in order to accommodate the transforming bond? Over what length scales do these adjustments occur?

Adolf and Ediger recently introduced a new technique for addressing these general questions of cooperativity [79]. Cooperativity in conformational transitions is examined by taking a snapshot of the atomic positions on a section of a chain just before and just after a conformational transition. By averaging over thousands of transitions, the average atomic displacements accompanying a conformational transition can be determined. Mathematically, the quantity

$$\Delta \vec{r}_i = \langle |\vec{r}_i(t = \tau_{trans} + \Delta t) - \vec{r}_i(t = \tau_{trans} - \Delta t)| \rangle \tag{8}$$

is calculated. Here $\vec{r}_i$ represents the coordinates of the i-th atom along the chain. The index "i" is the position of an atom along the chain contour relative to the position of the transforming bond. τ_{trans} is the time when the transforming bond crosses the torsional potential barrier and Δt is a variable which determines how near the transition state the snapshots are taken. Typically Δt is about 0.5 ps (the average time between transitions for polyethylene at 425 K is 55 ps).

The solid lines in Figs. 8 and 9 show Eq. (8) for polyethylene and polyisoprene (T = 425 K). As expected, the two atoms which define the transforming bond show the greatest change in position during the short time interval about τ_{trans}. The atomic position distortion drops off quickly as atoms farther from the transforming bond are considered. Six atoms along the chain contour from the transforming bond, the displacement is no greater than 20 atoms away. This baseline reflects thermal motions which occur in any 1 ps interval regardless of whether a transition has occurred nearby.

An analysis similar to Eq. (8) can be carried out for any degree of freedom in the simulation [24]. Figure 10 shows the average changes in torsional angles

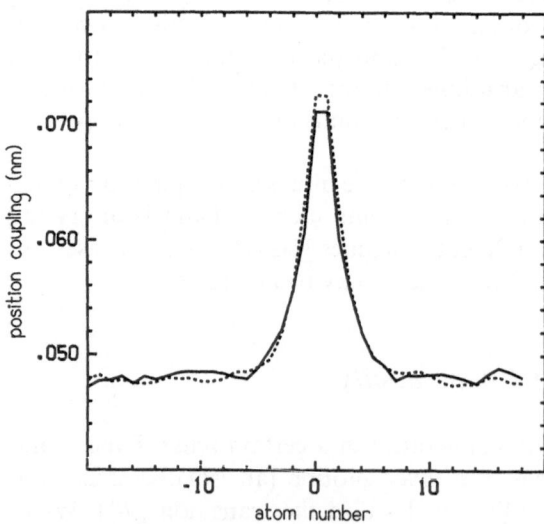

Fig. 8. Distortion of atomic positions accompanying a conformational transition in simulated polyethylene (Eq. (8)): ——, all transitions; – – –, cooperative transitions pairs only [79]. The conformational transition occurs between atoms 0 and 1. In each case, the distortion is limited to a few repeat units

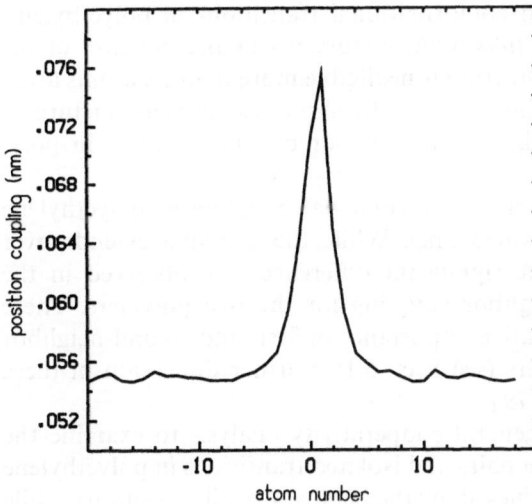

Fig. 9. Distortion of atomic positions accompanying a conformational transition in simulated polyisoprene (Eq. (8)) [24]. The conformational transition occurs between atoms 0 and 1. The distortion is limited to a few repeat units

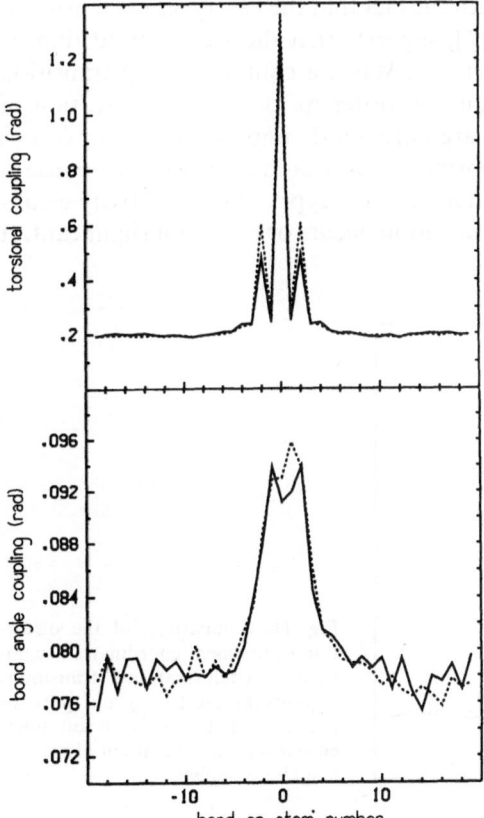

Fig. 10. Distortions of torsion angles (*top panel*) and bond angles (*bottom panel*) accompanying a conformational transition in simulated polyethylene: ——, all transitions; ---, cooperative transitions [79]. For the torsional modes, the transition occurs at bond 0; for the bond angles the transition occurs between atoms 0 and 1. The distortions are limited to a few repeat units. The bond angle coupling is small in magnitude (about 1°)

and bond angles associated with conformational transitions in polyethylene. Both panels of the figure show maximum distortions at the position of the conformational transition; the distortion is negligible more than six atoms away from the transforming bond. Thus Figs. 8–10 show a consistent picture of conformational transitions localized to about four backbone atoms in polyethylene and polyisoprene.

Figure 11 shows a comparison of the torsional coupling in polyethylene (solid line) and polyisoprene (dashed line). While the coupling extends over generally the same length scales, significant differences are observed in the amount of first- and second-neighbor coupling for the two polymers. These results are consistent with the relative importance of first- and second-neighbor transitions for these two polymers (see Sect. 5.1). Further discussion of these differences are available [63, 64, 79].

Adolf and Ediger used this general cooperativity analysis to examine the behavior of cooperative transition pairs and isolated transitions in polyethylene [79]. In Figs. 8 and 10, the solid lines show the behavior of all transitions, while the dashed line shows the behavior of only the cooperative transition pairs. Surprisingly, very little difference is observed. These figures indicate that very similar motions are involved whether transitions occur in an isolated fashion or as cooperative pairs. This, together with the fact that the two types of transitions have the same activation energies [92], suggests that the second transition be regarded as an event of minor importance. When a conformational transition occurs, many degrees of freedom adjust in order to localize the distortion of atomic coordinates. Torsional angles are particularly important in this process. Some fraction of the time a neighboring torsion adjusts enough to cause a second conformational transition. In terms of the types of cooperativity examined in Figs. 8–11, whether a second transition occurs or not is not significant. It

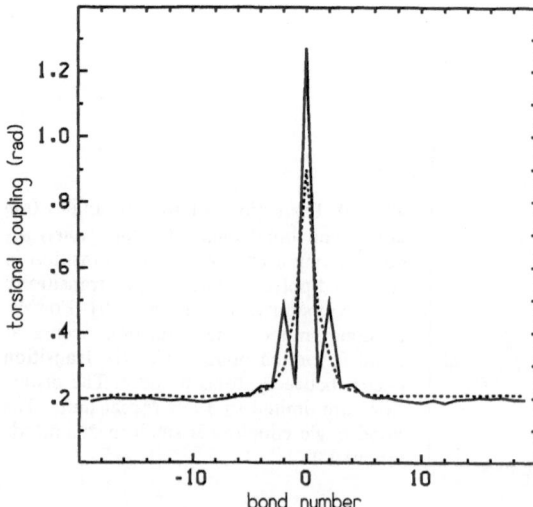

Fig. 11. Comparison of the distortion of torsional coordinates accompanying conformational transitions in polyethylene (——) and polyisoprene (– – –) [19]. A significant difference is seen in the amount of second-neighbor coupling

may provide more insight to view the displacements of the neighboring torsional angles as a continuous process instead of involving only a small number of RIS. Figure 5 provides a view of the type of torsional adjustment likely to be important in localizing conformational transitions on fast time scales. As discussed in Sect. 4.7, torsional adjustments of more than $25°$ can occur in one ps.

Recently Bahar et al. examined the motion of a freely rotating chain with tetrahedral bond angles using a new analytical approach [93, 94]. With this technique, a given dihedral angle in a short chain is forced to change while the response of the atoms on the chain is constrained so that the sum of squared atomic displacements is minimized. These authors averaged over a large number of starting configurations and calculated quantities similar to those displayed in Figs. 8 and 11. Since this is a zero temperature technique, the calculated quantities did not contain a thermal background like the one present in the figures. However, with this exception, the results of their calculations were in qualitative agreement with those shown in Figs. 8 and 11. Some ambiguity clouds this comparison since these results were obtained on different systems using techniques with different assumptions. Nevertheless, the similarity of the results possibly indicates that the height of the torsional barrier does not play an important role in determining the length scale of localization for conformational transitions.

The conclusion that conformational transitions in polyethylene and polyisoprene are localized to about four carbon atoms along the backbone is qualitatively supported from two additional sources. Fixman performed BD simulations of alkane-like chains with fixed $90°$ bond angles [27]. He observed that the conformational transition rates in a long chain were only 2–3 times smaller than those expected for an isolated four atom molecule. He interpreted this effect as meaning that only a few extra atoms (in addition to the four atom segment) were involved in a conformational transition in the middle of a long chain. Results from dielectric [95] and optical [96] experiments have been interpreted as indicating that the fundamental kinetic unit for motions in polyisoprene is somewhere in the range of one to three repeat units.

We conclude this section with two comments. First, the type of analysis in this section is more general than schemes based on the RIS approximation. Yet even with this approach, the event which draws our attention is a transition from one RIS to another. It may be possible to devise a scheme of characterizing the length scale of cooperativity in local motions without utilizing conformational transitions. The motivation for this is threefold: conformational transitions are not generally experimentally observable, they are difficult to define unambiguously, and in different chains different angular rotations are required to accomplish a transition. Perhaps the decay of cross-correlation functions of local vectors should be used. A related method has been used by Bahar et al. [94].

The second comment involves a new method for calculating chain dynamics known as the dynamic rotational isomeric state (DRIS) approach [97–100].

This technique is the dynamic analogue of the powerful RIS theory for calculating equilibrium properties. Since much of the computation is performed analytically, DRIS calculations can be performed for much longer times than a stochastic or MD simulation. The question is this: to what extent can a DRIS calculation correctly reproduce the *local* dynamics of a polymer chain? Zuniga et al. have argued that aspects of their MD results can be well accounted for by a model which completely ignored cooperativity [42]. This argument might indicate that the DRIS approach would be an accurate description of local dynamics. In contrast, the results discussed in this section have emphasized the importance of slight adjustments in torsional and bending coordinates in localizing conformational transitions. These degrees of freedom are missing in the DRIS approach. It is an open question whether features such as those shown in Figs. 8 and 9 can be correctly calculated within the DRIS framework.

5.3 Possible Connection Between Cooperativity and T_g

The length scale over which conformational transitions are localized is an interesting issue from a fundamental perspective. It is possible that this length scale is also related to important macroscopic properties of polymers such as the glass transition temperature T_g. The relationship between T_g and polymer structure is one of long standing interest in polymer science. It is not currently understood at a fundamental level.

The possible connection between the length scale of a conformational transition and T_g can be stated as follows. T_g is independent of molecular weight (for high molecular weights) and clearly a kinetic phenomenon as observed in the laboratory. Thus the time scales of local dynamics are certainly central to the glass transition. In the broadest terms, one expects that the time required for a conformational transition in a bulk polymer would depend upon some sort of an activation energy and upon the size of the unit which must move. The size factor would depend both upon the volume per repeat unit and upon the length scale over which cooperative motions are localized. Thus one possible contributing factor in the difference in T_gs for two different polymeric materials is that they require cooperativity on different length scales in order to accomplish a conformational transition. Computer simulations provide a way to characterize the mechanism of local conformational transitions in different polymers, and thus can assist in testing this picture.

6 Future Directions

This review has illustrated the important insights into the conformational dynamics of polymer chains which have been obtained from stochastic simulations. Stochastic simulations occupy an interesting position relative to completely atomistic MD simulations. MD simulations will generally be more accurate and complete. Yet, for the foreseeable future, there will be systems of interest which can only be simulated with the computationally more efficient stochastic approach.

An important goal for the near future is to understand under what conditions a stochastic simulation will provide a reasonably realistic description of local segmental dynamics. This will depend both on the question asked and the polymer simulated. Polymer chains with large, rigid side groups such as polystyrene will probably be better approximated using stochastic methods than very flexible polymers with small side groups such as polyisoprene or polybutadiene. For very flexible polymers, it is possible that stochastic methods will accurately describe the relative time scales of various local processes even if some error is expected in absolute time scales. Questions about the structural representation of polymers in simulations also need to be addressed. Systematic comparisons will reveal the circumstances in which united atom approximations can be safely utilized.

The question of realism has not been adequately addressed by those performing either stochastic or MD simulations. This issue should be addressed primarily by comparison to experiments. Careful comparisons of various simulation techniques using the same potentials can also play a role.

Computer simulations will play an important role in furthering our fundamental understanding of structure/property relationships. In order for this to occur, realistic simulations of a wider variety of polymer structures are needed on time scales long enough to observe conformational transitions.

Acknowledgements. Our work with BD simulations has been supported by the National Science Foundation (DMR Polymers Program, DMR-9123238). MDE acknowledges many helpful conversations with Eugene Helfand over several years. We thank Dan Gisser, Neil Moe, and Juan de Pablo for their careful reading of this review. MDE thanks the Alfred P. Sloan Foundation for a research fellowship. Figures 3–5 and 8–11 are reprinted with permission from the American Chemical Society. Figures 6 and 7 are reprinted with permission from the American Institute of Physics.

References

1. Amelar S, Krahn JR, Hermann KC, Morris RL, Lodge TP (1991) Spectrochim Acta Rev 14: 379
2. Gisser DJ, Ediger MD (1992) Macromolecules 25: 1284
3. Lodge TP (1993) J Phys Chem 97: 1480
4. Ediger MD (1991) Ann Rev Phys Chem 42: 225
5. Heatley F (1986) Ann Rep NMR Spect 17: 179
6. Heatley F (1979) Prog NMR Spectroscopy 13: 47
7. Helfand E (1985) J Polym Sci: Polym Symp 73: 39
8. Stockmayer WH (1973) Pure Appl Chem, Suppl Macromol Chem 8: 379
9. See for example: Chirico G, Langowski J (1992) Macromolecules 25: 769; Bitsanis I, Davis HT, Tirrell M (1990) Macromolecules 23: 1157; Cascales JJ, Navarro S, Garcia de la Torre J (1992) Macromolecules 25: 3574; Allison SA, Nambi P (1992) Macromolecules 25: 759; Fixman M (1991) J Chem Phys 95: 1410

10. See other contributions in this volume
11. van Gunsteren WF, Berendsen HJC (1990) Angew Chem — Int Ed Eng 29: 992
12. McQuarrie DA (1976) Statistical mechanics. Harper Collins, New York, Chapter 20.
13. Allen MP, Tildesley DJ (1987) Computer simulation of liquids. Oxford University Press, Oxford. Chapter 9.
14. Fixman M (1978) J Chem Phys 69: 1527
15. Pastor RW, Karplus M (1989) J Chem Phys 91: 211
16. Xiang T, Liu F, Grant DM (1991) J Chem Phys 95: 7576
17. Xiang T, Liu F, Grant DM (1991) J Chem Phys 94: 4463
18. Helfand E (1979) Bell Sys Tech J 58: 2289
19. Greenside HS, Helfand E (1981) Bell Sys Tech J 60: 1927
20. Fixman M (1986) Macromolecules 19: 1195
21. van Gunsteren WF, Berendsen HJC (1988) Mol Sim 1: 173
22. van Gunsteren WF, Berendsen HJC (1982) Mol Phys 45: 637
23. van Gunsteren WF, Berendsen HJC (1977) Mol Phys 34: 1311
24. Adolf DB, Ediger MD (1991) Macromolecules 24: 5834
25. Bird RB, Hassager O, Armstrong RC, Curtiss CF (1977) Dynamics of polymeric liquids: Volume II, Kinetic theory. Wiley, New York.
26. Brown D, Clarke JHR (1987) One aspect of this problem has been discussed in relation to small molecules. J Chem Phys 86: 6446
27. Fixman M (1978) J Chem Phys 69: 1538
28. Weber TA, Helfand E (1983) J Phys Chem 87: 2881
29. Woessner DE (1962) J Chem Phys 36: 1
30. Abragam A (1961) The Principles of nuclear magnetism. Clarendon, Oxford
31. Glowinkowski S, Gisser DJ, Ediger MD (1990) Macromolecules 23: 3520
32. Evans GT (1983) in: Molecular-Based Study of Fluids, Adv in Chem Ser 204 (eds) Haile JM, Mansoori GA, p. 423, ACS, Washington
33. Grant DM, Mayne CL, Liu F, Xiang TX (1991) Chem Rev 91: 1591
34. Fuson MM, Brown MS, Grant DM, Evans GT (1985) J Am Chem Soc 107: 6695
35. Brown MS, Grant DM, Horton WJ, Mayne CL, Evans GT (1985) J Am Chem Soc 107: 6698
36. Fuson MM, Grant DM (1988) Macromolecules 21: 944
37. Fuson MM, Anderson DJ, Lui F, Grant DM (1991) Macromolecules 24: 2594
38. Fuson MM, Miller JB (1993) Macromolecules 26: 3218
39. Helfand E, Wasserman ZR, Weber TA (1980) Macromolecules 13: 526
40. Brown D, Clarke JHR (1990) J Chem Phys 93: 4117
41. Jones AA, NMR and polymer chain dynamics (unpublished manuscript)
42. Zuniga I, Bahar I, Dodge R, Mattice WL (1991) J Chem Phys 95: 5348
43. de Gennes PG (1979) Scaling concepts in polymer physics. Cornell University Press, Ithaca. p. 167–72
44. Schatzki TF (1962) J Polym Sci 57: 496
45. Morawetz H (1989) J Lumin 43: 59
46. Liao TP, Morawetz H (1980) Macromolecules 13: 1228
47. Adolf DB, Ediger M (unpublished results)
48. See reference 12, Chapter 13
49. Yun-Yu S, Wang L, van Gunsteren WF (1988) Mol Sim 1: 369
50. Depner M, Schurmann BL, Auriemma F (1991) Mol Phys 74: 715
51. Widmalm G, Pastor RW (1992) J Chem Soc Far Trans 88: 1747
52. Gisser DJ (1992) Ph.D. Thesis, University of Wisconsin
53. Adolf DB, Ediger MD, Kitano T, Ito K (1992) Macromolecules 25: 867
54. Waldow DA, Ediger MD, Yamaguchi Y, Matsushita Y, Noda I (1991) Macromolecules 24: 3147
55. Lang MC, Laupretre F, Noel C, Monnerie L (1979) J Chem Soc Far Trans 2 75: 349
56. Lui KJ, Anderson JE (1970) Macromolecules 3: 163
57. De Loof H, Harvey SC, Segrest JP, Pastor RW (1991) Biochemistry 30: 2099
58. Bagchi B, Oxtoby DW (1983) J Chem Phys 78: 2735
59. Darinskii AA, Gotlib YY, Lyulin AV, Klushin LK, Neyelov IN (1990) Polym Sci USSR 32: 2192
60. Pear MR, Weiner JH (1980) J Chem Phys 72: 3939
61. Adolf DB (1991) Ph.D. Thesis, University of Wisconsin
62. Gronski W, Murayama N (1978) Makromol Chem 179: 1509
63. Skolnick J (1981) Macromolecules 14: 646

64. Rey A, Kolinski A, Skolnick J, Levine YK (1992) J Chem Phys 97: 1240
65. Gronski W (1977) Makromol Chem 178: 2949
66. Dejean de la Batie R, Laupretre F, Monnerie L (1989) Macromolecules 22: 122
67. Gisser DJ, Glowinkowski S, Ediger MD (1991) Macromolecules 24: 4270
68. Note that the labels on two lines (bond and out-of-plane) in Table II of reference 24 are switched.
69. Takeuchi H, Roe R (1991) J Chem Phys 94: 7446
70. Volterra V, Bucaro JA, Litovitz TA (1971) Ber Bunsenges Phys Chem 75: 309
71. Ivanov EN (1964) Sov Phys JETP 18: 1041
72. Anderson JE (1972) Far Symp Chem Soc 6: 82
73. van Gunsteren WF, Karplus M (1982) Macromolecules 15: 1528
74. Gottlieb M, Bird RB (1976) J Chem Phys 65: 2467
75. Reference 25, p. 633–4.
76. Helfand E, Wasserman ZR, Weber TA, Skolnick J, Runnels JH (1981) J Chem Phys 75: 4441
77. Perico A (1989) Acc Chem Res 22: 336
78. Helfand E (1984) Science 226: 647
79. Adolf DB, Ediger MD (1992) Macromolecules 25: 1074
80. Hall CK, Halfand E (1982) J Chem Phys 77: 3275
81. Bendler JT, Yaris R (1978) Macromolecules 11: 650
82. See for example references 80 and 81.
83. Hu Y, Fleming G, Freed K, Perico A (1991) Chem Phys 158: 395
84. Hyde PD, Waldow DA, Ediger MD, Kitano T, Ito K (1986) Macromolecules 19: 2533
85. Helfand E (1971) J Chem Phys 54: 4651
86. Zhan Y, Mattice WL (1992) Macromolecules 25: 4078
87. Dodge R, Mattice WL (1991) Macromolecules 24: 2709
88. Zhan Y, Mattice WL (1992) Macromolecules 25: 1554
89. Zhan Y, Mattice WL (1992) J Chem Phys 96: 3279
90. Sozzani P, Bovey FA, Schilling FC (1991) Macromolecules 24: 6764
91. Tonelli AE (1990) Macromolecules 23: 3134
92. This can be inferred from information in reference 39
93. Bahar I, Erman B, Monnerie L (1992) Macromolecules 25: 6309
94. Bahar I, Erman B, Monnerie L (1992) Macromolecules 25: 6315
95. Adachi K, Imanishi Y, Kotaka T (1989) J Chem Soc, Far Soc I 88: 1083
96. Hyde PD, Ediger MD (1990) J Chem Phys 92: 1036
97. Bahar I, Erman B (1987) Macromolecules 20: 2310
98. Bahar I, Erman B (1988) J Chem Phys 88: 1228
99. Bahar I, Erman B, Monnerie L (1989) Macromolecules 22: 431
100. Bahar I, Erman B, Monnerie L (1990) Macromolecules 23: 1174

Received: June 1993

MD Simulation Study of Glass Transition and Short Time Dynamics in Polymer Liquids

Ryong-Joon Roe
Department of Materials Science and Engineering, University of Cincinnati,
Cincinnati, OH 45221, USA

Molecular dynamics simulations have been performed on models of polyethylene (and n-alkane) liquids at their realistic bulk densities. The united atom approximation has been adopted. On stepwise cooling of the system under a constant pressure, the temperature coefficient of specific volume changes abruptly at a temperature mimicking the glass transition phenomenon observed in laboratory. The results of the simulation runs, lasting for the order of nanoseconds, were analyzed to investigate the short time dynamics of bond reorientation. The distribution of bond reorientation angle is much broader than is expected from a rotational diffusion with a single diffusion coefficient. Similarly the time-correlation function of bond reorientation is non-exponential and can be fitted well by the stretched exponential function. Attempts to explain these behaviors by assuming a superposition of rotational diffusion processes is met with many contradictions. Explanation is instead offered on the basis of the observed anisotropy of the bond reorientation motion, i.e., the finding that the chain axis reorients much more slowly than a vector attached perpendicular to the chain axis. Such anisotropy results from the intramolecular and intermolecular constraints to the motion of the chain imposed by the segments in the neighborhood.

1 Introduction

The study of static and dynamic properties of polymer liquids and glasses is one of the central topics of research in polymer science. A variety of experimental and theoretical techniques have been mobilized for this purpose. In recent years the techniques of computer simulations are increasingly finding application toward this goal. All the various methods of simulating polymer molecules on computers, such as Monte Carlo, molecular mechanics, Brownian dynamics, and molecular dynamics simulation techniques have been utilized. Early works relied mostly on Monte Carlo techniques applied to schematic models of polymer liquid built on a lattice. With increased capabilities of computers available in recent years, the use of more computationally intensive methods has become feasible, allowing simulations of more realistic, off-lattice models of bulk polymers by Brownian and molecular dynamics simulation techniques as well as by Monte Carlo methods.

In the simulation of off-lattice, atomistic models of polymers, one has to make a choice in the degree of realism that is to be incorporated in the model. For example, one may adopt an all atom model, with all the vibrational and torsional potentials as well as non-bonded interactions fully represented and partial charges on every atom explicitly specified. Such a detailed model would of course be computationally more intensive. One may, on the other hand, choose a much more simplified model, e.g., a spring-and-bead model in which a spherical particle represents monomeric units and interacts with bonded neighbors through a spring-like potential and with non-bonded neighbors through a repulsive potential. There are many possible models with intermediate degrees of realism. The choice of the model to use is dictated to a large part by the objective of the study. If the purpose is to determine accurate numerical constants of some physical properties of a specific polymer, for example T_g and cohesive energy density, then there is no alternative but to use the most detailed atomistic model of the polymer. If, on the other hand, the purpose is to derive a general relationship applicable to a class of polymers, it is feasible, and often desirable, to use a more 'generic' model of polymer in which varying degrees of approximations are incorporated to enhance computational efficiency.

In the study described below, we have been examining the structure and dynamic properties of polyethylene and n-alkane liquids by means of a model that incorporates the united atom approximation. It takes about 0.01 s on a Cray Y-MP to integrate the Newtonian equations of motion for all the 500 CH_2 united atoms comprising the system. With the time step of 10 fs taken for successive integrations, this means that about 20 min of Cray cpu time is required to make a simulation run lasting 1 ns. We find that of the 0.01 s required for a single round of integrations, more than 70% of the cpu time is spent in computing the non-bonded interaction potential. If we had adopted a spring-and-bead model, we would have gained only marginally in the computational efficiency, unless we simplified or eliminated altogether the non-bonded interaction potential. On the other hand, if we adopted an all atom model, the number density of atoms would have increased by a factor of three, the number of neighbor pairs within a given cutoff distance would have increased by a factor of nine, and therefore the cpu time required for the simulation of the same duration would have increased by almost an order of magnitude. Since our purpose is to gain understanding of the general characteristics of short time chain motion in polymer liquids and not to evaluate numeric values of specific properties of polyethylene-like liquids, the gain by an order of magnitude in the computational efficiency fully justifies the use of the united atom approximation. On the other hand, any physical constant we derive, for example the glass transition temperature, is not to be expected to agree exactly with experimental data, and should be viewed with some caution until one gains sufficient understanding of the errors introduced by the united atom approximation. One may be tempted to speed up the computation further by eliminating the non-bonded interaction altogether. Studies of this kind were made in the past. Without the non-bonded interactions, however, the system behaves essentially as a collection of isolated phantom chains, and its behavior will miss the important effect due to intermolecular interactions.

In our study of polyethylene-like systems, we normally ran a simulation lasting up to 1 ns, although on occasions we ran it up to 5 ns. Among the various types of motions probed by experimental techniques, many greatly exceed such a time scale. For example, the study of translational diffusion of long chain molecules or the evaluation of viscosity of polymer liquid, even at very high temperatures, requires a simulation much longer than a nanosecond duration. We can investigate only the short time dynamics of chain segments moving within the local environment, such as the conformational transitions, reorientation of a bond or of a segment consisting of a few bonds, fluctuation of local free volume (unoccupied space between atoms), etc. Fortunately many of the dynamic properties of amorphous polymers, probed by such widely-used techniques as NMR, light scattering, and dielectric relaxation, can be attributed to such local motions, and the result obtained by simulation can indeed be compared with experimental data.

The glass transition temperature may be regarded as the temperature at which the translational diffusion ceases. Often the temperature at which the liquid viscosity exceeds 10^{13} poise is defined as the glass transition temperature. Our simulation cannot be utilized to determine the glass transition temperature according to this definition. Glass transition temperatures of polymers were, however, originally determined mostly by means of dilatometry, i.e., from the change in the temperature coefficient of specific volume. On lowering the temperature, the rearrangement of atomic packing required to accommodate the increase in density involves only a local motion of atoms and segments, and can be probed by our simulation. Our results described below indeed suggest that the type of MD simulation we have performed could be adopted as a means of estimating the glass transition temperature of as-yet unsynthesized polymers, if the model is made sufficiently realistic (and if the effect of cooling rate differences is properly taken into account).

2 The Model and the Method of MD Simulation

The model employed to mimic n-alkane and polyethylene molecules has been described previously and the results obtained from its use have been published in a number of previous papers [1–9]. It consists of sequences of spherical CH_2 segments of mass 14 amu; chain ends (CH_3) and the monatomic homolog (CH_4) are also represented by the same pseudo-atoms. The segments along the chain are subject to the potentials E_b for bond stretching, E_θ for bond angle bending, and E_ϕ for torsional rotation according to

$$E_b = (1/2)\,k_b(1 - l_0)^2 \tag{1}$$

$$E_\theta = (1/2)\,k_\theta(\cos\theta - \cos\theta_0)^2 \tag{2}$$

$$E_\phi = k_\phi \sum_{n=0}^{5} a_n \cos^n\phi. \tag{3}$$

Non-bonded interactions between segments in different chains and between segments separated by more than three bonds along the chain backbone are given by a truncated and shifted Lennard-Jonnes 6–12 potential function

$$E_{nb} = 4\varepsilon\left[\left(\frac{\sigma}{r}\right)^{12} - \left(\frac{\sigma}{r}\right)^{6}\right] + C \quad r < r_c$$

$$\tag{4}$$

$$= 0 \qquad\qquad\qquad r > r_c.$$

Values of the parameter used are listed in Table 1. For comparison purposes we have also investigated, to a limited extent, a so-called freely-rotating chain model, which is identical to the above polyethylene model except that k_ϕ is set equal to zero, i.e., the torsional barrier to conformational transitions is eliminated.

The simulated system consists of a cubic box containing 500 CH_2 segments in most cases (and 600 or 2000 segments in some cases). Periodic boundary conditions in the three Cartesian coordinate directions are imposed as usual to eliminate surface effects. The 500 CH_2 segments are divided into chains of x segments, x ranging from 1 to 500. In the case of x = 500, for example, the system contains only one molecule of 500 CH_2 segments and its replicas. In the infinite chain length model, an additional periodic condition ties one end of a 600 segment chain, spanning three cubic boxes, to the image of its other end, so

Table 1. Constants used for potential functions

l_0	0.152 nm
$\cos\theta_0$	-0.3333
k_b	$3.46 \times 10^7\,\mathrm{J\,nm^{-2}\,mol^{-1}}$
k_θ	$5 \times 10^5\,\mathrm{J\,mol^{-1}}$
k_ϕ	$9 \times 10^3\,\mathrm{J\,mol^{-1}}$
a_0	1
a_1	1.3100
a_2	-1.4140
a_3	-0.3297
a_4	2.8280
a_5	-3.3943
ε	$500\,\mathrm{J\,mol^{-1}}$
σ	0.38 nm
r_c	0.57 nm

that no chain ends are present in the system, in accordance with a scheme first used by Weber and Helfand [10]. In this infinite chain model, therefore, the end-to-end distance is equal to three times the box edge length, and changes with the temperature, not according to the trans-gauche population redistribution but rather simply in accordance with the change in the specific volume. The local chain dynamics we are investigating is unlikely to be perturbed to a significant extent by the long range effect that arises, in the case of $x = 500$, from the chain interacting with its own images or from the constraint, in the case of the infinite chain model, in the end-to-end distance.

The Newtonian equations of motion for individual segments are integrated numerically using the Verlet algorithm [11] with a time step of 10 fs (or 6 fs in the case of infinite chain model) at constant energy and volume. The temperature T is evaluated from the average kinetic energy per atom, and the pressure by

$$pV = NT + (1/3)W \tag{5}$$

with

$$W = \sum_{i<j} r_{ij} f_{ij}. \tag{6}$$

In calculating the virial W, only the forces f_{ij} acting between atomic centers (bond stretching and non-bonded interaction) are included [1]. Within a given simulation run both T and p fluctuate considerably [1] and their averages are evaluated. The average pressure thus evaluated, for a system at a given temperature and specific volume, is affected sensitively by the truncation limit of the non-bonded potential function used [12] or whether or not the united atom approximation is adopted [13]. Most of the results reported here have been obtained at a nominal average pressure of 1.36 kbar. Although algorithms for constant T and constant p are available, their effect on the dynamic trajectory is not always clear, and we have instead chosen to use the microcanonical ensemble method. Very occasionally the system is found to undergo crystallization as revealed by a rapid change in temperature and pressure and by an onset of non-vanishing values of average orientation of all the bonds in the system. Since we are interested in the behavior of amorphous polymer, such a crystallizing system is immediately discarded and the simulation is restarted from a fresh set of atomic configurations.

Different strategies have been used by different workers in producing the starting set of atomic coordinates. Some use [14] the rotational isomeric states model [15] to generate polymer chains having the correct trans-gauche probabilities, or rely on a method [16] very similar to the Monte Carlo technique to eliminate improbable chain conformations. Our method does not require any program specifically written for the generation of the initial configurations, but instead simply makes use of our main molecular dynamics simulation program. We start from polymer chain configurations produced arbitrarily, either arranged entirely randomly or on a regular lattice. The MD run is then carried out at a high temperature and at a much reduced density, with the force constants reduced, and particularly with the repulsive part of the Lennard–Jones potential substantially reduced. As the atoms move according to the equation of motion, overlapping atomic coordinates are eliminated on their own and the chain configurations gradually settle down to low-energy, high-probability states. The force constants are then restored in steps to their proper values, and the temperature and density are also returned to their desired starting conditions.

3 Glass Transition Temperature

On lowering the temperature of the simulated system stepwise, some of its properties exhibit fairly abrupt changes around a temperature in a way that characterizes the glass transition in laboratory liquid systems. Figure 1 shows how the specific volume, evaluated for systems comprising chains of length 125 and 500 respectively, changes as the temperature is lowered in steps under a constant pressure. At each temperature the simulation was performed for a 100 ps duration. By noting the changes in the slope the glass transition temperature can be identified to within a few degrees. The transition temperatures determined in this way are plotted in Fig. 2 as a function of chain length x. The result for the monatomic liquid ($x = 1$) is the one obtained by Fox and Andersen [17]. The T_g initially increases rapidly with x, and then apparently approaches an asymptote at large x, in agreement with what one expects on the basis of laboratory experiments. When the chain is made significantly more flexible by eliminating the barrier to torsional rotation (the freely-rotating chain model), the T_g obtained is reduced, as shown in Fig. 3, to around 40 K, which appears to be independent of the chain length ($x = 10$, 20, and infinity) [1,7].

At around the temperature where the temperature coefficient of specific volume undergoes an abrupt change, other properties of the system are also seen

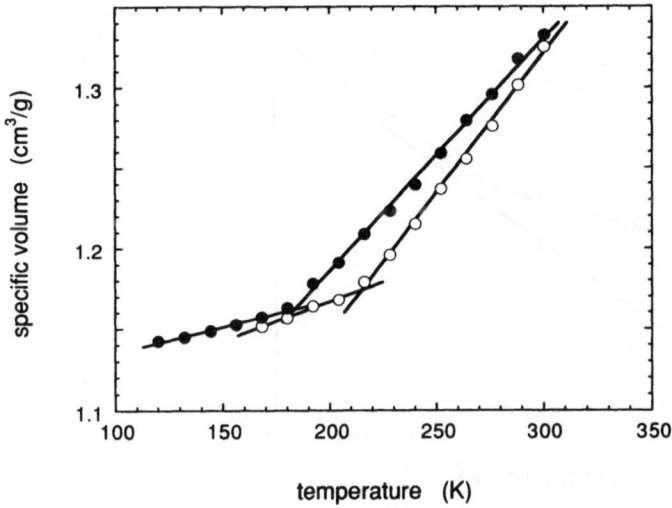

Fig. 1. Specific volume against temperature obtained [9] during stepwise cooling runs for x = 125 (●) and 500 (○)

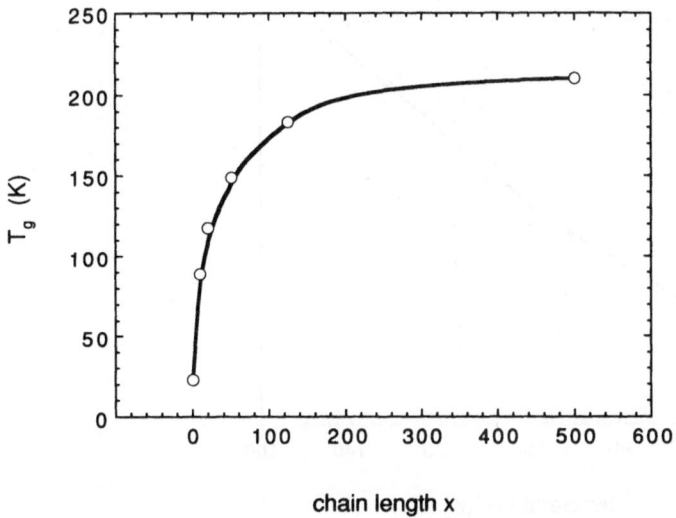

Fig. 2. Dependence of T_g on chain length [9]. The value for x = 1 comes from the work of Fox and Andersen [15]

to exhibit changes characteristic of glass transition. Figure 4 shows the dependence of the internal energy per CH_2 on temperature, obtained on stepwise cooling [1]. One can clearly recognize a break in the temperature coefficient at about the same temperature at which the thermal expansion coefficient changes abruptly, although the transition exhibited by the former is much weaker (as

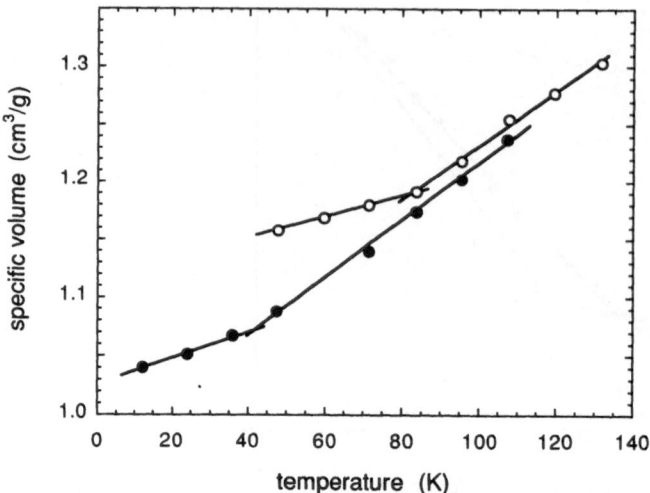

Fig. 3. Specific volume against temperature obtained [1] during stepwise cooling for x = 10. The *open circles* are for chains subject to full torsional potential, whereas the *filled circles* are for flexible chains with their torsional potential switched off

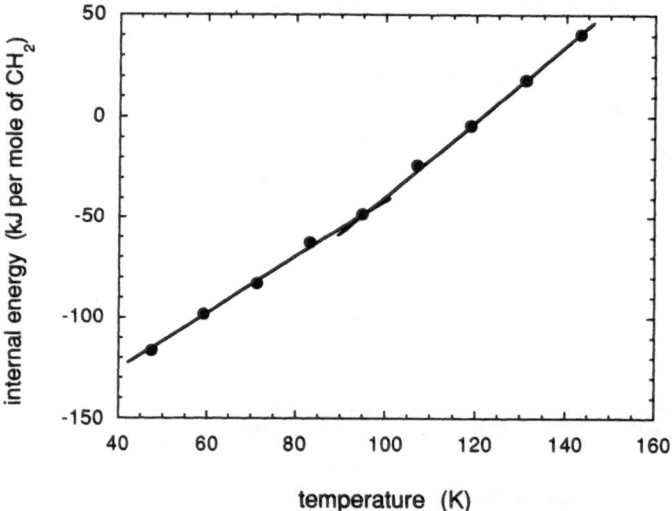

Fig. 4. Internal energy against temperature obtained [1] during stepwise cooling for x = 10. Change in the temperature coefficient is seen to occur at about the same temperature where the specific volume against temperature similarly changes its slope

Fox and Andersen [17] also noted previously in their study of the monatomic system). Figure 5 shows the self-diffusion coefficient D evaluated [1] from the mean square displacement of all CH_2 units in a system consisting of chains of length 10. D appears to decrease to negligibly small values as the temperature is lowered toward the volumetrically-determined T_g, although it is not quite clear

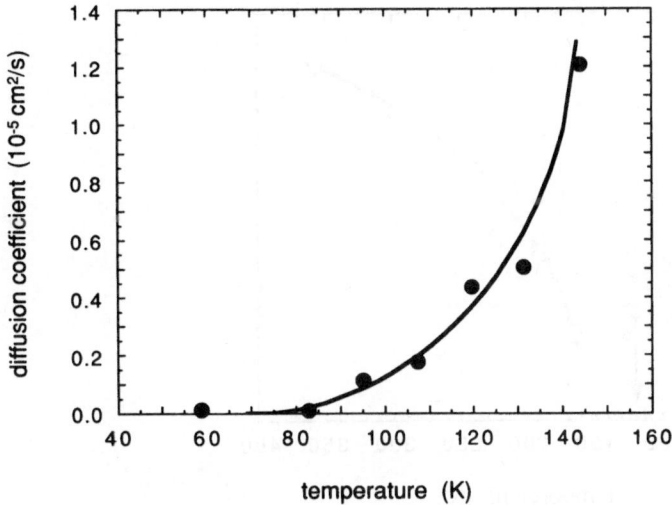

Fig. 5. Segmental diffusion coefficient against temperature obtained [1] during stepwise cooling for x = 10

whether the segmental diffusion has totally ceased at the T_g. Figure 6 gives the variation [1] with temperature of characteristic ratio $\langle r^2 \rangle / nl^2$ found for chains of x = 10, and shows that the polymer chain conformation becomes frozen as the temperature is lowered beyond T_g. This of course reflects the fact that the trans-gauche conformational transition rate decreases rapidly as the temperature is lowered toward the T_g, as shown in Fig. 7 [6].

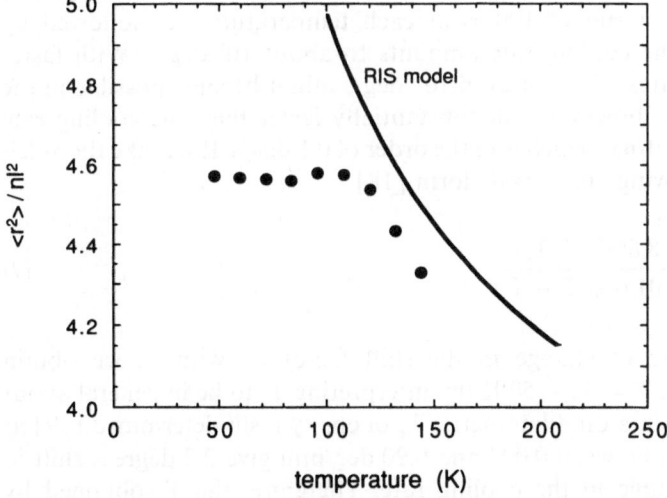

Fig. 6. Characteristic ratio $\langle r^2 \rangle / nl^2$ of chain dimension against temperature obtained [1] for system of x = 10. The *solid curve* gives the values calculated by the rotational isomeric states model

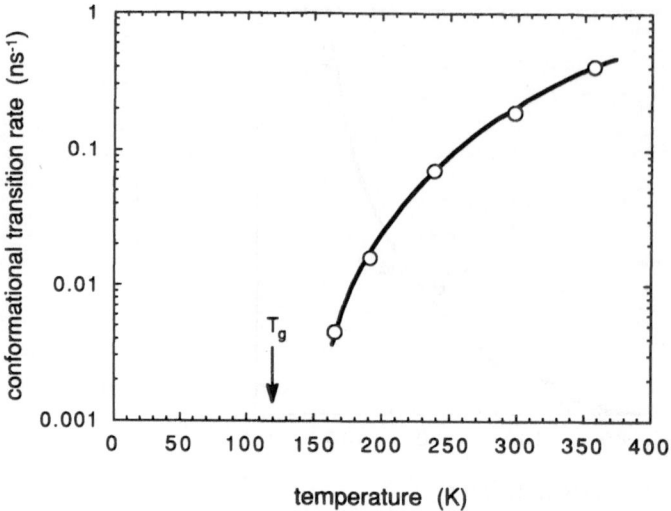

Fig. 7. Torsional barrier crossing rate against temperature observed [6] during stepwise cooling for system of x = 20

The results described above illustrate that a molecular dynamics simulation may eventually be able to serve as a tool to estimate the glass transition temperature of a new polymer as yet unsynthesized. Before this can be realized routinely, the computational speed of available computers has to increase by an order or two of magnitude so that much more realistic, all atom models of complicated polymers can be handled. Even then, the question of cooling rate (or, equivalently, the affordable length of simulation runs) will remain. In our simulation, in which a run of 100 ps at each temperature was followed by a cooling by 12 K, the cooling rate amounts to about 10^8 deg/s. With faster computers, cooling rates of 10^7 or even 10^6 deg/s might become possible in the near future. They are, however, still substantially faster than the cooling rate practiced in the laboratory, which is of the order of 0.1 deg/s. If we take the WLF equation in the following "universal" form [18]

$$\log_{10} a_T = -\frac{8.86(T - T_s)}{101.6 + T - T_s} \qquad (7)$$

and evaluate the rate of change in the shift factor a_T with T, we obtain $\partial T / \partial \log_{10} a_T = 2.95$ at $T = T_s - 50$ K (by interpreting T_s to be in general about 50 K above T_g). The data on dilatometric T_g of epoxy resin determined [19] at different cooling rates between 0.003 and 0.90 deg/min give 2.2 degrees shift in T_g per decade of change in the cooling rate. Therefore, the T_g obtained by simulation is expected in general to be about 20 degrees higher than the T_g obtained in the laboratory by dilatometry or calorimetry.

4 Static Structure

Before discussing the dynamic properties of our simulated systems, it is useful to describe briefly some aspects of equilibrium structure that have been evaluated. Any explanation of the dynamics properties has to be based on a firm understanding of the equilibrium structure. In this connection, we will touch upon two topics: the free volume distribution and the short-range order.

4.1 Free Volume Distribution

The term free volume is often used in explaining various phenomena associated with polymer liquids and glasses. It enjoys the advantage of being a physical concept that can easily be comprehended and visualized, but suffers from the lack of a precise definition. Its usefulness in a way stems from its ambiguity, in allowing its meaning to be subtly altered depending on the phenomenon at hand that is to be explained. When one attempts to evaluate free volume from simulation results, this latitude disappears and one is forced to choose a definition unambiguously. One of the more rational way of defining free volume, in the case of systems consisting of atoms with soft repulsive potentials such as ours, would be to evaluate the integral [20,21] of $\exp[-\phi(r)/kT]$ over positions r around an atom, where $\phi(r)$ is the potential energy acted on by surrounding atoms. A much simpler approach is to replace the atom with a hard sphere, and to evaluate the unoccupied cavity that is left in the space surrounded by neighboring hard spheres.

In our study [4] of cavity size distribution a sphere of diameter D is prescribed on every atomic center, and the cavity is evaluated as the pocket of contiguous space unoccupied by any of these spheres. The "exclusion diameter" D has to be equated to the van der Waals diameter D_{vdW} of a CH_2 unit for the evaluation of the total cavity volume. When diffusion of small gas molecules through polymers is of interest, the unoccupied space available for inclusion of the small molecule can be evaluated by setting D equal to $D_p + D_{vdW}$, where D_p is the diameter of the small probe molecule. The results of the cavity volume analysis were therefore presented [4] as a function of the "exclusion diameter" D varying over a range of values.

Some salient results obtained from our analysis of cavity-volume sizes are given below. For more details, the reader is referred to the original article [4]. Figure 8 gives the distribution of cavities according to their size, as evaluated for a system of x = 20 at 60 K. Here the four plots show how the distribution changes when the exclusion diameter D is varied from 1.2 times to 1.8 times the Lennard–Jones parameter r* (= 0.38 nm). The distribution is very broad, ranging from very fragmentary, little cavities to fairly large, inter-connected regions, and the overall feature suggests a tendency toward a bimodal distribution of sizes. With increase in the exclusion diameter D the average size of the

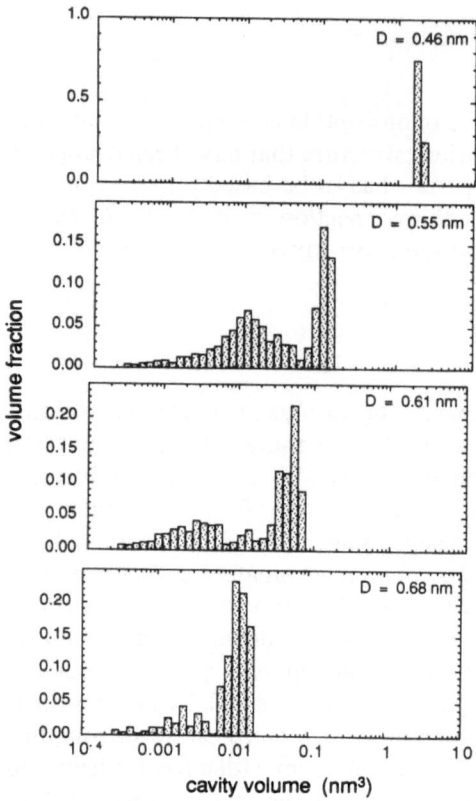

Fig. 8. Distribution [4] of cavities according to their size at T = 60 K for system of x = 20. The value of the exclusion diameter D assumed in each case is indicated

cavities decreases, but this bimodal character remains unchanged. Some idea of the shape of these cavities can be gleaned from the surface/volume ratio, which was found [4] to vary from unity (round, compact shape) to more than four (elongated and irregular shape).

These cavities are not static, even below T_g, and constantly change their shape and size as the molecules themselves undergo local motions. A glimpse into the residual local motions still active in the glassy state can be gained from Fig. 9, where cavities present at succeeding times separated by 5 ps intervals were evaluated for D = 0.68 nm at 60 K, and are shown as projections on the x–y plane (the squares being the projection of the basic MD cube of edge length approximately 2.3 nm). Three different cavities are identified and are distinguished by different shadings. The time-lapse projections show that at a temperature about 60 degrees below the T_g a large-amplitude, low-frequency motion of molecules still persists and leads to a waxing and waning of the cavities. The result suggests that a glass is far from being a frozen, immobile structure retaining only local vibrations, but is rather capable of large scale cooperative motions. A further, detailed analysis is obviously required to unravel the nature of such cooperative motions.

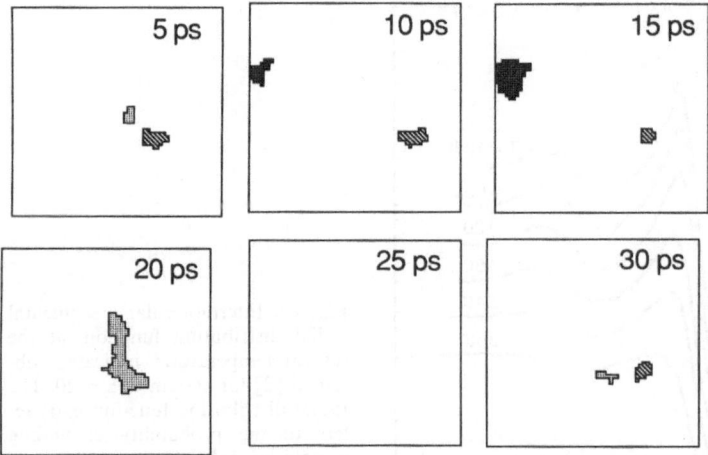

Fig. 9. Projections on the X–Y plane of cavities evaluated [4] at T = 60 K and D = 0.68 nm at successive time frames separated by 5 ps

4.2 Short Range Order

The question of whether polymer liquids possess short range order much beyond what is normally observed with small molecule liquids has been vigorously debated until some years ago. On the basis of observations obtained by electron microscopy and X-ray and electron scattering methods, suggestions [22–24] were made that ordered domains, variously called nodules, bundles, meander arrays, etc., of size in the range of 10 nm exist in amorphous polymers. The analysis of data obtained by depolarized light scattering, on the other hand, led to the conclusion [25–27] that only a limited degree of short range order, comparable to that observed in ordinary liquid, is present in polymer liquids and glasses. More recent works by optical [28–30] and deuterium NMR [31,32] techniques on deformed melts and networks also give evidence to the presence of short range orientational coupling.

With polymer liquids, the radial distribution function that can be derived from wide-angle X-ray diffraction is dominated by the pair correlation among segments belonging to the same chain, and the extent of the *intermolecular* pair correlation, that is of interest, is difficult to be discerned. From the simulation results one can extract the radial distribution function characterizing only the correlation between pairs of segments belonging to different chains. The result obtained [2] with a system of x = 20 is given in Fig. 10. What is plotted here is not the atom–atom pair correlation but rather a correlation between the centers of mass of two short subchains, each consisting of four CH_2 units. An atom–atom pair correlation, or correlation between pairs of subchains of length 2, 3, 4, etc., are found [2] to give very similar correlation functions. In Fig. 10

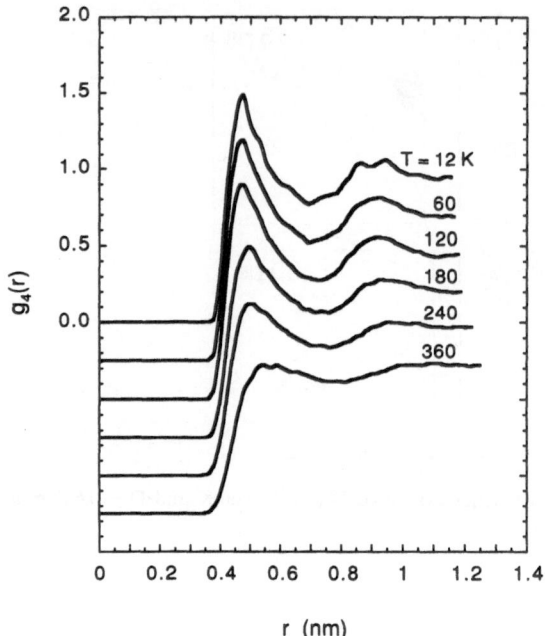

Fig. 10. Intermolecular segmental radial distribution function, at the several temperatures indicated, obtained [2] for system of $x = 20$. The radial distribution function $g_4(r)$ refers to the probability of finding a pair of subchains consisting of 4 CH_2 units and each residing on different chains. Successive curves are displayed vertically for legibility

both the first and second neighbor shells are clearly displayed, although with decreasing temperature the peaks grow progressively. Below the T_g (~ 120 K) the peaks continue to become sharper with fall in temperature, again indicating that even below T_g reorganization of the local structure continues through short range displacement of atoms.

With polymers, as well as with liquids of non-spherical molecules, the extent of short range order present is demonstrated more clearly by the degree of spatial orientational correlation exhibited among neighboring segments, subchains, or molecules. We characterize [2] the spatial orientation correlation by means of

$$P_2(r) = (1/2)(3\langle\cos^2\theta_{ij}\rangle - 1) \tag{8}$$

where θ_{ij} is the angle between the end-to-end vectors of subchain i and j, each consisting of four atoms, and r is the distance between the centers of mass of these two subchains. To be more meaningful, this orientation correlation function $P_2(r)$ is further multiplied by the probability $g(r)$ of finding the subchain pair at separation r, and, in Fig. 11, $P_2(r)g(r)$ is plotted against r for several temperatures. Clearly a tendency for parallel alignment of neighboring segments persists at least up to the second neighbor shell, and this tendency becomes stronger as the temperature is lowered. The increasing orientation correlation below the T_g again suggests that a sufficient short-range mobility of segments exists in the glassy state to allow local reorganization of segmental packing with changes in temperature.

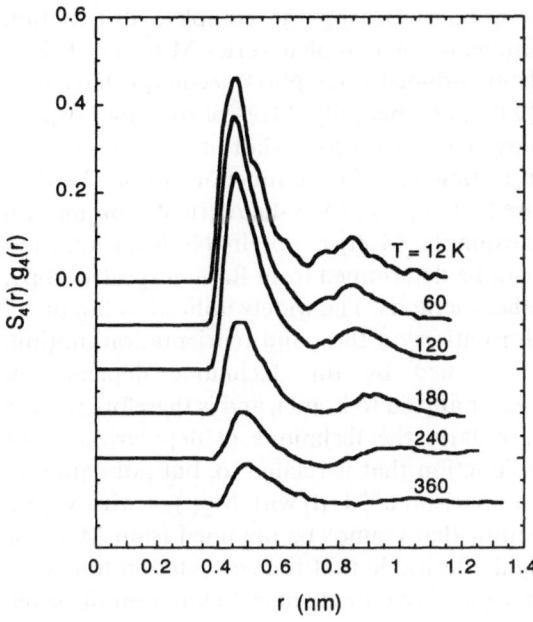

Fig. 11. Static orientation correlation $S_4(r)$ weighted by the probability $g_4(r)$ of finding subchains at separation r, obtained [2] for system of $x = 20$ at various temperatures indicated. Successive curves are displaced vertically for legibility

5 Bond Reorientation Dynamics

5.1 Reorientation Angle Distribution and Time Correlation Functions

Among the various types of local motions that take place in polymer liquids, the change in the orientation of individual bonds with time is one that can be readily evaluated from the molecular dynamics simulations. The bond reorientation motions are also amenable to experimental measurement by means of a number of techniques. We investigate the bond reorientation dynamics by evaluating three types of quantities, the distribution $W(\theta, t)$ of reorientation angle θ that a bond has undergone during a time interval t, and the time correlation functions

$$M_1(t) = \langle \mathbf{u}(0) \cdot \mathbf{u}(t) \rangle \tag{9}$$

$$M_2(t) = (1/2)(3 \langle [\mathbf{u}(0) \cdot \mathbf{u}(t)]^2 \rangle - 1) \tag{10}$$

where $\mathbf{u}(t)$ is a unit vector affixed to the bond, and the product $\mathbf{u}(0) \cdot \mathbf{u}(t)$ is equal to the cosine of the reorientation angle θ. The time correlation functions are the moments of the distribution function $W(\theta, t)$ and can be obtained from the latter by

$$M_i(t) = \int P_i(\cos \theta) W(\theta, t) d\theta \tag{11}$$

where $P_i(x)$ is the Legendre polynomial. Having the complete distribution function $W(\theta,t)$ is equivalent to knowing the complete series $M_i(t)$, $i = 1, 2, 3,$ .... Recent advances [33,34] in the two-dimensional NMR technique have now opened the way to determine $W(\theta,t)$ experimentally. Most of the other experimental techniques that are employed to investigate short time dynamics of polymers, on the other hand, lead to time correlation functions of one kind or another [35,36]. A number of these techniques probe the reorientation motion of bonds in polymer chains. For example, $M_1(t)$ is obtainable from infra-red band shape analysis, while $M_2(t)$ can be determined from Raman spectroscopy, fluorescence depolarization and other methods. The widely utilized technique of dielectric relaxation also gives information on the bond reorientation motion. The time correlation function obtained by this technique depends on $\langle \mathbf{u}_i(0) \cdot \mathbf{u}_j(t) \rangle$, where $\mathbf{u}_i$ is the unit vector affixed to bond i, and is therefore related to, but not identical to, $M_1(t)$. Similarly the technique of depolarized light scattering gives a time correlation function that is related to, but not equal to, $M_2(t)$. As will be seen below, the comparison of $M_1(t)$ with $M_2(t)$ or with $W(\theta,t)$ provides much additional information that cannot be obtained from $M_1(t)$ or $M_2(t)$ alone, and it is therefore highly desirable that time correlation functions be determined experimentally by means of two or more techniques on the same material under closely similar conditions.

Typical examples of the reorientation angle distribution function $W(\theta,t)$ are given in Fig. 12, where results obtained [9] with the system of x = 500 at 300 K are plotted at several values of t. $W(\theta,t)$ is defined so that the probability of finding the reorientation angle between θ and $\theta + d\theta$ after time interval t is given by $W(\theta,t)d\theta$, and it approaches $W(\theta, \infty) = (1/2)\sin\theta$ at long t. The time correlation functions $M_1(t)$ and $M_2(t)$ obtained from the same simulation run are given

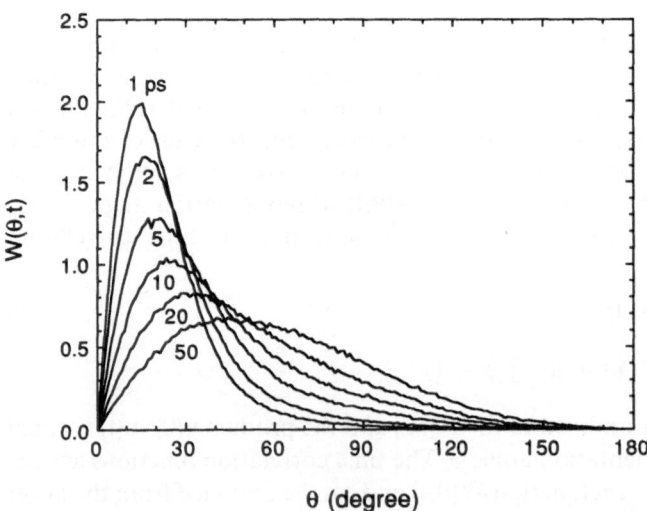

Fig. 12. Bond reorientation angle distribution $W(\theta,t)$ obtained [9] for system of x = 500 at 300 K. The time interval t allowed for reorientation is indicated for each curve

in Fig. 13. The correlation functions are obviously much broader than an exponential function. They can in fact be fitted well by means of the stretched exponential or the Kohlrausch-Williams-Watts function [37]

$$M_i(t) = \exp[-(t/\tau_i)^{\beta_i}] \qquad (12)$$

as shown by the dotted curves in Fig. 13. The values of β_1 and β_2 used for the fit are 0.40 and 0.50 respectively, both in the range of values usually found with experimentally determined time-correlation functions. The effective relaxation time τ_1 and τ_2 can be evaluated by noting the time at which the correlation functions decay to 1/e in Fig. 13. Another measure of the time scale of decay of correlation that is commonly used is the average correlation time $\langle\tau\rangle$ defined by

$$\langle\tau\rangle_i = \int M_i(t)\,dt \qquad (13)$$

which, when M(t) is represented by the stretched exponential function, Eq. (12), is equal to

$$\langle\tau\rangle_i = (\tau_i/\beta_i)\,\Gamma(1/\beta_i) \qquad (14)$$

where Γ is the gamma function. In this article the relaxation time τ_i always refers to the time at which the time-correlation function decays to 1/e.

Table 2 gives the values of τ_1 and τ_2 evaluated [9] for systems consisting of chains of different lengths as a function of temperature. The relaxation times depend somewhat [9] on where the bond being considered is located along the chain, and the values given in Table 2 are those evaluated for bonds in the middle of the chain unless otherwise noted. The ratio τ_1/τ_2 is often regarded as an indicator of the mechanism of reorientation motion [38]. For example, if the

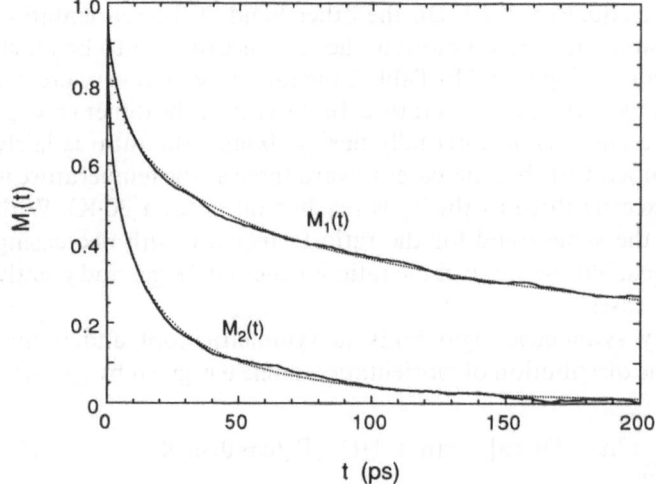

Fig. 13. Bond reorientation time-correlation function, $M_1(t)$ and $M_2(t)$, obtained for system of x = 500 at 300 K. The *dotted curves* are the fit by the KWW function

Table 2. τ_1 and τ_2 as a function of chain length and temperature[a]

Chain length	Temperature (K)	τ_1(ps)	τ_2(ps)	τ_1/τ_2
x = 2	360	0.219	0.134	1.63
	300	0.230	0.140	1.64
	240	0.274	0.166	1.65
	180	0.328	0.200	1.71
	120	0.431	0.241	1.79
	60	1.317	0.439	3.00
	36	7.50	1.785	4.20
4	360	0.480	0.234	2.05
	240	0.651	0.289	2.25
	180	0.921	0.332	2.77
	120	1.658	0.438	3.78
	60	10.90	1.688	6.46
10	360	2.94	0.874	3.36
	300	4.66	1.246	3.74
	240	7.56	1.889	4.00
	180	26.1	4.17	6.27
	150	80.4	8.74	9.20
20	360	8.52	1.93	4.42
	300	17.04	3.86	4.42
	240	17.29	3.90	4.43
125	300	82.1	10.2	8.04
500	300	101.0	10.8	9.35

[a] τ_1 and τ_2 are those evaluated for the center bond, except that, in the case of x = 125 and 500, they are averaged over 80% of the bonds in the interior part of the chain

motion of a rigid body consists of a succession of small angle reorientations as to allow description of the reorientation motion as a rotational diffusion, the ratio τ_1/τ_2 is expected to be equal to three. On the other hand, if the reorientation occurs by a discrete, large-angle jump motion, the ratio is expected to be much smaller, the limiting value being unity. In Table 2 the ratio is seen to vary greatly depending on the chain length and temperature. In the case of the dimer (x = 2), which is dumbbell-like and has no internally flexible bonds, the ratio is fairly small (1.63) at high temperature, but increases toward three as the temperature is lowered and finally exceeds three as the T_g is reached (at around 36 K). With longer chain systems the same trend for the ratio to increase with decreasing temperature is displayed, but the value of the ratio is generally larger and greatly exceeds three in most cases.

For a cylindrically symmetric rigid body (a symmetric top) undergoing rotational diffusion, the distribution of reorientation angle θ is given by [38–41]

$$W_D(\theta,t) = \frac{1}{2} \sum_{n=0}^{\infty} (2n + 1)\exp[-n(n + 1)Dt]P_n(\cos\theta)\sin\theta \qquad (15)$$

where D is the rotational diffusion coefficient and $P_n(x)$ is the Legendre polynomial. Substitution of Eq. (15) for $W(\theta,t)$ in Eq. (11) gives the time-correlation functions

$$M_1(t) = e^{-2Dt} \tag{16}$$

and

$$M_2(t) = e^{-6Dt} \tag{17}$$

indicating that τ_1 and τ_2 are to be given by 1/2D and 1/6D respectively. The behavior of bond reorientation in our simulated system deviates from what is expected of rotational diffusion in a number of ways. The time-correlation functions, as given in Fig. 13, deviate from a single exponential decay. The ratio τ_1/τ_2 exceeds three in most cases. The observed distribution $W(\theta,t)$ cannot be fitted by the distribution $W_D(\theta,t)$, given in Eq. (15), that is expected of rotational diffusion. In Fig. 14, two of the curves in Fig. 12 are replotted, and are compared with the corresponding curves calculated by Eq. (15) with the values of D chosen to match the peak angles. Clearly the observed distribution is much broader than is expected from a rotational diffusion.

The way the observed distributions $W(\theta,t)$ deviate from the theoretical distribution $W_D(\theta,t)$ predicted from a rotational diffusion can also be illustrated by the results presented in Fig. 15. Here the angles θ_{max} at which the observed curves $W(\theta,t)$, such as those in Fig. 12, exhibit the maximum are plotted as a function of time. Also plotted as the solid curve in Fig. 15 is the angle

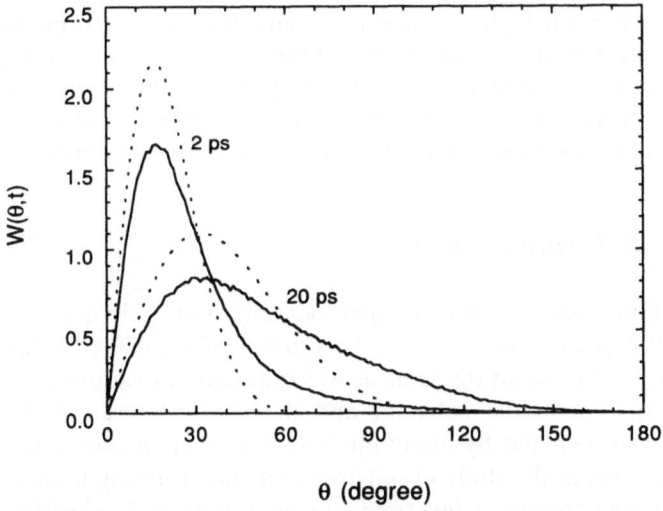

Fig. 14. Comparison of observed bond reorientation angle distribution $W(\theta,t)$ (*solid curves*) with theoretical curves $W_D(\theta,t)$ (*dotted curves*) calculated on the assumption of a rotational diffusion with a single diffusion coefficient. The rotational diffusion coefficients used for the calculation are 0.02 and 0.008 ps^{-1} for the case of t equal to 2 and 20 ps, respectively. These D values are chosen to make the calculated peak angles match the observed ones

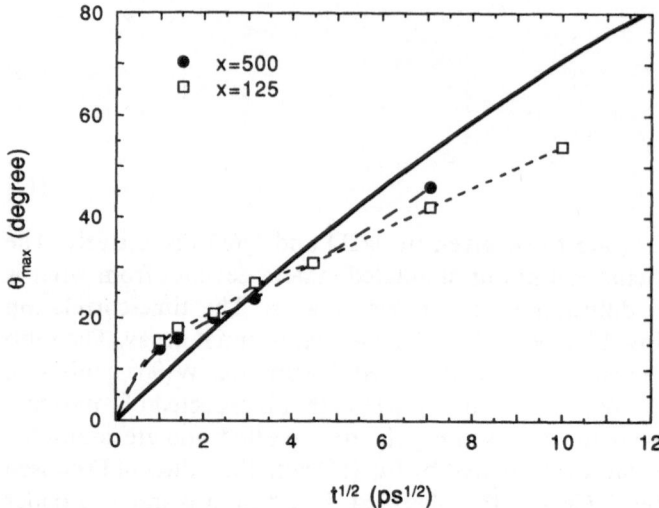

Fig. 15. The angle θ_{max} at which the distribution $W(\theta,t)$ exhibits a maximum is plotted against t. The points are those observed with systems of $x = 500$ and $x = 125$, respectively, at 300 K. The **thick solid** line gives the results calculated from the theoretical expression $W_D(\theta,t)$ for a rotational diffusion process

θ_{max} calculated from Eq. (15). In this calculation the rotational diffusion coefficient D is arbitrarily given a numerical value 0.01 ps^{-1} to make the calculated curve coincide roughly with the results evaluated from the simulation data. If we had chosen a value of D much higher in order to make the calculated curve match the first observed point at 1 ps, all the rest of the observed points would have fallen much below the calculated curve. One may therefore describe the observed reorientation process as resembling the rotational diffusion but slowing down after the initial period in comparison to the ideal rotational diffusion.

5.2 Distribution of Relaxation Times

The time-correlation functions determined experimentally with polymers are never known to decay exponentially, but can be fitted in most cases by the KWW function with a β value about the same as was found in our simulations. The reason for the non-exponential behavior and the physical basis of the KWW function have been debated by many for long, and remain one of the central unresolved problems in the study of polymer and glass-forming liquids. The non-exponential time-correlation functions are often regarded as arising from superpositions of exponential processes each with a different relaxation time. Similarly, one may try to explain the broad reorientation angle distribution $W(\theta,t)$, such as those shown in Figs. 12 and 14, as arising from a superposition of rotational diffusions with a spectrum of diffusion coefficients. As the physical basis for such an assumption it has been often suggested [42–44] that the individual elements (chemical bonds in the present case) that undergo relaxation might be embedded in the local environments that fluctuate from place to place and from moment to moment, and that the relaxation time spectrum that can be derived from the observed time-correlation function reflects the heterogeneity of the local environment.

Any attempt to explain our result of bond reorientation dynamics on the basis of superposition of rotational diffusion processes encounters a contradiction. If such an explanation was to be valid, the same spectrum of D had to be able to explain the shape of the observed $M_1(t)$ and $M_2(t)$ functions and the broad nature of the reorientation angle distribution $W(\theta,t)$ at the same time. The spectrum $g(\tau)$ of correlation time τ or the equivalent spectrum $g(D)$ of the rotational diffusion coefficient D can be evaluated from the correlation functions by means of a numerical procedure such as CONTIN [45]. When the correlation function can be represented by an analytical function, the spectrum can be obtained more conveniently by means of inverse Laplace transformation. In the case of a KWW function, with τ and β characterizing the function as given in Eq. (12), $g(D)$ can be calculated by [46]

$$g(D) = \frac{\tau}{\pi} \int_0^\infty \exp[-x\tau u - u^\beta \cos \pi\beta] \sin(u^\beta \sin \pi\beta) \, du \qquad (18)$$

where x is either 2D or 6D depending on whether it is $M_1(t)$ or $M_2(t)$ that is fitted by the KWW function. From the KWW fits to $M_1(t)$ and $M_2(t)$ shown in Fig. 13, the spectra $g(D)$ have been evaluated [9] by means of Eq. (18), and the results are plotted in Fig. 16. The spectra obtained from $M_1(t)$ and $M_2(t)$ respectively are not identical with each other. If we had used a numerical procedure to derive $g(D)$, instead of relying on the KWW function whose fit is good but not perfect, the spectra obtained would have differed in detail from the ones shown in Fig. 16, but the fact would still have stood that the two spectra derived from $M_1(t)$ and $M_2(t)$ respectively do not agree with each other.

Moreover, neither one of the two spectra shown in Fig. 16 is able to explain the shape of the observed $W(\theta,t)$ either. This is demonstrated in Fig. 17, where the spectrum $g_2(D)$ obtained from $M_2(t)$ as shown in Fig. 16 was used to construct a composite distribution $W(\theta,t)$ according to

$$W(\theta,t) = \int W_D(\theta,t) \, g(D) \, dD \qquad (19)$$

where $W_D(\theta,t)$ is the ideal distribution given by Eq. (15) for rotational diffusion with a single diffusion coefficient D. The fit of the constructed distribution to the observed one is not good; if we had used $g_1(D)$ in Fig. 16 derived from $M_1(t)$, the agreement would have been even worse. To achieve good fits to the observed $W(\theta,t)$s at different t, one in fact has to assume a spectrum $g(D,t)$ which itself changes and becomes broader with increasing time. It is difficult to say whether the notion of a time-dependent spectrum of diffusion coefficients or correlation time can be given a meaningful physical interpretation.

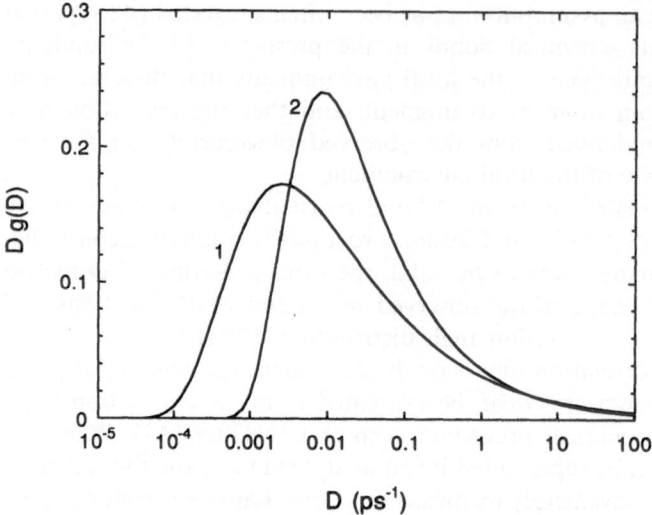

Fig. 16. Dg(D) plotted against D, where g(D) is the spectrum of rotational diffusion coefficient D evaluated from the assumption that nonexponential time-correlation functions arise from superposition of rotational diffusions with different Ds. *Curves 1 and 2* were obtained from the KWW function fit to $M_1(t)$ and $M_2(t)$, respectively, shown in Fig. 13

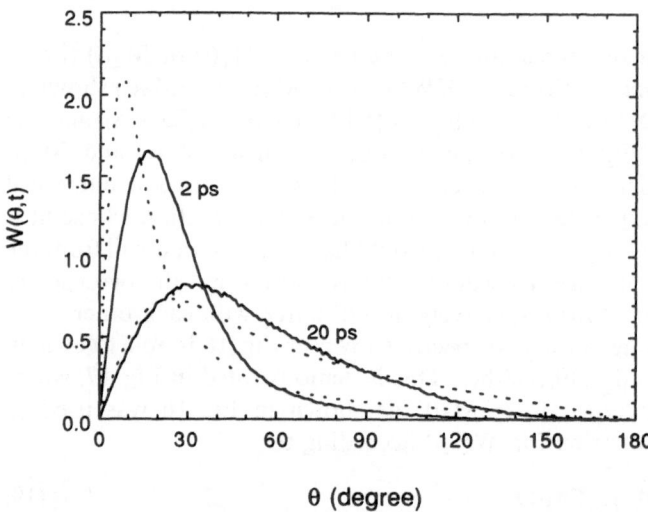

Fig. 17. Comparison of observed bond reorientation angle distribution $W(\theta,t)$ (*solid curves*) with curves (*dotted curves*) calculated on the assumption that the distributions arise from superposition of rotational diffusions. The spectrum g(D) of diffusion coefficient obtained from analysis of $M_2(t)$ and shown as *curve 2 in Fig. 16* is used

The difficulty of explaining our results by the assumption of superpositions of rotational diffusions can be pointed out again in another way. As was mentioned earlier and as is evident from Eqs. (16) and (17), the ratio τ_1/τ_2 should be equal to three if the rotational diffusion mechanism is valid. In our results given in Table 2, however, the ratio in most cases exceeds three. The ratio larger than three cannot be reconciled with the rotational diffusion even when the existence of a spectrum of diffusion coefficient is assumed. In such cases, the time correlation functions have to be given by

$$M_1(t) = \int e^{-2Dt} g(D) \, dD \tag{20}$$

and

$$M_2(t) = \int e^{-6Dt} g(D) \, dD \tag{21}$$

where $g(D)$ is the assumed spectrum of the rotational diffusion coefficient. If Eqs. (20) and (21) are to be obeyed simultaneously, however, the ratio τ_1/τ_2 has to be equal to three no matter what the $g(D)$ is. This can be shown as follows. From the definition of τ_i being the time at which $M_i(t)$ decays to $1/e$, one can write

$$1/e = \int \exp(-2D\tau_1) g(D) \, dD = \int \exp(-6D\tau_2) g(D) \, dD. \tag{22}$$

The second equality in the above is valid only if τ_1 equals to three times τ_2.

5.3 Anisotropy of Bond Reorientation Motion

The results given above clearly show that the bond reorientation motion cannot be regarded either as a rotational diffusion with a single diffusion coefficient or as a superposition of rotational diffusions with many different diffusion coefficients. Yet it is obvious that the bond reorientation occurs not through large angle jumps but rather through accumulation of many small angle motions. The broad angular distribution at all t exhibited by the observed $W(\theta, t)$ gives evidence to this view. We thus have to regard the reorientation motion as being similar to rotational diffusion but somewhat modified from it due to the complex nature of the polymer liquid environment. The formalism of the rotational diffusion would be exact only if a rigid body undergoes a succession of truly random reorientations. A polymer segment cannot undergo random motions because of the constraints imposed by the rest of the chain connected covalently to the bond as well as those imposed by the chain segments present in the neighborhood. We may thus regard the degree of departure from the true rotational diffusion, as represented by the departure of the observed τ_1/τ_2 ratio from three, as revealing the extent of constraints imposed on the bond reorientation motion at any given circumstance. If we look at Table 2, the ratio increases as the chain length increases at any given temperature, in agreement with the expectation that the intramolecular constraint is more severe with longer chains. The ratio in Table 2 is also observed to increase, for any given chain length, as the temperature is lowered toward the T_g, and this may be taken to indicate that at lower temperatures a motion of a bond has to be accompanied by cooperative motions of more segments in the neighborhood extending wider.

To place the above speculative considerations on a more quantitative basis, we now investigate the anisotropy of the bond reorientation motion. A pure rotational diffusion, resulting from random motions, should produce reorientations into all directions with equal probabilities. In contrast to this, the intramolecular and intermolecular constraints imposed to the reorientation of a bond are not isotropic and depends on the structure of the chain and the non-bonded neighbors in the vicinity. To investigate the anisotropy of the local motion, we now evaluate the time correlation functions $M_1(t)$ and $M_2(t)$ of the reorientation of a vector u that is not necessarily coincident with a bond but rather varies in all directions. As illustrated in Fig. 18, the direction of u is specified by Θ and Φ with respect to the local chain structure defined by two successive bonds. From the time correlation functions evaluated for the reorientation of u, the relaxation times τ_1 and τ_2 were determined as a function of Θ and Φ, and Fig. 19 gives the results obtained [9] for central segments in the system of chain length 500 at 300 K. An appreciable degree of anisotropy is seen in these plots. The longest relaxation time is seen when $\Theta = 0°$, that is when vector u is directed toward the chain axis, and the shortest relaxation time is exhibited when $\Theta = 90°$ and $\Phi = 90°$, that is when vector u is perpendicular both to the chain axis and to the zig-zag plane of the carbon bonds. The results indicate that the change in the direction of the chain axis takes place much more slowly than the rotation of the chain around its own local axis.

The anisotropy of segmental motion exhibited in Fig. 19 may arise, as noted above, either from the intramolecular or from the intermolecular constraint to the rotational motion. The anisotropy of orientational correlation decay was indeed noted already by Weber and Helfand [47] in their Brownian dynamics simulation of polyethylene of infinite chain length. Their orientational time-correlation function of the chord vector ($\Theta = 0°$) decayed much more slowly than those of either the bisector vector ($\Theta = 0°$, $\Phi = 0°$) or the out-of-plane vector ($\Theta = 0°$, $\Phi = 90°$). What they modeled was a phantom chain having no non-bonded (Lennard–Jones) interaction terms. Their model was thus subjected to no *inter*molecular constraints nor long range *intra*molecular interactions, and therefore the observed anisotropy was the consequences of the burden exerted by their immediate bonded neighbors. Bahar and Erman [48], using their dynamic rotational isomeric states model [49], studied the dynamic orientation

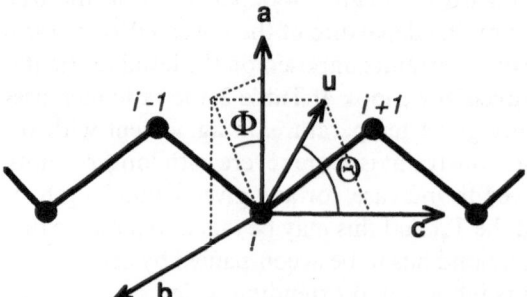

Fig. 18. Definition of Θ and Φ specifying the orientation of vector u in relation to the local structure of the chain

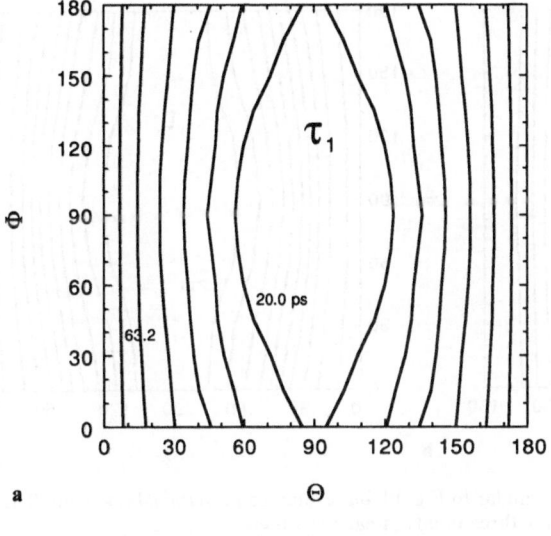

a

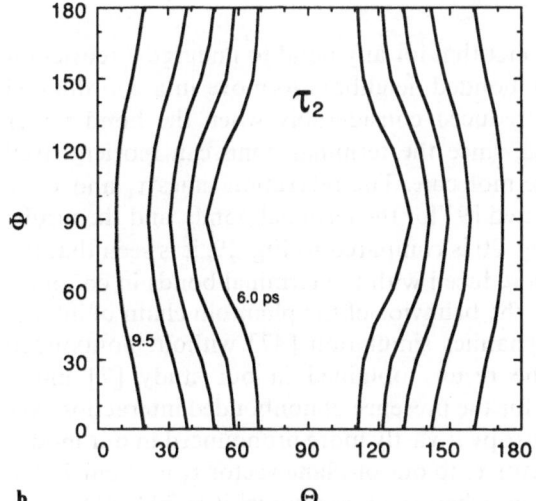

b

Fig. 19a,b. Dependence of reorientation relaxation times, τ_1 and τ_2, on the direction of vector **u** attached to *central* segments of chains of length 500 at 300 K. (The relaxation times shown are averages over ten bonds in the middle of the chain.) The direction of vector **u** is given by Θ and Φ in relation to the local structure of the chain. The τ_i values for the innermost and outermost contour lines are given on the plots. The successive contour lines denote levels increasing by a factor of $10^{0.1}$ ($= 1.259$) in the case of τ_1 and $10^{0.05}$ ($= 1.122$) in the case of τ_2

correlation of an *n*-alkane chain, and noted a similar anisotropy in the decay of orientational correlation depending on the direction of the vector in relation to the chain axis. Such an anisotropy was found even though the model considered the behavior of a single isolated alkane chain that is not subject to the influence of any neighboring chains.

In contrast, in this study modeling ensembles of chains at a realistic bulk density, any constraint to the reorientational motion of a bond is possibly exerted not only by the covalently bonded neighbors but also by the nonbonded neighbors. Our results in fact suggest that the *intermolecular* interaction is more important than the *intramolecular* interaction in causing the anisotropy. This conclusion is based on the following two observations. First, the intramolecular

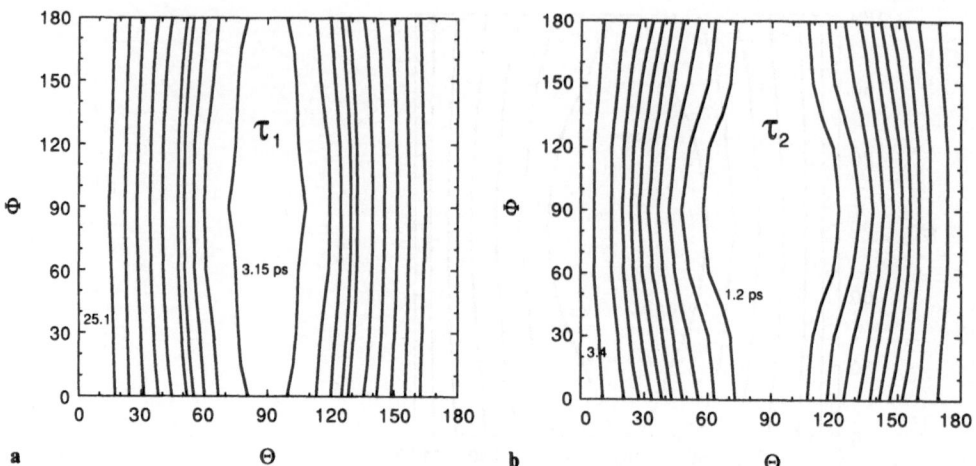

Fig. 20a,b. Contour maps of τ_1 and τ_2, similar to Fig. 19, but evaluated for *terminal* segments. The relaxation times shown are averages over three bonds at each chain end

constraint arises mainly from the fact that for any bond to undergo a reorientation it has to drag the covalently bonded neighbors to move in a coordinated manner. Such a burden will be reduced considerably when the bond under consideration is at the chain end, since the terminal bond can reorient itself without disturbing the rest of the molecule. The relaxation times τ_1 and τ_2 as a function of Θ and Φ were evaluated [9] for the terminal bonds, and the results are presented in Fig. 20. When Fig. 20 is compared to Fig. 19, it is seen that the extent of anisotropy has not been reduced with the terminal bonds in comparison to the central bonds. Second, the behavior of the phantom chain of infinite chain length in the Brownian dynamics simulation [47] without nonbonded interaction was compared to the results obtained in our study [7] under a closely similar condition except for the presence of nonbonded interaction. We found [7] that the degree of anisotropy is vastly more pronounced in our model. For example, the ratio of cord vector τ_1 to out-of-plane vector τ_1 is about 7–8 in the Brownian dynamics simulation, whereas in our model it is 300–400.

The finding that the bond reorientation motion is highly anisotropic and this anisotropy arises more from the *intermolecular* constraint due to the presence of the nonbonded neighbors has prompted us to propose the following pictorial model [7] to describe the local chain motion in bulk polymer liquids. In the melt a polymer chain can be considered to be confined to a "pipe" formed by its neighbors. Within the "pipe" the chain is tumbling rapidly around the local chain axis, while the "pipe", hence the chain axis itself, slowly changes its direction as the surrounding chains undergo relaxation. This picture of a "pipe" should be clearly distinguished from that of a "tube" proposed in the reptation model [50] of diffusion and flow in polymer liquids. The "pipe" and the "tube" differ from each other by at least an order of magnitude both in their spatial and temporal scales.

5.4 Decomposition of Bond Reorientation Motion into Two Components

The large disparity in the relaxation times between the chain axis and the vectors perpendicular to it suggests that the reorientation motion could be decomposed into two, more or less independent, components. The one involving rotation around the chain axis is represented by angle $\psi(t)$, while the other denoting the reorientation of the chain axis itself is represented by $\chi(t)$. Suppose that the local frame, defined by two successive bonds as in Fig. 18, is represented by the three axes $\mathbf{a}(0)$, $\mathbf{b}(0)$, and $\mathbf{c}(0)$ at time $t = 0$. The axes change their directions after a duration t to $\mathbf{a}(t)$, $\mathbf{b}(t)$, and $\mathbf{c}(t)$, as shown in Fig. 21b. One of the components of this motion is the change in the direction of the chain axis by $\chi(t)$, without any rotation around it, to bring the axes into $\mathbf{a}^*$, $\mathbf{b}^*$, and $\mathbf{c}^*$ shown in Fig. 21c. Next, the azimuthal rotation by $\psi(t)$ around the chain axis $\mathbf{c}^*$ [$=\mathbf{c}(t)$] brings $\mathbf{a}^*$ and $\mathbf{b}^*$ to coincide with $\mathbf{a}(t)$ and $\mathbf{b}(t)$, respectively, as in Fig. 21d. The time correlation functions can then be defined, in analogy to Eqs. (9) and (10), by

$$M_1(t) = \langle \cos \chi(t) \rangle \tag{23}$$

$$M_2(t) = (1/2)(3 \langle \cos^2 \chi(t) \rangle - 1) \tag{24}$$

and similarly with respect to $\psi(t)$. Figure 22 gives the time correlation functions for $\chi(t)$, evaluated with the polyethylene model and the freely-rotating chain model of infinite chain length, and Fig. 23 gives the same for $\psi(t)$. In Fig. 22 it is clearly seen that the chain axis, after an initial, fairly rapid, loss of correlation,

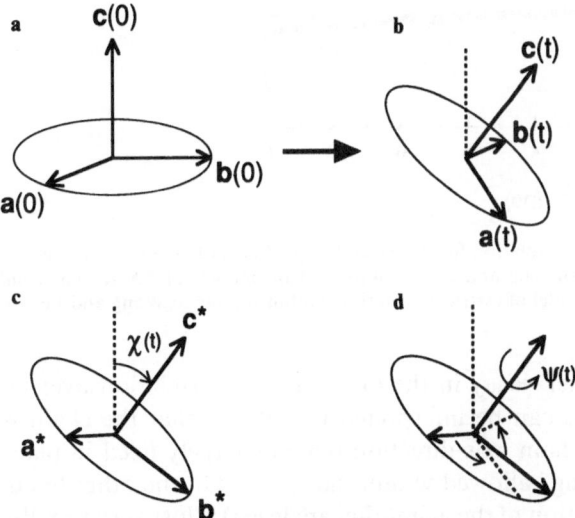

Fig. 21a–d. Schematic drawing to illustrate how the reorientation of the local frame, represented by **a**, **b**, and **c** axes, is decomposed into two components, one representing axial reorientation by $\chi(t)$ and the other azimuthal rotation by $\chi(t)$

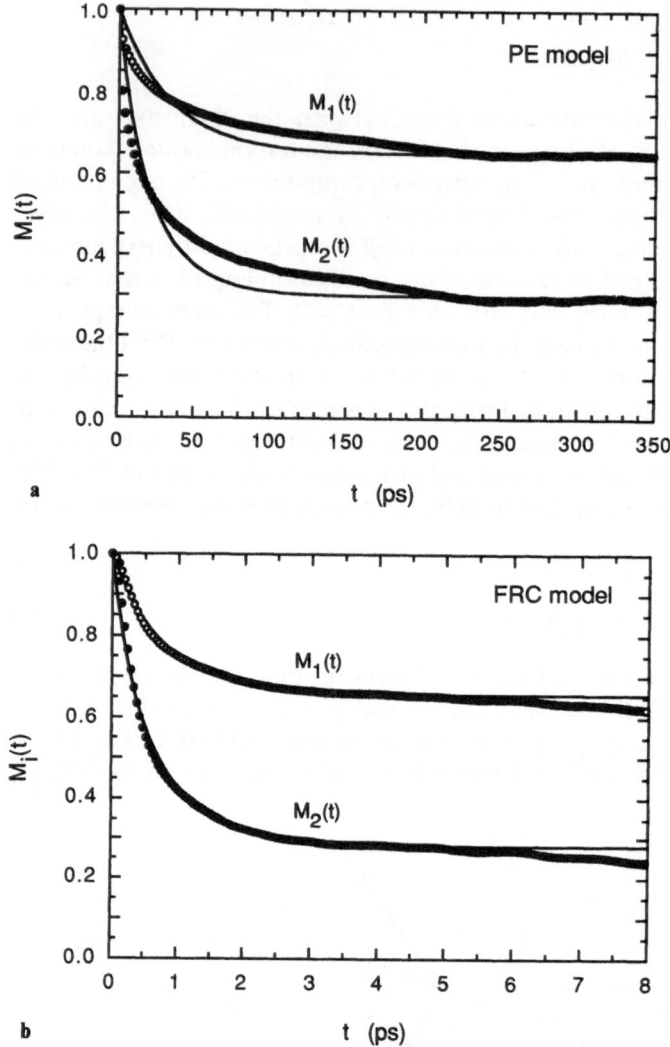

Fig. 22a,b. The *points* are the time-correlation functions $M_1(t)$ and $M_2(t)$ of chain axis reorientation angle $\chi(t)$ obtained with polyethylene and freely-rotating chain models at 300 K. The *solid curves* are the fit by the theoretical model of restricted rotational diffusion due to Wang and Pecora [44]

suffers only a very slow further decay in the orientational correlation over an extended period of time. This can be interpreted to indicate that the chain is trapped in a "pipe", and its chain axis direction remains largely fixed in place except for some local "wiggling" allowed within the "pipe." On the other hand, Fig. 23 shows that the correlation of the azimuthal angle $\varphi(t)$ is lost very rapidly. (Note that, in Fig. 23, $M_2(t)$ does not decay to zero because, for an azimuthal angle that ranges from 0 to 2π on a plane, $\langle \cos^2 \varphi(t) \rangle \to 1/2$ and $M_2(t) \to 1/4$ as $t \to \infty$.)

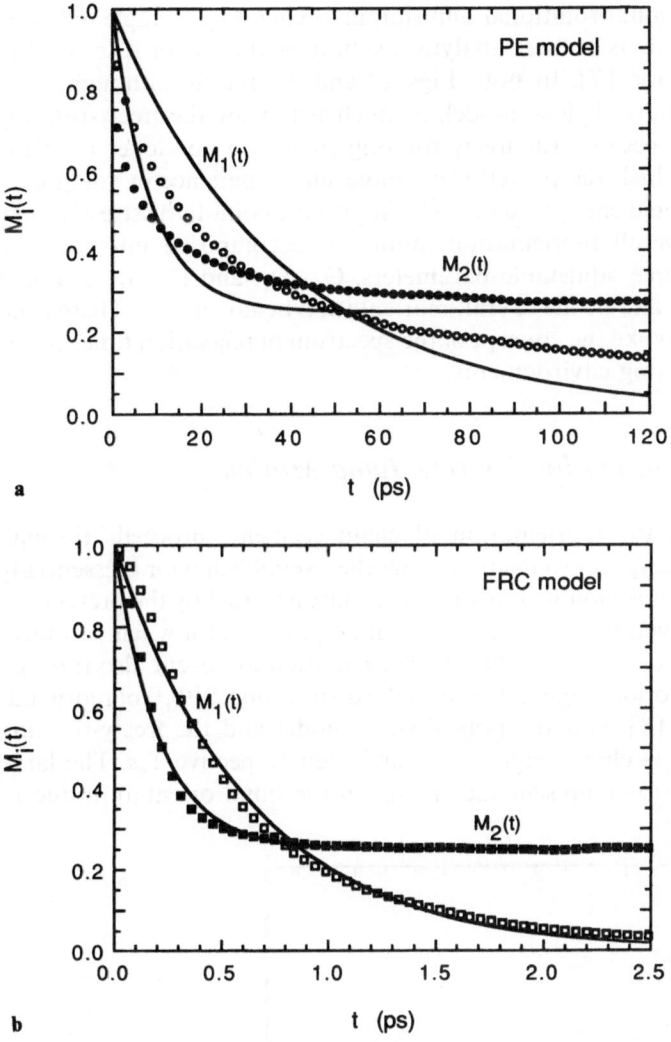

Fig. 23a,b. The *points* are the time-correlation functions $M_1(t)$ and $M_2(t)$ of azimuthal rotation angle $\psi(t)$ obtained with polyethylene and freely-rotating chain models at 300 K. The *solid curves* are the fit by the theoretical expression for rotational diffusion in two-dimension

Wang and Pecora [51] earlier presented an analytical solution applicable to a restricted rotational diffusion model, in which a rigid rod undergoing a rotational diffusion is allowed to change its direction only up to an angle Θ_0 and not beyond. The solution contains two parameters, Θ_0 and the diffusion coefficient D_r. In Fig. 22 the solid lines are those fitted by the analytic equations using nearly identical numerical values of Θ_0 and D_r for both $M_1(t)$ and $M_2(t)$ in either the polyethylene or the freely-rotating chain model. Similarly in Fig. 23 the observed time correlation functions are fitted by analytical solutions for a simple two-dimensional rotational diffusion involving only a single diffusion coefficient D_c. For details of these analytical solutions, the reader is referred to the original publication [7]. In both Figs. 22 and 23, the fit, although fairly satisfactory for the polyethylene model, is much better for the freely-rotating chain model. This is because the freely-rotating chains do not suffer from the degree of jerkiness which the polyethylene molecules experience in going over a trans-gauche torsional energy barrier. The important point to be stressed here is that, once the overall reorientation motion is decomposed into the two components, only three adjustable parameters, Θ_0, D_r, and D_c, are required altogether to explain and fit the overall reorientation behavior well. There is no longer any need to invoke the concept of the spectrum of relaxation times or the heterogeneity of relaxing environments.

5.5 Bond Reorientation by Discrete Jump Motion

As discussed above, the reorientation of chain segments proceeds through a succession of small angle movements, so that the overall behavior is essentially diffusional, although it is modified by the constraints imposed by the presence of the neighbors. However, under some conditions, especially at low temperatures approaching the T_g, it is observed that the reorientation can occur also through a large-angle jump motion. Figure 24 shows the distribution $W(\theta,t)$ of reorientation angle obtained [8] with the polyethylene model and the freely-rotating chain model of infinite chain length at around their respective T_gs. The large peaks at around 10–15° represent the change in the bond orientation due to

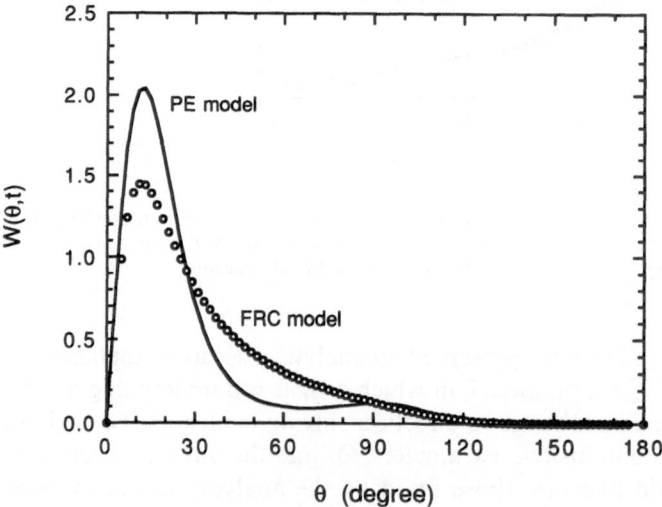

Fig. 24. Bond reorientation angle distribution $W(\theta,t)$ obtained at T_g for the polyethylene model (*solid curve*) is compared against the distribution obtained for the freely-rotating chain model (*open circles*) β at its own T_g

libration of the bonds within their cages, but of interest here is the small second peak at around 90° exhibited by the polyethylene model. The position of the second peak does not shift with time, suggesting that it does not represent a reorientation resulting from a succession of smaller angle displacements. The presence of a small secondary peak at around 70° was noted in the distributions $W(\theta,t)$ obtained [6] with alkane chain models of chain length 20 and smaller. Figure 24 shows that such a secondary peak is not exhibited by the freely-rotating chain model, and this suggests that the motion leading to the discrete jump is associated with the trans-gauche conformational transition. As the temperature is increased, the presence of the secondary peak becomes less noticeable, since the height of the secondary peak barely increases, while the tails of the main peak in $W(\theta,t)$ due to the diffusional motion spreads out to higher angles. This suggests that the activation energy for the discrete motion might be lower than the activation energy for the diffusional reorientation. The detailed behavior of such discrete jump motion and its mechanism are topics that require further study. We stress here that the presence of the two different mechanisms of bond reorientation has been revealed by the study of the distribution $W(\theta,t)$ itself, and the discrete jump motion would have been totally missed if we had relied on the study of time-correlation functions alone.

6 Summary

Molecular dynamics simulations have been performed on models of polyethylene (and *n*-alkane) liquids at their realistic bulk densities. The united atom approximation has been adopted. For comparison purposes a freely-rotating chain model, which is identical to the polyethylene model except that the energy barrier to the torsional rotation is absent, has also been studied. With these models it is feasible, without straining the available computational resources, to make simulation runs lasting in the order of nanoseconds in real time. Such a length of run is sufficient for investigation of properties that depend on the local motion of bonds and short sequences of segments. The local dynamic behavior observed in this study reflects the consequences of both *intramolecular* constraints exerted by the segments along the chain and *intermolecular* constraints exerted by other chains in the neighborhood. This has to be contrasted to the results of other studies in which a single isolated chain or a chain without nonbonded interactions is investigated.

When the temperature of the simulated system is lowered stepwise while maintaining a constant pressure, it has been found that the temperature coefficient of specific volume is found to exhibit an abrupt change at a temperature which we identify as the glass transition temperature. Around this temperature other properties of the system undergo changes that are characteristic of glass forming polymer liquids in laboratory experiments. The internal energy shows a break in temperature coefficient, the segmental self-diffusion coefficient diminishes to a negligible value, the trans-gauche conformational transition rate is reduced drastically, and the characteristic ratio of the polymer chain becomes frozen. These results lend support to the hope that the glass transition temperature of a new polymer yet unsynthesized could be estimated in the future. In attempting such an estimation, the large disparity in the cooling rate between simulation and experiment has to be kept in mind.

As a preliminary to the discussion of local dynamics, some aspects of the static structure of the simulated systems are discussed. The distribution of free volume has been investigated by noting the sizes of the cavities that are left unoccupied when the surrounding atoms are replaced by hard spheres. The distribution is very broad and exhibits a bimodal character. Even below the glass transition temperature these cavities change their sizes and shapes continually as the polymer chains engage in cooperative local motions. The extent of the short-range order present is revealed by the evaluation of the radial distribution function and the spatial orientation correlation of neighboring short subchains. The extent of the short range order steadily increases as the temperature is lowered through and below the glass transition temperature.

The chain local dynamics is investigated mainly through the evaluation of the bond reorientation dynamics. For this purpose we evaluate the distribution $W(\theta, t)$ of the bond reorientation angle θ as well as the time-correlation functions $M_1(t) = \langle \cos \theta \rangle$ and $M_2(t) = \frac{1}{2}(3 \langle \cos^2 \theta \rangle - 1)$. The observed distributions $W(\theta, t)$ are much broader than those expected from a rotational diffusion motion. Similarly, the observed time-correlation functions are much broader than a single exponential. They can be approximated well by the stretched exponential (or KWW function) with β equal to 0.4–0.5, which are in the range of values often found experimentally. When the relaxation times τ_1 and τ_2 are evaluated from $M_1(t)$ and $M_2(t)$, the ratio τ_1/τ_2 in most cases greatly exceeds the three that is expected of a rotational diffusion. Any attempt to explain such behavior by assuming a superposition of rotational diffusion processes, each having a different diffusion coefficient, is met with various kinds of contradictions.

The explanation is, rather, sought by investigating the anisotropy of the bond reorientation motion. When a vector is affixed to the local structure of a segment defined by two successive bonds and its direction in relation to the local structure is varied, the relaxation times τ_1 and τ_2 are found to be longest when the vector is along the chain axis, and shortest when it is perpendicular to the chain axis. Such an anisotropy of the bond reorientation motions arises from the constraints imposed by neighboring segments. We present evidence indicating that, of such constraints, those arising from *intermolecular*, nonbonded interactions are more important than those arising from *intramolecular* interactions with segments along the chain. In view of the anisotropy, the bond reorientation motion can be better understood if the overall reorientation is decomposed into two components, one denoting the slow reorientation of the chain axis and the other denoting the fast rotation of the chain around its local axis. It is shown that the chain axis reorientation can be described by the model of a restricted rotational diffusion with two adjustable parameters and the rotation around its own chain axis is well represented by a two-dimensional rotational diffusion with only one adjustable parameter. These findings lead to the proposal of a pictorial description of the local chain motions: in the melt the polymer chain can be considered to be confined to a "pipe" formed by its neighbors, and within the "pipe" the chain is tumbling rapidly around the chain axis, while the "pipe" itself slowly changes its direction as the surrounding chains undergo relaxation.

Examination of the reorientation angle distribution $W(\theta, t)$ obtained at T_g exhibits a secondary peak at around $\theta = 90°$ that does not change its angular position with time. This is interpreted to indicate the presence of a second reorientation mechanism by which the bond undergoes a large angle jump. Such a revelation of the presence of the second mechanism would not have been possible if only the time-correlation functions, and not the full distributions, had been evaluated. This provides an example in which the molecular modeling offers information that cannot easily be obtained from experiment.

Acknowledgment. This work was supported in part by NSF Grant DMR8909232. Most of the results reported here were obtained in collaboration with Drs. D. Rigby, H. Takeuchi, and H. Furuya, with the use of Cray YMPs at the Pittsburgh Supercomputing Center and the Ohio Supercomputing Center.

References

1. Rigby D, Roe RJ (1987) J Chem Phys 87: 7285
2. Rigby D, Roe RJ (1988) J Chem Phys 89: 5280
3. Rigby D, Roe RJ (1989) J Macromolecules 22: 2259
4. Rigby D, Roe RJ (1990) J Macromolecules 23: 5312
5. Takeuchi H, Roe RJ, Mark JE (1990) J Chem Phys 93: 9042
6. Rigby D, Roe RJ (1991) In: Roe RJ (ed) Computer simulation of polymers. Prentice Hall, Englewood Cliffs, NJ, p 79
7. Takeuchi H, Roe RJ (1991) J Chem Phys 94: 7446
8. Takeuchi H, Roe RJ (1991) J Chem Phys 94: 7458
9. Roe RJ, Rigby D, Furuya H, Takeuchi H (1992) Comp Polymer Sci 2: 32
10. Weber TA, Helfand E (1979) J Chem Phys 71: 4760
11. Verlet L (1967) Phys Rev 159: 98
12. Pant PVK, Boyd RH (1992) Macromolecules 25: 494
13. Toxvaerd S (1990) J Chem Phys 93: 4290
14. Theodorou DN, Suter UW (1985) Macromolecules 18: 1467
15. Flory PJ (1969) Statistical mechanics of chain molecules. Wiley-Interscience, New York
16. Clarke JHR, Brown D (1989) Mol Sim 3: 27
17. Fox H, Andersen HC (1984) J Phys Chem 88: 4019
18. Ferry JD (1980) Viscoelastic properties of polymers, 3rd edition. Wiley, New York
19. Bero CA, Plazek DJ (1991) J Polymer Sci, Part B, Polymer Phys 29: 39
20. Ichimura I, Ogita N, Ueda A (1978) Japan J Phys Soc 45: 252
21. Hiwatari Y (1982) J Chem Phys 76: 5502
22. Geil PH (1976) J Macromol Sci, Phys Ed 12: 173
23. Wang CS, Yeh GSY (1978) J Macromol Sci, Phys Ed 15: 107

24. Pechhold WR, Grossman HP (1979) Faraday Disc Chem Soc 68: 58
25. Flory PJ (1979) Faraday Disc Chem Soc 68: 14
26. Patterson GD, Kennedy AP, Latham JP (1977) Macromolecules 10: 667
27. Fischer EW, Strobl GR, Dettenmaier M, Stamm M, Steidle H (1979) Faraday Disc Chem Soc 68: 26
28. Jarry JP, Monnerie L (1978) J Polymer Sci, Polymer Phys Ed 16: 443
29. Thulstrup EW, Michl J (1982) J Am Chem Soc 104: 5594
30. Kornfield JA, Fuller GG, Pearson DS (1989) Macromolecules 22: 1334
31. Jacobi MM, Stadler R, Gronski W (1986) Macromolecules 19: 2884
32. Sotta P, Deloche B, Herz J, Lapp A, Durand D, Rabadeux J-C (1987) Macromolecules 20: 2769
33. Schmidt C, Wefing S, Blümich B, Spiess HW (1986) Chem Phys Lett 130: 84
34. Wefing S, Kaufmann S, Spiess HW (1988) J Chem Phys 89: 1234
35. Böttcher CJF, Bordewijk P (1978) Theory of electric polarization, Vol II, Elsevier, New York, Chap 11
36. Williams G (1979) Chem Soc Rev 7: 89
37. Williams G, Watts DC (1977) Trans Faraday Soc 66: 80
38. Ivanov EN (1963) Zh Eksp Teor Fiz 45: 1509 [Sov Phys JETP (1964) 18: 1041]
39. Roberts PH, Ursell DD (1960) Proc Roy Soc London A252: 317
40. Cukier RI, Lakatos-Lindenberg K (1972) J Chem Phys 57: 3427
41. Cukier RI (1974) J Chem Phys 60: 734
42. Anderson JE, Ullman R (1967) J Chem Phys 47: 2178
43. Silescu H (1971) J Chem Phys 54: 2110
44. Kovacs AJ, Aklonis JJ, Hutchinson JM, Ramos AR (1979) J Polymer Sci, Polymer Phys Ed 17: 1097
45. Provencher SW (1982) Comp Phys Comm 27: 213
46. Lindsey CP, Patterson GD (1980) J Chem Phys 73: 3348
47. Weber TA, Helfand E (1983) J Phys Chem 87: 2881
48. Bahar I, Erman B (1988) J Chem Phys 88: 1228
49. Bahar I, Erman B (1987) Macromolecules 20: 1368
50. Doi M, Edwards F (1986) The theory of polymer dynamics. Clarendon, Oxford
51. Wang CC, Pecora R (1980) J Chem Phys 72: 5333

Received: June 1993

Effect of Molecular Structure on Local Chain Dynamics:
Analytical Approaches and Computational Methods

I. Bahar[1], B. Erman[1] and L. Monnerie[2]
[1] Polymer Research Center and School of Engineering, Bogazici University, and TUBITAK Advanced Polymeric Materials Research Center, Bebek 80815, Istanbul, Turkey
[2] Ecole Supérieure de Physique et de Chimie Industrielles de Paris, PCSM, 10 Rue Vauquelin, Cedex 05, Paris 75231, France

Analytical and numerical approaches for the treatment of local chain dynamics are reviewed. Two main approaches are considered. First the dynamic rotational isomeric state (DRIS) formalism which has been developed as the dynamic counterpart of the classical rotational isomeric state theory of chain statistics is recapitulated and compared with other analytical treatments of chain conformational dynamics. The DRIS model is based on the solution of the multivariate master equation describing the time evolution of discrete chain configurations. The limitations and implications of the formalism are discussed. The real conformational and structural characteristics of the chain are rigorously included in the DRIS formalism. This feature makes it particularly suitable for application to specific polymer chains. However, the cooperative motion of relatively long chain segments is considered in the DRIS formalism through a mean field approximation, only. This shortcoming is overcome by a newly developed approach, referred to as the cooperative dynamics model. The latter takes account of the restrictions imposed on the mechanism of conformational transitions by chain connectivity and environmental frictional resistance in addition to those from internal conformational barriers. Comparison of both DRIS and the cooperative kinematics approaches with Brownian simulations indicates that these approaches may be advantageously used for predicting the time evolution of bond rotameric states, the distribution of angular reorientations of bonds in the neighborhood of a rotating one, the types of coupled transitions which are most strongly favored by the particular chain structure. This type of information which would otherwise be extracted from the statistical analysis of long trajectories generated by Brownian or molecular dynamics simulations is readily obtainable from either the DRIS or the cooperative kinematics theories.

Advances in Polymer Science, Vol. 116
© Springer-Verlag Berlin Heidelberg 1994

1 Introduction

1.1 General Perspective

The dynamics of a polymer chain results from the coordinated motions of its atoms. Depending on the length and time scale of observation, the dynamics may be analyzed in two broad regimes: (1) The Rouse-Zimm [1, 2] and (2) the local dynamics regimes. The former involves large scale motions of the chain. The smallest distinguishable length scale in this regime is that of a subchain that contains a sufficiently large number of bonds so that its end-to-end separation exhibits Gaussian distribution. The configurational statistics of polymers indicates that the number of bonds comprising such a subchain must be above fifty in a typical flexible polymer chain [3]. The mean-square end-to-end separation of the subchains and their mobility are the two relevant structural parameters which determine the equilibrium and dynamic behavior of the chain in this regime. The mean-square end-to-end separation underlies the elastic response of the chain whereas the mobility controls the time scale of motion. In this respect, fastest relaxational modes belonging to the Rouse-Zimm regime are those involving the cooperative reorganization of individual Gaussian subchains, and cover time scales of 10^{-7} s or more in solution. Dynamic processes encompassing length and time scales below these ranges would be identified as participating in local dynamics regime.

Evidence from various experimental techniques leads to the following information: (i) High frequency relaxational motions are localized along the chain and their characteristic correlation times are independent of the molecular weight M_w of polymers for chains exceeding ~ 200 bonds in solution and ~ 30 bonds in the melt. This is in contrast to the behavior of the slowest modes of the Rouse-Zimm model which scale as M_w^2. (ii) The mechanism and time scale of local relaxation depend on the identity of the specific chain. As a result, the structural and conformational details of investigated polymers need be considered. (iii) Activation energies associated with local motions are below 15 kJ/mol, in general. This value is of the order of the activation energy for a single isomeric rotation of a given backbone bond. This picture is incompatible with the crankshaft model of local conformational transitions in which two or more bond rotations occur at a time, and hence higher activation energies are required. Instead, the experimental observation supports isolated single-bond motions accompanied by small spatial rearrangements of the neighboring bonds along the chain. (iv) The orientational correlation functions for chains, in dilute solution as well as in the bulk state, show marked departures from the classical Debye type single exponential decay obtained from the relaxation of isolated dipoles. The occurrence of a nonexponential relaxation, resulting from the linear connectivity of the polymer chain, eventually enhanced by intermolecular effects, is central to the phenomenon of local chain dynamics.

Local motions may be broadly classified into three groups: (i) High frequency motions such as bond stretching, bond-angle bending and small-amplitude oscillations about rotational energy minima of backbone bonds. (ii) Side chain motions. (iii) Jumps or discrete transitions of backbone bonds from one rotational isomeric state to another, referred to as 'rotameric transitions'. In the present review, the stochastic process of rotameric transitions between rotamers is considered (i) within the framework of the dynamic rotational isomeric state (DRIS) formalism [4–10], and (ii) with the aid of a kinematic model [11–13] incorporating the long-range connectivity of the chain, referred to hereafter as the cooperative kinematics (CK) model. An alternative stochastic treatment of chain dynamics resorting to the RIS approximation has been independently performed by Ferrarini, Moro and Nordio [14].

The DRIS model has first been presented in a systematic way by Jernigan [4] and later developed in several papers by Bahar, Erman and Monnerie [5–10]. This is an extension of the classical rotational isomeric state model (RIS) [3] of chain statistics in which specific conformational and structural properties are rigorously considered. Short-range conformational energetics are emphasized in this approach. Here successive transitions of bond torsional states from one rotational isomeric minima to another, is assumed to be the only dynamic process governing local motions, thus neglecting the relaxation of other degrees of freedom of the chain such as bond stretching, bond angle bending and torsional librations. Interdependence of backbone torsional states and transitions is limited to first neighbors along the chain. In contrast to this Markov chain approach, the newly developed CK model systematically accounts for the restrictions imposed by the long-range connectivity of the chain and the frictional resistance of the environment [11, 12]. This approach is based on the minimization of energy spent against the environment during the conformational rearrangements of a chain. This process necessitates the cooperative response of all bonds to external perturbations. Motions engendering large amplitude swinging of the chain segments surrounding a rotating bond are automatically impeded in the CK model by concerted distortion of bond rotational states as well as translation and reorientation of the chain as a whole. In view of these considerations, the DRIS approach seems more appropriate for dilute polymer systems in which environmental effects might be negligibly small compared to internal barriers to rotations. The CK model, on the other hand is suitably applicable to the dynamics of polymer chains in dense media.

Clearly, the use of detailed molecular level structural and conformational information is essential in analyzing motions belonging to local dynamics regime. The most complete theoretical approach of studying local chain dynamics is the molecular dynamics (MD) simulation method in which all components of a given polymeric system are considered at the atomistic level. MD simulations are accurate to the extent that the energy and length parameters and functions used therein yield a realistic representation of interatomic interactions, and the validity and realism of these parameters may be accepted to be sufficiently well established for most cases. However, due to the large number of degrees of freedom which needs be handled in this approach, MD simulations are suitable only for the investigation of relatively short chains and/or can be extended only to time scales up to tenths of nanoseconds. Brownian dynamics on the other hand present a computationally more efficient tool inasmuch as the solvent is taken therein as a continuum and its role is mimicked by Gaussian distributed white noise. However, the attainment of the timescale of experimental observables and the establishment of statistically reliable relationships still necessitate considerable computational times and occasionally the adoption of simplified models has been resorted to, such as unified atoms approximating the collective behavior of groups of atoms in the backbone. The DRIS and CK models and methods presented here, being based on assumptions that render the problem analytically tractable, are less precise than those two numerical approaches. Nevertheless, they present the advantage of gaining an understanding of the mechanism and basic factors controlling the stochastic process of local relaxation in specific polymers without recourse to extensive simulations. In particular, as will be demonstrated below, several characteristics of local chain dynamics, such as the anisotropy of the motion, the temperature dependence, the stretched exponential decay of correlations at intermediate times, the specific types of torsional coupling between neighboring bonds, the range of orientational correlations along the chain, which have been partly clarified at the expense of extensive simulations, may be readily reproduced with rather small computational effort by using the DRIS and CK models.

The paper is organized as follows. Experimental and theoretical work related to local chain dynamics are outlined in this section, with emphasis on experimentally observable dynamic properties which lend themselves to quantitative estimation by the two present models. This is not a complete presentation but a summary of some techniques and observations relevant to our theoretical treatment, only. In Sect. 2, the stochastic theory of conformational transitions and the DRIS approach are presented. The master equation formalism constitutes the basic mathematical approach of the DRIS model, and is outlined as an alternative form of the Chapman-Kolmogorov equation defining Markov processes. Its relation to Fokker-Planck equation is briefly retraced in Appendix 5.1. The DRIS model and assumptions for an isolated polymer chain are presented in the same section with the matrix multiplication method as an efficient tool to compute correlation functions associated with the stochastic process of rotameric transitions. The DRIS predictions are compared with those of other theories, the results from Brownian dynamics simulations and experimental measurements in the second part of this section. The resistance to motion imposed by either chain connectivity and/or intermolecular interactions may be approximately incorporated into the DRIS formalism by adopting an effective frictional coefficient increasing with the size of the reorienting segment. Also, a bistate model may account for the heterogeneity of the environment as a first approximation. The implementation of these refinements in the DRIS formalism for a sound interpretation of the behavior of polymers in the bulk state, is presented in Sect. 2.2. Yet, a more rigorous approach describing the cooperative relaxation of a chain segment in the presence of constraints is the CK formalism which is presented in Sect. 3. A critical discussion of the achievements on the one hand, and limitations on the other, of both approaches is given in Sect. 4, and future prospects for a more complete and realistic analysis of local chain dynamics are presented.

1.2 Experiments on Local Orientational Dynamics

Local dynamics of polymers are probed by spectroscopic techniques such as time-resolved fluorescence anisotropy, nuclear magnetic resonance, electron spin resonance, quasielastic neutron scattering, Rayleigh scattering depolarization, Raman line shape analysis, ultrasonic relaxation and high frequency dielectric relaxation. Here, some general aspects of these techniques are summarized. The reader is referred to the reviews by Monnerie [15], Heatley [16, 17] Spiess [18] and more recently by Ediger [19] and several references cited therein, for a more detailed presentation of the experimental techniques and studies on local chain dynamics. Orientational autocorrelation functions (OACF) and/or their spectral densities are in general probed in these experiments. The OACF associated with the motion of a given unit vector $\mathbf{m}(\tau)$ may be expressed in the general form

$$M_i(\tau) = \langle P_i(\mathbf{m}(0) \cdot \mathbf{m}(\tau)) \rangle \ . \tag{1}$$

Here P_i is the Legendre polynomial of order i, the angular brackets refer to the ensemble average over all conformational transitions undergone within the time interval τ by the chain segment to which $\mathbf{m}$ is rigidly affixed. The subscript i assumes the values 1 and 2 in the respective cases of first and second OACFs.

In time-resolved fluorescence anisotropy studies the orientational motion of a fluorescent label rigidly affixed to a point on the chain, or that of a probe dissolved in the system, are studied. These experiments yield directly the time decay of the second orientational autocorrelation function, $M_2(\tau)$, associated with the unit vector $\mathbf{m}(\tau)$ along the transition moment of the chromophore at time τ. $M_2(\tau)$ is given by

$$M_2(\tau) = (3/2) \langle [\mathbf{m}(\tau_0) \cdot \mathbf{m}(\tau_0 + \tau)]^2 \rangle - 1/2 \ . \tag{2}$$

Here, the brackets refer to an ensemble average at a given time τ, or alternatively a time average over various initial times τ_0. A correlation time t_{corr} for the investigated motion may be deduced from the integral of $M_2(\tau)$, according to

$$t_{corr} = \int_0^\infty \frac{[M_2(\tau) - M_2(\infty)]}{[M_2(0) - M_2(\infty)]} \, d\tau \ . \tag{3}$$

$M_2(\tau)$ decays from unity to zero at long times when the vector $\mathbf{m}(\tau)$ is viewed from a laboratory-fixed frame. In the case of a chain-embedded local frame with respect to which $M_2(\tau)$ is expressed, $M_2(\infty)$ equates to a finite asymptotic value due to the absence of isotropic overall tumbling superposed on the segmental motion. Local chain motions are detected in these experiments inasmuch as the dimensions of the labels or the probes are of the order of the size of a chain repeat unit. In particular, holographic grating experiments have proven suitable to probe sub-picosecond motions in polymer chains.

Nuclear magnetic resonance spectroscopy is another powerful technique for probing local chain dynamics. With this technique, the spectral density of local relaxational frequencies, spin-lattice relaxation times and orientational correlation times are characterized. Here, the spectral density function is the Fourier transform of the OACF associated with the investigated internuclear vector, given as

$$J(\omega) \equiv \int_0^\infty M_2(\tau) \exp\{i\omega\tau\}\, d\tau. \tag{4}$$

The internuclear vector is usually taken along the direction of the ^{13}C–H bond in ^{13}C-NMR, and in the direction of the proton pair in ^{1}H-NMR, assuming a purely ^{1}H–^{1}H dipolar relaxation mechanism. The spin-lattice relaxation times, T_{1C} and T_{1H} measured in ^{13}C–H and ^{1}H-NMR respectively, may be expressed in terms of $J(\omega)$ as

$$1/T_{1C} = [n_0(h/2\pi)^2\, \gamma_C^2\, \gamma_H^2/10 r_{CH}^6]$$

$$\times [J(\omega_H - \omega_C) + 3J(\omega_C) + 6J(\omega_H + \omega_C)] \tag{5}$$

and

$$1/T_{1H} = [3(h/2\pi)^2\, \gamma_H^4/10 r_{HH}^6][J(\omega_H) + 4J(2\omega_H)]. \tag{6}$$

Here γ_C and γ_H are the gyromagnetic ratios, ω_H and ω_C are the resonance frequencies of the hydrogen and carbon nuclei, h is the Planck constant, r_{CH} is the internuclear distance, and n_0 denotes the number of protons contributing to the dipolar interaction with the ^{13}C nucleus. Along similar lines, orientational correlation times relating to local motions may be determined by measurements of electron spin resonance. The recently developed two-dimensional (2D) exchange NMR method applied to polymers has clarified many details of molecular reorientation geometry. Typically, local motions with correlation times above microseconds, in systems not far above the glass transition temperature, have been examined with the 2D-NMR technique.

Dielectric spectroscopy is one of the earlier techniques for probing local chain dynamics as reviewed by Williams [20]. This technique measures the time-dependent complex dielectric constant $\varepsilon^*(\tau)$ which is then transformed by the use of linear response theory into the dipole–dipole correlation function $\Phi(\tau)$, according to

$$\frac{\varepsilon^*(\tau) - \varepsilon(\infty)}{\varepsilon(0) - \varepsilon(\infty)} = L[d\Phi(\tau)/d\tau]. \tag{7}$$

Here $\varepsilon(0)$ and $\varepsilon(\infty)$ are the static and infinite time dielectric constants, and L denotes the Laplace transform. The normalized response function $\Phi(\tau)$ is expressed in terms of the instantaneous dipole vectors μ_i of the molecules as

$$\Phi(\tau) = \frac{\sum_i \sum_k M_1^{ik}(\tau) - M_1^{ik}(\infty)}{\sum_i \sum_k M_1^{ik}(0) - M_1^{ik}(\infty)} \tag{8}$$

with

$$M_1^{ik}(\tau) \equiv \langle \mu_i(\tau_0) \cdot \mu_k(\tau_0 + \tau) \rangle . \tag{9}$$

$M_1^{ik}(\tau)$ is the first OACF associated with the dipole moment vectors μ_i and μ_k. The major contribution to $\Phi(\tau)$ is due to autocorrelation functions (i = k) in the summations, while cross-correlations (i $\neq$ k) play a secondary role, their effect diminishing rapidly with increasing separation of dipoles along the chain. The summations in Eq. (8) include all pairs of interacting dipoles within the window of observation. The possibility of measuring the relaxation of dipoles with components perpendicular to the chain contour [21] makes this technique particularly useful for the study of local motions.

1.3 Theoretical Approaches to Local Dynamics

The emphasis on the role of torsional barriers in local dynamics goes back to the work of Kuhn in 1945, where the apparent 'stiffening' of chains at high frequencies, due to contributions from internal energy barriers was originally recognized [22, 23], and referred to as 'internal friction'. Based on the idea of internal friction, the local dynamics of an isolated chain has been investigated in several papers [24–29]. The number of Kuhn segments above which Rouse dynamics is applicable and below which the effects of internal energy barriers have to be considered is discussed by deGennes [30]. Incorporation of torsional energy barriers into local chain dynamics by Allegra through the use of the Langevin equation [31–33] may be cited as an important step in the development of the theory, where the intramolecular potential of polyethylene (PE) is approximated by a quadratic function. The Langevin theory was further improved [34–36] and applied to the analysis of the dynamics of atactic polystyrene [37] and poly(dimethyl siloxane) [38]. Although these studies recognize the rotational mobility of backbone bonds as a major factor controlling local motions, the interdependence of bond conformations, resulting from chain connectivity, is not accounted for.

The near neighbor interdependence along the chain, which is of fundamental importance in prescribing the type of mechanism of local motions, is first considered by Glauber in his study of time dependent statistics of the one-dimensional array of a two-spin state Ising model [39]. As described in more detail below, the time-dependent correlation of spin states exhibits a strong departure from single exponential decay. Results of Glauber, like those obtained in a later study by Shore and Zwanzig [40] for one-dimensional polymer chains with perpendicular dipoles, separate the relaxation interval into three regions. The short and long time regions exhibit almost a single exponential decay while the intermediate region of the spectrum shows a nonexponential behavior, apparently resulting from the combined contribution of a distribution of relaxation frequencies. The single exponential decay predicted at short times ($\leqslant 10$ ps) might be attributed to the effect of highest frequency motions, like small amplitude oscillations in bond lengths and angles, or weak librations about bond torsional minima, which do not necessitate a cooperative reorganization of the chain but involve a single or rather narrow range of relaxational modes. Likewise, the theoretically predicted single exponential behavior at long times reflects the overall molecular diffusion as a rigid body and is beyond the scope of the present work. The intermediate range, which covers several decades of time involving both the Rouse-Zimm and local dynamics regimes described above, will be of interest here.

Consequences of the three-dimensional structure of the chains were first studied by Helfand [41] by considering specific conformational changes in short sequences of bonds and their effects on the remaining portions of the chain called the 'tails'. Conformational changes are classified into three types according to the effect they have on the tails. Type I changes are those which leave the tails undisturbed. These are the crankshaft motions of Shatzki [42] and Boyer [43]. They involve simultaneous rotational transition of more than a single bond. The three-bond and four-bond motions on tetrahedral lattice introduced by Monnerie et al. [44–46] belong to this category. In Type II motions, asserted to be a highly probable type of transitions, one of the tails undergoes a translational motion as a result of the conformational transition(s) that take(s) place in the central sequence. These motions involve correlated sequential transitions between next-to-nearest neighbors along the chain, as will be described below. In Type III motions, the transition of the bond in question results in a change of the orientation of one tail with respect to the other.

The transition mechanisms described briefly in the preceding paragraph form the basis of conformational jump models. The bond orientational autocorrelation function resulting from conformational jumps has been investigated by various analytical methods. Monnerie and collaborators used three-bond and four-bond motions on tetrahedral lattice and derived a Markovian master equation for the orientational probability and the autocorrelation of a bond along the backbone [44–47]. Jones and Stockmayer [48] treated the master equation in matrix form and obtained the eigenvalue solution for the OACF. Bendler and Yaris [49] transformed the master equation into a one-dimensional diffusion equation for an infinitely long chain. Hall and Helfand solved the master equation for conformations using Pauli spin matrices in terms of modified Bessel functions [50]. All of these theoretical studies as well as others [51–54] contain the essence of conformational transitions, mainly the near-neighbor interdependence associated with chain connectivity. They are nevertheless of limited applicability since they do not incorporate the spatial struc-

tural features of real polymer chains. The DRIS scheme brings an improvement
to the study of local chain dynamics by accounting for the effects of specific
conformational and structural properties of polymers.

A comprehensive master equation approach to the problem of conformational kinetics of polymers, incorporating real structural properties of polymers,
has been performed by Moro and collaborators [14, 55–63]. In their studies, the
effects of torsional energy barriers are incorporated on local chain dynamics by
applying the projection operator formalism to the Smoluchowski equation to
obtain discrete master equations. The theory of Moro et al. is based on the
master equation formalism and Kramers' rate theory [64]. Kramers' rate
theory, together with the free volume theory, has previously been used by
Mashimo in the analysis of local chain dynamics to estimate the relaxation time
for local conformational transitions in a polymer chain [65]. The master
equation formalism and the Kramers' multidimensional rate theory forms the
basis of the DRIS model and is described in detail below. The model developed
by Moro and collaborators and the DRIS approach of the present authors show
similarities in many respects. The present review concentrates on the DRIS
model of the present authors.

An exact assessment of local configurational dynamics in the presence of
constraints imposed by the tails, which has also been studied by several other
authors [66–69], is possible by Brownian dynamics (BD) and molecular dynamics (MD) simulations [70]. In general, trajectories of bond dihedral angles
resulting from simulations exhibit the occurrence of discrete jumps between
rotameric states, interrupted by relatively longer residence periods at rotameric
minima [71, 72]. This behavior is illustrated in Fig. 1, where a typical dihedral
angle history of the central bond is displayed for a chain of 50 bonds, as
observed during a 1.2 ns BD simulation at 400 K [73]. The preference for
rotational angles centered within $\pm 30°$ fluctuations about the rotational
isomeric states $\phi = 0°$, $-120°$, and $+120°$ is clearly apparent from those
trajectories. Bond rotations are accommodated either by the cooperative small

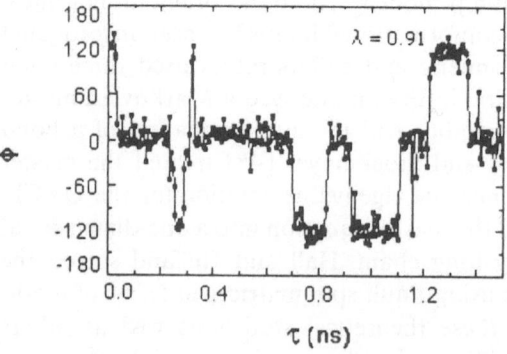

Fig. 1. An example trajectory of 1.2 ns for the dihedral angle ϕ of the central bonds in a polyethylene chain of 50 bonds, resulting from BD simulations at 400 K. The ends of the chain are held fixed
at an extension ratio $\lambda = 0.91$, where λ is defined as the ratio $r/\langle r^2 \rangle^{1/2}$ of the end-to-end distance r to
the unperturbed chain length

amplitude rearrangements of several topological neighbors, or by a correlated isomeric jump induced in the close neighborhood of the rotating bond, thus highly localizing the motion. Correlated rotameric jumps succeed the original motion within very short time intervals. The strength and type of correlations depend however on the specific geometry and conformational nature of the chain. In polyethylene chains, strong correlations between torsional motions of next-to-nearest neighboring bonds are observed. These are negligibly small in polyisoprene, apparently due to the departure of bond angles from approximately tetrahedral geometry, and the heterogeneity of the conformational behavior of backbone bonds [74, 75]. On the other hand, in poly(dialkyl siloxanes), the rotational states and transitions of the pair of bonds centered about oxygen atoms are strongly coupled. Secondly, there is a totally different level of coupling between adjacent monomeric units, extending between bonds i and i + 3, which covers timescales of about three orders of magnitude slower than the first type of correlated motions [72]. These observations lend support to the adoption of specific structural and conformational features of polymer chains for a thorough understanding or realistic estimation of local dynamic properties. BD and MD simulations performed with realistic potentials serve this purpose. These two techniques are reviewed in this volume [76, 77]. However, due to computational speed and memory constraints, the simulated time and length scales usually fall below those observed in experiments. This is the main motivation underlying the development of the DRIS and CK formalisms.

2 Stochastic Theory of Conformational Transitions and Dynamic Rotational Isomeric State (DRIS) Approach

2.1 Theoretical Background

2.1.1 Markov Processes and Master Equation Formalism

We consider the discrete state space $\{\Omega\} \equiv \{\Omega_\alpha, \Omega_\beta, \ldots, \Omega_{\nu^{N-2}}\}$ of the ν^{N-2} rotational isomeric configurations of a chain of N bonds having ν states accessible to each bond. The stochastic process of $\nu^{N-2} \times \nu^{N-2}$ transitions between those configurations is the object of study. In the following, the terms "states" and "system" will be used interchangeably for "configurations" and "chain", respectively.

The Markov assumption applies in the sense that the state of the system at a given time $\tau_0 + \tau$, is prescribed entirely by its state at time τ_0. Accordingly, the probability $P(\Omega_\alpha, \tau_0 + \tau)$ of state Ω_α at time $\tau_0 + \tau$ is written in terms of the time-delayed conditional probability $C(\Omega_\alpha, \tau_0 + \tau | \Omega_\beta, \tau_0)$ of state Ω_α at time $\tau_0 + \tau$ given state Ω_β at time τ_0

$$P(\Omega_\alpha, \tau_0 + \tau) = \sum_\beta C(\Omega_\alpha, \tau_0 + \tau | \Omega_\beta, \tau_0) P(\Omega_\beta, \tau_0) \qquad (10)$$

where the summation is carried over all original states Ω_β. $C(\Omega_\alpha, \tau_0 + \tau | \Omega_\beta, \tau_0)$ is also referred to as the transition probability of passage from Ω_β to Ω_α during the time range $(\tau_0, \tau_0 + \tau)$.

Following the hierarchy of equations describing distribution functions of high orders in Markov processes, the time-delayed conditional probability of state Ω_α at time $\tau_0 + \tau + \tau'$ given state Ω_β at time τ_0 is given by the Chapman-Kolmogorov equation [78]

$$C(\Omega_\alpha, \tau_0 + \tau + \tau' | \Omega_\beta, \tau_0) = \sum_\gamma C(\Omega_\alpha, \tau_0 + \tau + \tau' | \Omega_\gamma, \tau_0 + \tau)$$

$$\times C(\Omega_\gamma, \tau_0 + \tau | \Omega_\beta, \tau_0) . \tag{11}$$

Here the summation is performed over all intermediate configurations. For stationary processes, which occur at equilibrium in the absence of any external perturbation, the choice of the time origin is immaterial. The state at time $\tau_0 + \tau$ depends upon the elapsed time τ only, such that knowledge of the transition probabilities for a given time interval τ together with the equilibrium distribution of probabilities is sufficient to completely specify the state of the system. Equations (10) and (11) may be conveniently rewritten in this case as

$$P_\tau(\Omega_\alpha) = \sum_\gamma C_\tau(\Omega_\alpha | \Omega_\gamma) P_0(\Omega_\gamma) \tag{12}$$

and

$$C_{\tau+\tau'}(\Omega_\alpha | \Omega_\beta) = \sum_\gamma C_{\tau'}(\Omega_\alpha | \Omega_\gamma) C_\tau(\Omega_\gamma | \Omega_\beta) \tag{13}$$

respectively. Here the conditional probability for the transition $\Omega_\beta \to \Omega_\alpha$ within the time interval τ is denoted by $C_\tau(\Omega_\alpha | \Omega_\beta)$ and equilibrium probabilities are indicated with the subscript zero. $C_\tau(\Omega_\alpha | \Omega_\beta)$ may be conveniently expressed as the ij-th element of a conditional probability matrix C_τ of order v^{N-2}. Likewise, the probabilities of occurrence of the configurations Ω_α, $1 \leqslant \alpha \leqslant v^{N-2}$, may be organized as the elements of the probability vector $P(\Omega)$. For stationary processes, $P(\Omega)$ being conserved and equal to the equilibrium array $P_0(\Omega)$ at any τ, it is appropriate to omit the subscript τ in $P(\Omega)$.

For small τ', $C_{\tau'}(\Omega_\alpha | \Omega_\gamma)$ may be written as [78]

$$C_{\tau'}(\Omega_\alpha | \Omega_\gamma) = (1 - a_0(\Omega_\gamma) \tau') \, \delta(\Omega_\alpha, \Omega_\gamma) + \tau' A(\Omega_\alpha | \Omega_\gamma) + O(\tau') \tag{14}$$

where $A(\Omega_\alpha | \Omega_\gamma)$ is the *transition probability per unit time* from Ω_γ to Ω_α, $\delta(\Omega_\alpha, \Omega_\gamma)$ is the Kronecker delta which is equal to one for $\Omega_\alpha = \Omega_\gamma$ and vanishes for $\Omega_\alpha \neq \Omega_\gamma$. $O(\tau')$ represents terms which are higher order in τ' and thus tends to zero at small times. The coefficient $a_0(\Omega_\gamma)$ is given by

$$a_0(\Omega_\gamma) = \sum_\gamma A(\Omega_\alpha | \Omega_\gamma) . \tag{15}$$

The first term on the right hand side of Eq. (14) yields the probability that no transition occurs at time τ'. This term operates for the case $\Omega_\alpha = \Omega_\gamma$. The second term, which is the only term surviving for $\Omega_\alpha \neq \Omega_\gamma$, is a linear approximation to $C_{\tau'}(\Omega_\alpha | \Omega_\gamma)$, becoming exact as τ' approaches zero. In parallel with the conditional probability matrix $\mathbf{C}_\tau$, it will prove mathematically convenient to define a transition rate matrix $\mathbf{A}$ of order v^{N-2} whose element $\alpha\gamma$ is $A(\Omega_\alpha | \Omega_\gamma)$, for $\alpha \neq \gamma$. Substitution of Eq. (14) into Eq. (13) leads to

$$
\begin{aligned}
C_{\tau+\tau'}(\Omega_\alpha | \Omega_\beta) = &(1 - a_0(\Omega_\alpha)\tau')C_\tau(\Omega_\alpha | \Omega_\beta) \\
&+ \tau' \sum_\gamma A(\Omega_\alpha | \Omega_\gamma)C_\tau(\Omega_\gamma | \Omega_\beta)
\end{aligned}
\tag{16}
$$

which, upon dividing by τ' and rearranging, reduces to the *master equation*

$$
\begin{aligned}
\partial C_\tau(\Omega_\alpha | \Omega_\beta)/\partial\tau = \sum_\gamma \{ &A(\Omega_\alpha | \Omega_\gamma)C_\tau(\Omega_\gamma | \Omega_\beta) \\
&- A(\Omega_\gamma | \Omega_\alpha)C_\tau(\Omega_\alpha | \Omega_\beta)\}
\end{aligned}
\tag{17}
$$

in the limit as $\tau' \to 0$. $a_0(\Omega_\alpha)$ in Eq. (16) has been replaced by using Eq. (15) in obtaining the master equation. The master equation may be rewritten in the same form as Eq. (17) by replacing all of the conditional probabilities in the latter with the corresponding joint probabilities upon multiplication of both sides of the equality by the equilibrium probability $P_0(\Omega_\beta)$ of state (Ω_β). Alternatively, the identification of a given original configuration (Ω_β) may be totally removed, and consequently Eq. (17) becomes [79]

$$
\partial P(\Omega_\alpha)/\partial\tau = \sum_\gamma \{ A(\Omega_\alpha | \Omega_\gamma)P(\Omega_\gamma) - A(\Omega_\gamma | \Omega_\alpha)P(\Omega_\alpha)\}
\tag{18}
$$

which may be organized in matrix form as

$$
\partial P(\Omega)/\partial\tau = \mathbf{A} P(\Omega) .
\tag{19}
$$

From comparison of Eqs. (18) and (19) it is clear that the α-th diagonal element of the transition rate matrix $\mathbf{A}$ is defined by

$$
A(\Omega_\alpha | \Omega_\alpha) = -a_0(\Omega_\alpha) = -\sum_\gamma A(\Omega_\gamma | \Omega_\alpha)
\tag{20}
$$

thus representing the total rate of escape from state Ω_α, whereas the off-diagonal elements are simply $A(\Omega_\alpha | \Omega_\gamma)$ as delineated above. From the principle of detailed balance, $A(\Omega_\alpha | \Omega_\gamma)$ reads

$$
A(\Omega_\alpha | \Omega_\gamma) = A(\Omega_\gamma | \Omega_\alpha)P(\Omega_\alpha)/P(\Omega_\gamma)
\tag{21}
$$

where the instantaneous probabilities $P(\Omega_\gamma)$ and $P(\Omega_\alpha)$ remain unchanged and equal to the equilibrium probabilities $P_0(\Omega_\gamma)$ and $P_0(\Omega_\alpha)$ for stationary processes.

The master equation, Eq. (19), constitutes the basic differential equation which will be solved for the description of the stochastics of local chain motions in the DRIS formalism. It is noted that any isolated physical system can be described as a Markov process by proper introduction of all microscopic variables [78]. Also, the above derivation demonstrates that the master equation is simply an equivalent form of the Chapman-Kolmogorov equation for Markov processes, as explained in more detail by van Kampen [78]. In view of these arguments, we might project that the dynamics of a given chain may be described with high degree of accuracy provided that the rates adopted in the matrix **A** give a realistic account of the cooperative nature of local conformational transitions. In principle, if those rates are deduced by methods such as molecular dynamics or Brownian dynamics simulations of sufficiently long chain segments, application of the master equation formalism could yield information on a large number of dynamic properties involving time regimes not readily accessible by present computational techniques.

The formal solution to the master equation, Eq. (19) reads

$$\mathbf{P}(\Omega) = \exp\{\mathbf{A}\tau\} \, \mathbf{P}_0(\Omega) = \mathbf{B}\exp\{\Lambda\tau\} \, \mathbf{B}^{-1} \, \mathbf{P}_0(\Omega) \tag{22}$$

where Λ is the diagonal matrix of the eigenvalues of **A**, **B** is the matrix of the corresponding eigenvectors, and $\mathbf{B}^{-1}$ is the inverse of **B**. One of the eigenvalues of **A** is equal to zero and ensures that the equilibrium distribution is approached at long times. All other eigenvalues of **A** are real and negative, since a symmetric matrix **S** may be constructed from **A** by defining its elements to be [79]

$$S(\Omega_\alpha, \Omega_\gamma) = [\mathbf{P}_0(\Omega_\alpha)]^{1/2} \, A(\Omega_\alpha \,|\, \Omega_\gamma) \, [\mathbf{P}_0(\Omega_\gamma)]^{-1/2} \, . \tag{23}$$

From the physical meaning of Eq. (22), it is noted that the time-delayed conditional probability matrix $\mathbf{C}_\tau$ is equal to the term multiplying $\mathbf{P}_0(\Omega)$ on the right hand side of the equality, i.e.,

$$\mathbf{C}_\tau = \mathbf{B}\exp\{\Lambda\tau\} \, \mathbf{B}^{-1} \, . \tag{24}$$

Knowledge of $\mathbf{C}_\tau$, together with the equilibrium probabilities, $\mathbf{P}_0(\Omega)$, gives a complete description of the stochastics of conformational transitions. Any transient property $\langle f(\tau) \rangle$ depending on conformational transitions may be readily evaluated by using the elements of $\mathbf{C}_\tau$ and $\mathbf{P}_0(\Omega)$ in

$$\langle f(\tau) \rangle = \sum_\alpha \sum_\beta C_\tau(\Omega_\alpha \,|\, \Omega_\beta) \, \mathbf{P}_0(\Omega_\beta) \, f(\Omega_\alpha; \Omega_\beta) \, . \tag{25}$$

Here $f(\Omega_\alpha; \Omega_\beta)$ is the value assumed by the investigated property when the transition takes place from configuration Ω_β to configuration Ω_α, and the summations are performed over the whole configurational space. We note that the product $C_\tau(\Omega_\alpha \,|\, \Omega_\beta) \, \mathbf{P}_0(\Omega_\beta)$ represents the joint probability of occurrence of configurations Ω_β and Ω_α with a time delay of τ. Using Eq. (24), Eq. (25) may be rewritten as

$$\langle f(\tau) \rangle = \sum_\gamma k_\gamma \exp\{\lambda_\gamma \tau\} \tag{26}$$

where λ_γ is the γ-th eigenvalue of Λ. The amplitude factor k_γ is a static quantity, expressed as a function of the specific investigated property, the equilibrium conformational distribution of the chain and the particular elements of $\mathbf{B}$ and $\mathbf{B}^{-1}$ as

$$k_\gamma = \sum_\alpha \sum_\beta B_{\alpha\gamma} [\mathbf{B}^{-1}]_{\gamma\beta} P_0(\Omega_\beta) f(\Omega_\alpha; \Omega_\beta) . \tag{27}$$

Following Eq. (26), a distribution of relaxational modes, with frequencies equal to λ_γ, contributes additivity to the macroscopically observed transient property $\langle f(\tau) \rangle$. $\langle f(\tau) \rangle$ may be conveniently replaced by the above described correlation functions associated with specific vectorial quantities, depending on the experimental system, or on the investigated quantity. This representation is particularly suitable for taking the Fourier transform and expressing the spectral density of $\langle f(\tau) \rangle = M_2(\tau)$, for example, as a sum of Lorentzians

$$J(\omega) = - \sum_\gamma k_\gamma \lambda_\gamma / (\omega^2 + \lambda_\gamma^2) . \tag{28}$$

Equations (25)–(28) are fundamental in establishing the correspondence between theory and experiments.

In view of the fact that a large number of theoretical treatments of chain dynamics adopt the Fokker-Planck equation as the starting point, it might be informative to consider briefly the connection between the master equation formalism and the Fokker-Planck equation before proceeding further with the details of the former. The Fokker-Planck equation may be obtained as an approximation to the master equation, as first derived by Planck [78]. The derivation relies on a Taylor series expansion of the configurational probability distribution function $P(y)$ up to second order in the continuous random variable y. The reader is referred to the Appendix for more details. Gardiner adopts another approach in which the master equation is shown to be a particular case of the so-called differential Chapman-Kolmogorov equation [80], which also leads to Fokker-Planck and Langevin equations under suitable simplifications.

2.1.2 Dynamic Rotational Isomeric State Model and Assumptions

Following the conventional rotational isomeric state approach [3, 81] of chain equilibrium statistics, bond dihedral angles are assumed to be the only variables defining a given conformation Ω_α. Thus, among the $3(N + 1)$ degrees of freedom of a chain of $N + 1$ backbone atoms indexed from 0 to N, as shown in Fig. 2, those associated with bond stretching and bond angle bending are frozen, as well as the external degrees of freedom accounting for the rigid body translation and reorientation of the chain in space. The torsional rotations ϕ_i of internal bonds, with indices $2 \leqslant i \leqslant N - 1$, are the only relevant structural parameters characterizing chain configuration. Thus, a given configuration is represented by a sequence of dihedral angles as $\Omega_\alpha = \{\phi_2, \phi_3, \phi_4, \ldots, \phi_{N-1}\}$. The torsions of the terminal bonds ϕ_1 and ϕ_N determine the absolute orientation of the chain in space, which is inconsequential in internal dynamics of the chain. It will prove mathematically convenient to define a laboratory-fixed coordinate system OXYZ to fix the position and the orientation of the chain in space, and bond-based coordinate systems $x_i y_i z_i$ as reference for local motions and for frame transformations, as commonly used in chain statistics. The origin of the i-th bond-based frame coincides with atom $i - 1$. The x_i-axis is directed along the bond vector 1_i, y_i makes an acute angle with the prolongation of x_{i-1}, lying in the plane of bonds $i - 1$ and i, and z_i completes a right-handed system.

Discrete values corresponding to rotational isomeric minima will be adopted for torsional angles, and fluctuations about the potential energy well are neglected. The reader is referred to the recent work of Moro [59] for the effects of small angle torsional librations on conformational dynamics. Rotational angle of $\phi_i = 0$ for bond i, leading to planar geometry of bonds $i - 1$, i and $i + 1$, defines the trans (t) state. Torsions of $\pm 120°$ from planar geometry give the gauche$^\pm$ (g$^\pm$) states. These three rotational states of bonds may be conveniently indexed as $t \equiv 1$, $g^+ \equiv 2$ and $g^- \equiv 3$. Using this notation, a given chain conformation may be written as $\{2, 2, 1, 3, \ldots, 1\}$, for example. Conformational transitions occur via collective rotational jumps of backbone bonds between isomeric states. Single bond rotation is assumed to occur at a given time. This assumption does not preclude the possibility of coupled, consecutive jumps occurring within infinitesimally short time intervals, when a high energy isomeric state is occasionally visited.

In general, local conformational rearrangements extend over a finite number of bonds, referred to as a kinetic unit. The size of the kinetic unit is not precisely defined, but depends on the specific structural characteristics of the chain. The orientational motion of a given vectorial quantity **m**, rigidly affixed to the n-th bond-based frame along the kinetic unit, is conveniently expressed relative to the bond-based frame appended to the first bond of the kinetic unit. A schematic drawing is presented in Fig. 3. The kinetic unit is shown by the portion of the curve between points A and B. The remaining portions of the chain, shown by the dotted lines, are referred to as the tails and are assumed to be subject to

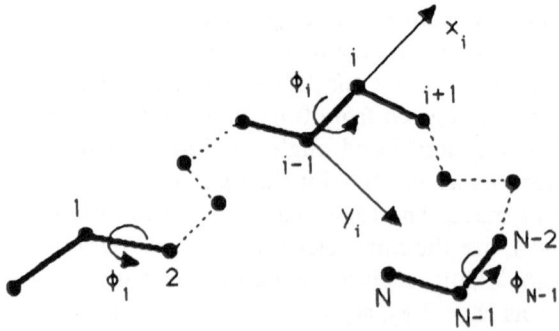

Fig. 2. An instantaneous configuration of a chain of N backbone bonds. Bond lengths and angles are fixed. The only degree of freedom is torsional rotation ϕ_i about bonds, i, $0 \leqslant i \leqslant N$, as indicated by *curved arrows*. A bond-based coordinate system associated with the i-th bond is shown

independent rotational kinematics, thus not affecting the conformational dynamics of the kinetic unit. Bonds belonging to the kinetic unit, on the other hand, are coupled to each other in the following two senses. First, there is a short-range intramolecular coupling, in general limited to first neighboring bonds along the chain, and related to the torsional energetics of interdependent bonds. Markov chain assumption has proven to be a valid approach for the treatment of chain statistics and will be adopted in the DRIS scheme as well. Second, there is a longer-range coupling between all of the bonds in the kinetic unit, such that conformational transitions leading to larger amplitude displacements in space are impeded by stronger frictional resistance. This second effect will be introduced by a mean field approach, for the case of chains in the bulk state or in dense medium, but will be neglected in the case of isolated chains in dilute solutions. In a restrictive environment, the torsional transition of a given bond i in the kinetic unit, which induces a rotational motion of chain segment about an axis coinciding with the bond i itself, is not directly propagated in the kinetic unit but assumed to be gradually damped away from bond i. Accordingly, the effect of transitions about the i-th bond of **m** should depend on the location of i. Rotations about a bond closer to **m** should have more pronounced effect on **m**. This has been included into the DRIS model [10, 82] by choosing a position-dependent friction coefficient, first introduced by Paul and Mazo [83].

2.1.3 Transition Rate Matrices in the DRIS Formalism

In the DRIS model, the $\gamma\alpha$ element $A(\Omega_\gamma|\Omega_\alpha)$ of A represents the rate of transition from a set of rotational angles defining configuration Ω_α to another set corresponding to Ω_γ. The accurate estimation of these rates is critically important for establishing a quantitative relation between theory and experiments. As mentioned above, $A(\Omega_\gamma|\Omega_\alpha)$ for $\gamma \neq \alpha$ may be estimated from the

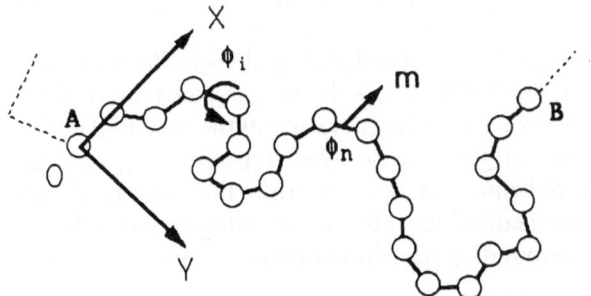

Fig. 3. The kinetic unit between atoms A and B on the chain. The *dashed lines* indicate the tails whose conformational dynamics is assumed to be uncoupled to that of the kinetic unit. Orientational motions in the kinetic unit are expressed with reference to the frame OXYZ appended to the first bond. Torsional rotation ϕ_i about bond i affect the orientation of the vector **m** affixed to bond n

initial slope of the transition probability $C(\Omega_\gamma | \Omega_\alpha)$, following Eq. (14), provided that $C(\Omega_\gamma | \Omega_\alpha)$ is determined by BD or MD simulations. Alternatively, $\mathbf{A}$ may be estimated from the kinetics of one or more bonds as follows. In the simplest case of bonds subject to independent conformational kinetics, the transition rate matrix $\mathbf{A}$ operating in a chain of N bonds is found from the convolution of the rate matrices $\mathbf{A}_i^{(1)}$ corresponding to individual rotatable bonds i, $2 \leqslant i \leqslant N - 1$, as

$$\mathbf{A} = (\mathbf{A}_2^{(1)} \otimes \mathbf{E} \otimes \mathbf{E} \ldots \otimes \mathbf{E}) + (\mathbf{E} \otimes \mathbf{A}_3^{(1)} \otimes \mathbf{E} \ldots \otimes \mathbf{E})$$

$$+ \ldots + (\mathbf{E} \otimes \mathbf{E} \otimes \mathbf{E} \ldots \otimes \mathbf{A}_{N-1}^{(1)}) . \tag{29}$$

Here $\mathbf{E}$ is the identity matrix of order v, and $\mathbf{A}_i^{(1)}$ is the rate matrix for the $v \times v$ rotational transitions of bond i, and $\otimes$ denotes the direct product. $\mathbf{A}_i^{(1)}$ obeys the master equation $\partial \mathbf{P}_i^{(1)} / \partial \tau = \mathbf{A}_i^{(1)} \mathbf{P}_i^{(1)}$, in which $\mathbf{P}_i^{(1)}$ represents the probability array of the v isomeric states accessible to the i-th bond, independently of the states of other bonds along the chain. Equation (29) readily follows from the time derivative of [4]

$$\mathbf{P}(\Omega) = \mathbf{P}_2^{(1)} \otimes \mathbf{P}_3^{(1)} \otimes \ldots \mathbf{P}_{N-2}^{(1)} \otimes \mathbf{P}_{N-1}^{(1)} \tag{30}$$

which applies for the case of independent bonds only.

The form of $\mathbf{A}_i^{(1)}$ depends on the particular conformational characteristics of the investigated chain. For polyethylene-like chains, with three states t, g^+ and g^- accessible to each bond, the kinetic scheme

$$g^+ \underset{r_2}{\overset{r_1}{\rightleftharpoons}} t \underset{r_2}{\overset{r_1}{\rightleftharpoons}} g^- \qquad \qquad \textbf{Scheme 1}$$

has been postulated on a large number of theoretical studies. r_1 and r_2 are the Arrhenius type rate constants associated with the indicated transitions. Expressions for the rate constants are written on the basis of the high friction limit of Kramers rate theory as

$$r_i = A_0 \exp \{ - E_{act}/RT \} = \frac{(\gamma^* \gamma)^{1/2}}{2\pi \zeta_{i, \, eff}} \exp \{ - E_{act}/RT \} \tag{31}$$

where E_{act} is the activation energy to be crossed during the specific transition, A_0 is referred to as the preexponential factor, R is the gas constant, T is the absolute temperature, and γ^* and γ refer to the curvature of the potential energy surface at the barrier and at the original energy well respectively. $\zeta_{i, \, eff}$ is the effective frictional coefficient. Scheme 1 is inferred from the conformational energetics of n-butane, and related equations rely on the assumption of independent backbone bonds. The corresponding transition rate matrix for bond i takes on the form

$$\mathbf{A}_i^{(1)} = \begin{bmatrix} -2r_1 & r_2 & r_2 \\ r_1 & -r_2 & 0 \\ r_1 & 0 & -r_2 \end{bmatrix} . \tag{32}$$

A more realistic yet mathematically tractable approach is that of a Markov chain model based on the pairwise interdependence of first neighboring bonds, in parallel with the one-dimensional Ising model of ferromagnetic phase transition. This approach requires precise knowledge of the kinetic scheme of transitions between rotameric states accessible to interdependent pairs of bonds. In this case, two-dimensional conformational energy contours constructed as a function of two consecutive dihedral angles [3] supply information on (i) the number and types of rotational isomeric states accessible to interdependent pairs of bonds, (ii) their equilibrium energies, and (iii) the path and the height of saddle to be crossed during rotational transitions. The latter yields the activation energy associated with specific transitions. For polyethylene, the kinetic scheme

$$
\begin{array}{ccccc}
g^+g^- & \underset{r_{-3}}{\overset{r_3}{\rightleftharpoons}} & g^+t & \underset{r_{-2}}{\overset{r_2}{\rightleftharpoons}} & g^+g^+ \\[4pt]
r_3 \big\Updownarrow r_{-3} & & r_1 \big\Updownarrow r_{-1} & & r_{-2} \big\Updownarrow r_2 \\[4pt]
tg^- & \underset{r_1}{\overset{r_{-1}}{\rightleftharpoons}} & tt & \underset{r_{-1}}{\overset{r_1}{\rightleftharpoons}} & tg^+ \\[4pt]
r_2 \big\Updownarrow r_{-2} & & r_1 \big\Updownarrow r_{-1} & & r_{-3} \big\Updownarrow r_3 \\[4pt]
g^-g^- & \underset{r_2}{\overset{r_{-2}}{\rightleftharpoons}} & g^-t & \underset{r_3}{\overset{r_{-3}}{\rightleftharpoons}} & g^-g^+
\end{array}
$$

Scheme 2

applies for pairwise interdependent bonds i and $j = i + 1$. The subscript ij may be omitted for chains with identical backbone bonds like polyethylene, but should otherwise be kept. The corresponding transition rate matrix $A_{ij}^{(2)}$ of order 9, is readily deduced by the proper substitution of the indicated rate constants: Considering the nine states $\Omega_\alpha = \{1, 1\}$, $\Omega_\beta = \{1, 2\}, \ldots$, etc. available in this reduced configurational space $\{\Omega\}$, the $\alpha\beta$ off-diagonal element $A_{ij}^{(2)}(\Omega_\alpha | \Omega_\beta)$ is the rate constant associated with the transition $\Omega_\beta \to \Omega_\alpha$, while the diagonal elements are evaluated from Eq. (20). In the case of chains composed of distinct pairs, it may prove useful to decompose $A_{ij}^{(2)}$ into two components, $A_i^{(2)}$ and $A_j^{(2)}$, representing the transition rate matrices of the left and right bonds of the pair, respectively, such that $A_i^{(2)} + A_j^{(2)} = A_{ij}^{(2)}$ [8, 84]. For a chain segment of N bonds, upon labeling the bonds with indices from 1 to N, the transition rate matrix becomes

$$
\mathbf{A} = [(\mathbf{A}_1^{(2)} + \tfrac{1}{2}\mathbf{A}_2^{(2)}) \otimes \mathbf{E} \otimes \mathbf{E} \ldots \otimes \mathbf{E}] + [\mathbf{E} \otimes \tfrac{1}{2}\mathbf{A}_{23}^{(2)} \otimes \mathbf{E} \ldots \otimes \mathbf{E}]
$$

$$
+ [\mathbf{E} \otimes \mathbf{E} \otimes \tfrac{1}{2}\mathbf{A}_{34}^{(2)} \ldots \otimes \mathbf{E}]
$$

$$
+ \ldots + [\mathbf{E} \otimes \mathbf{E} \otimes \mathbf{E} \ldots \otimes (\tfrac{1}{2}\mathbf{A}_{N-1}^{(2)} + \mathbf{A}_N^{(2)})] . \tag{33}
$$

The factor 1/2 prevents duplicate counting of the same bound by two consecutive terms in square brackets. End effects are taken into consideration by substituting the expressions $[A_1^{(2)} + A_2^{(2)}/2]$ and $[A_{N-1}^{(2)}/2 + A_N^{(2)}]$ for the transition rate matrices associated with terminal bonds, as previously performed for poly(oxyethylene) and polyisoprene [8, 84]. The reader is referred to these references for the explicit expressions for transition rate matrices operating in interdependent bond doublets or triplets, and their combinations to yield $\mathbf{A}$ for specific segment of N bonds.

Similarity transformation of the transition rate matrix $\mathbf{A}$ of a chain segment of N bonds, or the corresponding symmetric matrix $\mathbf{S}$, yields the set of eigenvalues and eigenvectors which, upon proper substitution in Eqs. (26)–(28), readily allow the computation of time-dependent properties or their counterparts in the Fourier transform domain. However, the inversion of a matrix of order v^N is required in this approach for a chain segment of N coupled bonds with v states accessible to each bond, which renders this method practically inapplicable beyond about $N \geqslant 9$. Instead, a computationally efficient matrix multiplication scheme, based on the pairwise interdependence of the stochastics of adjacent bonds, has recently been developed [85, 86], along the lines originally devised by Kramers and Wannier [87, 88] for the analysis of one-dimensional Ising model. This new matrix multiplication method associated with the DRIS formalism may be viewed as an extension of the conventional mathematical methods of equilibrium statistics, now including the time variable as well, and also incorporating real chain structural and conformational features. The main advantage is that it allows for the evaluation of various correlation functions in segments of any size, at the cost of negligibly small computation time, as will be outlined in the next subsection. Only the general approach and final expressions of the matrix multiplication scheme are presented here for brevity. The interested reader is referred to Appendix 5.2. and the original works on the subject for further details.

2.1.4 Matrix Multiplication Scheme for Conformational Stochastics of Markov Chains

The matrix multiplication scheme using statistical weights and generator matrices has been used extensively in conformational statistics of polymers, following the early studies of Lifson [89, 90], Nagai [91], and Birshtein and Ptitsyn [92]. This method is based on the formulation of a configurational partition function upon serial multiplication of statistical weight matrices associated with the rotational isomeric states of pairwise interdependent bonds. For, e.g., polyethylene-like chains, the statistical weight matrix for the pair of bonds $(i - 1, i)$ is

$$\mathbf{U}_i = \begin{bmatrix} \eta & \sigma & \sigma \\ 1 & \sigma\psi & \sigma\omega \\ 1 & \sigma\omega & \sigma\psi \end{bmatrix}. \tag{34}$$

Here, $\sigma = \exp(-E_\sigma/RT)$, $\omega = \exp(-E_\omega/RT)$, $\psi = \exp(-E_\psi/RT)$, $\eta = \exp(-E_\eta/RT)$ where E_σ is the equilibrium energy of the gauche$^\pm$ state in excess of the trans state, E_ω is the secondary interaction energy associated with the state

g^+g^- (or g^-g^+), and E_ψ and E_η are the secondary interaction energies for states g^+g^+ (or g^-g^-) and tt respectively. Analogous with this approach, the matrix multiplication scheme developed for treating conformational dynamics within the DRIS frame work stipulates the use of stochastic weight matrices associated with the conformational transitions between those states.

The stochastic weight matrices $V_i(\tau)$ are defined [85, 86] in terms of the elements of the joint probabilities $P_{ij}^{(2)}$ and $P_i^{(1)}$ obtained from the similarity transformation of $A_{ij}^{(2)}$ and $A_i^{(1)}$, respectively. We will denote these elements as (i) $p_i(\alpha\beta, \tau; \xi\eta, 0)$ for the transition $\{\xi, \eta\} \rightarrow \{\alpha, \beta\}$ of the pair of interdependent bonds $(i - 1, i)$ during the time interval τ, and (ii) $p_i(\alpha, \tau; \xi, 0))$ for the transition $\{\xi\} \rightarrow \{\alpha\}$ at time τ, of the single bond i, in the absence of any coupling to neighbors. Explicit expressions for $V_i(\tau)$ are given in Appendix 5.2. The transition partition function $Z_N(\tau)$ for a Markov chain of N bonds is evaluated from the serial multiplication of the stochastic weight matrices for a given τ according to

$$Z_N(\tau) = J^T \left[\prod_{i=2}^{N-1} V_i(\tau) \right] J \tag{35}$$

where $J \equiv \text{col}(1, 1, \ldots, 1)$ and the superscript T denotes the transpose. Accordingly, the a priori probability of specific transitions at a given time, such as the passage $\{\xi\} \rightarrow \{\alpha\}$ of bond j, may be computed from

$$p_j^*(\alpha, \tau; \xi, 0) = Z_N(\tau)^{-1} J^T \left[\prod_{i=2}^{j-1} V_i(\tau) \right] V_j^*(\tau) \left[\prod_{i=j+1}^{N-1} V_i(\tau) \right] J. \tag{36}$$

Here $V_j^*(\tau)$ is the stochastic weight matrix in which all elements are equated to zero except for those associated with the specific transition $\{\xi\} \rightarrow \{\alpha\}$ of bond j, indicated on the left hand side of the equality. Likewise, a priori probabilities $p_j^*(\alpha\beta, \tau; \xi\eta, 0)$ for the coupled rotations of adjacent pair of bonds $(j - 1, j)$ may be readily computed by an expression analogous to Eq. (36). In this latter case, $V_j^*(\tau)$ has only a single element, $v_j(\alpha\beta; \xi\eta)$, different from zero. A priori probabilities $p_j^*(\alpha, \tau; \xi, 0)$ and $p_j^*(\alpha\beta, \tau; \xi\eta, 0)$ differ from the respective joint probabilities $p_i(\alpha, \tau; \xi, 0)$ and $p_i(\alpha\beta, \tau; \xi\eta, 0)$ of isolated bonds or bond pairs, inasmuch as the former two incorporate the intrachain coupling due to connectivity and the perturbations induced by chain end effects. Calculations performed for $p_j^*(\alpha, \tau; \xi, 0)$ and $p_j^*(\alpha\beta, \tau; \xi\eta, 0)$ as a function of bond location j indicate an even-odd dependence of the a priori transition probabilities on bond serial order along the chain [86], in conformity with the well-known end effects of equilibrium statistics. This reflects the time-ensemble equivalence of conformational characteristics: accordingly, if a rotameric state has low equilibrium probability of occurrence, the period of time a given bond stays in that state is short, or vice versa.

Matrix multiplication method has been used as an efficient tool for the calculation of orientational auto- and cross-correlations for vectorial quantities such as dipole moments, transition moments etc., rigidly embedded in polymer chains [10, 84, 86]. Let us consider, for example, the orientational correlation function between two vectors **m** and **n** affixed to the i-th and j-th bond-based frames. Let us assume that $i \leqslant j$. The equality $i = j$ holds for the case of orientational autocorrelation. For a given transition from state $\{\Omega_\alpha\}$ at time 0 to state $\{\Omega_\beta\}$ at time τ, the scalar product $\mathbf{m}(0) \cdot \mathbf{n}(\tau)$ may be written as

$$\mathbf{m}(0) \cdot \mathbf{n}(t) = \mathbf{T}\{\Omega_\alpha\} \mathbf{m}^0 \cdot \mathbf{T}\{\Omega_\beta\} \mathbf{n}^0 = \mathbf{m}^{0\mathrm{T}}[\mathbf{T}^{\mathrm{T}}\{\Omega_\alpha\} \mathbf{T}\{\Omega_\beta\}] \mathbf{n}^0 \qquad (37)$$

where $\mathbf{m}^0$ and $\mathbf{n}^0$ are the vectors expressed in their local bond-based coordinate systems, and $\mathbf{T}\{\Omega_\alpha\}$ and $\mathbf{T}\{\Omega_\beta\}$ are the frame transformation operators expressing them in a common frame, say that associated with the first bond of the chain. The ensemble average of Eq. (37) over the $v^{j-2} \times v^{j-2}$ possible conformational transitions of the chain segment extending between bond 1 and j, omitting the rotations of bond 1 and j, is obtained [10, 86] upon an appeal to a theorem on direct products as

$$\langle \mathbf{m}(0) \cdot \mathbf{n}(\tau) \rangle = \mathbf{m}^{0\mathrm{T}}(\mathbf{E} \otimes \mathbf{F})(\mathbf{D}(\tau)^{\mathrm{T}} \otimes \mathbf{E})\mathbf{n}^0 . \qquad (38)$$

Here **E** is the identity matrix of order v, $\mathbf{F} \equiv \mathrm{row}(\mathbf{E})$, written in reading order, and the matrix $\mathbf{D}(\tau)$ incorporating the frame transformation matrices is given in Appendix 5.2.

2.2 Comparison of DRIS Results with Other Theories, Simulations and Experiments

2.2.1 Comparison with other Theoretical Approaches and Empirical Expressions

Here we will first compare the implication of the DRIS formalism with the predictions of the kinetic Ising model of Glauber [39] and discuss the results in relation to the theory of Shore and Zwanzig [40]. It should be noted that the model of Glauber is of fundamental importance, inasmuch as this is the first work in which the Markov character of a chain is rigorously considered and an analytical expression is provided for the time decay of correlation functions associated with a linear array of pairwise interdependent units. In the second part of this section, some empirical expressions previously proposed for describing OACFs, will be considered.

2.2.1.1 Comparison with the Kinetic Ising Model of Glauber

This model is based on the assumption that the spins of a linear array of N individual particles are stochastic functions $\sigma_i(\tau)$, $(i = 1, \ldots, N)$ and take on values of either $+ 1$ or $- 1$. The random transitions between these two states result from the interaction of the spins with an external agent. Intramolecular correlations are introduced by choosing the transition probabilities for any one spin to depend on the state of the neighboring spins. The master equation governing the time evolution of the system is given by the expression

$$\frac{d}{dt} p(\sigma_1, \ldots, \sigma_N \tau) = -\left[\sum_i w_i(\sigma_i) \right] p(\sigma_1, \ldots, \sigma_N \tau)$$

$$+ \sum_i w_i(-\sigma_i)] p(\sigma_1, \ldots, -\sigma_i, \ldots, \sigma_N \tau) . \quad (39)$$

Here, $p(\sigma_1, \ldots, \sigma_N \tau)$ is the joint probability of having the spin states $\sigma_1, \ldots, \sigma_N$ for the N spins at time τ and $w_i(\sigma_i)$ is the probability per unit time that the i-th spin flips from the value σ_i to $-\sigma_i$ while the others remain fixed. The first summation on the right hand side of Eq. (39) gives the rate of destruction of the configuration $\{\sigma_1, \ldots, \sigma_N\}$ and the second term gives the creation of this configuration by the transition of each spin i from the state $-\sigma_i$ to σ_i. Equation (39) is equivalent to Eq. (17) for the transition probabilities of bond isomeric states. Here, the master equation is rewritten for the joint probabilities of spin states, and $w_i(\sigma_i)$ is the counterpart of the elements of the rate matrix **A**.

The elements of the transition rate matrix **A** in the DRIS model are estimated from the multidimensional energy surface associated with the interdependent rotation of neighboring bonds using Kramers' rate theory, as described above. Accordingly, the probability of occurrence of a given isomeric state for a bond depends on the state of its first neighbors along the chain. Likewise, in the kinetic Ising model the transition rate $w_i(\sigma_i)$ of the i-th spin is assumed to be coupled to the state of its first neighbors by the relation

$$w_i(\sigma_i) = \tfrac{1}{2}\alpha\{1 - \tfrac{1}{2}\gamma\sigma_i(\sigma_{i-1} + \sigma_{i+1})\} . \quad (40)$$

Here, $w_i(\sigma_i)$ takes the value $\tfrac{1}{2}\alpha(1 - \gamma)$ if the i-th spin is parallel to both of its two neighboring spins, $\tfrac{1}{2}\alpha$ if it is parallel to one of the neighbors, and $\tfrac{1}{2}\alpha(1 + \gamma)$ if it is antiparallel to both of its neighbors. The parameter α determines the absolute value of transition rates in general. The parameter γ is a measure of the coupling between neighboring spins. A positive value for γ enhances parallel to alignment and a negative value favors antiparallel alignment. γ is related to the energy of alignment, J, by

$$\gamma = \tanh(2J/kT) \quad (41)$$

where k is the Boltzmann constant and T is the temperature.

For a system in thermal equilibrium, the time-delayed spin correlation function $M_1^{ik}(\tau)$ between the i-th and the k-th spins is given by Glauber as

$$M_1^{ik}(\tau) = \langle \sigma_i(0) \sigma_k(\tau) \rangle = \sum_{\{\sigma'\}}\sum_{\{\sigma\}} \sigma_i \sigma_k' \, p(\sigma_1' \sigma_2' \ldots \sigma_N', \tau; \sigma_1 \sigma_2 \ldots \sigma_N, 0)$$

$$(42)$$

$$= e^{-\alpha\tau} \sum_{m=-\infty}^{\infty} \eta^{|i-k+m|} I_m(\gamma\alpha\tau) . \quad (43)$$

Here, $p(\sigma'_1 \sigma'_2 \ldots \sigma'_N, \tau; \sigma_1 \sigma_2 \ldots \sigma_N, 0)$ is the joint probability of occurrence of conformations $\{\sigma\} \equiv \{\sigma_1, \sigma_2, \ldots, \sigma_N\}$ at time zero and $\{\sigma'\} \equiv \{\sigma'_1, \sigma'_2, \ldots, \sigma'_N\}$ at time τ. I_m is the modified Bessel function of order m and

$$\eta \equiv \tanh(J/kT) = \gamma^{-1} \left\{ 1 - (1 - \gamma^2)^{1/2} \right\} . \tag{44}$$

For $i = k$, $M_1^{ik}(\tau)$ represents the expectation value that the state of a given spin i remains unchanged during the time interval τ. Thus, $M_1^{ii}(\tau)$ will also be referred to as the conformational autocorrelation function. It is to be noted that the appearance of the modified Bessel function is common to several theoretical treatments of local orientation [39, 40, 50].

Autocorrelation decay functions calculated on the basis of Eq. (44) for different values of the intermolecular coupling parameter γ are presented in Fig. 4. The ordinate is $\log[- \ln M_1^{ii}(\tau)]$ and the abscissa is $\log(\tau)$, α being taken as unity. This choice of the coordinates allows direct comparison with the empirical stretched exponential or the Kohlrausch-Williams-Watt (KWW) function [93, 94]

$$M_i(\tau) = \exp\left\{ - (\tau/\tau_c)^\beta) \right\} . \tag{45}$$

Here τ_c is a characteristic correlation time and the exponent β is less than unity. The occurrence of a stretched exponential behavior is also pointed out by Ngai, Mashimo and Fytas [95] in a recent comparison of relaxation times for local

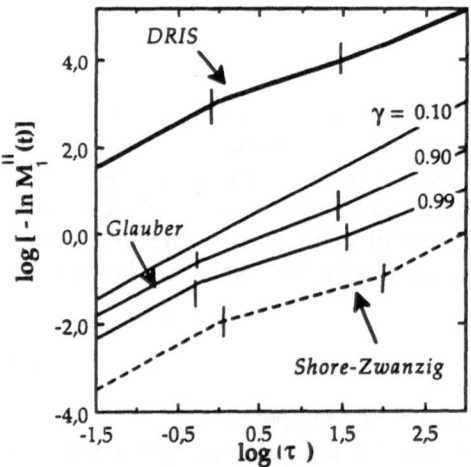

Fig. 4. Time dependence of the first OACF obtained from the indicated three different theories. The curves are arbitrarily shifted along the vertical axis for visual clarity. Three different degrees of intramolecular coupling are considered in the Glauber model, as indicated by the value of the parameter γ in each case. Three time regimes separated by *the vertical bars* are distinguished in the presence of internal coupling, yielding different values of stretched exponent, such that β is close to unity at short times, becomes $\leqslant 0.7$ at intermediate times, and increases again at terminal portions of the curves

segmental motion determined by photon correlation spectroscopy and viscoelastic measurements. For a fixed value of γ, Eq. (44) is evaluated at closely spaced times and the points obtained in this manner are joined by least squares straight lines. For $\gamma = 0.1$ a straight line is obtained indicating that the relaxation may be fitted by a single KWW function with $\beta = 0.998$ and $\tau_c = 1.01/\alpha$. However, as the degree of coupling between neighboring spin states is increased, the relaxation functions are observed to be best approximated by three lines of different slopes, indicating three distinct regimes. The curves obtained for $\gamma = 0.9$ and 0.99 obey this behavior, as illustrated by the middle three curves of various γ in Fig. 4. The curves are arbitrarily shifted along the vertical axis, isasmuch as the qualitative time dependence, rather than absolute values of correlations, is of interest here. The vertical bars on these curves indicate the boundaries of the regions with different slopes. For $\gamma = 0.9$, the straight lines have the slopes $\beta = 0.96$ for short times, $\beta = 0.72$ for intermediate times, and $\beta = 0.85$ for long times. For $\gamma = 0.99$, the respective exponents are $\beta = 0.98, 0.60$ and 0.73.

These results show that the initial relaxation region is relatively little affected by intramolecular correlations and β remains in the vicinity of unity, which corresponds to the relaxation of independent, isolated spins. The intermediate times, on the other hand, are strongly affected by near neighbor interdependence and a stretched exponential behavior, becoming increasingly more pronounced with the strength of intramolecular correlations, appears. The exponent is found to increase back to larger values at long times. This behavior is compatible with the picture of a few slowest relaxation modes surviving at long times, in contrast to the more cooperative mechanism operating at intermediate times.

The model of Glauber describes the relaxation due to large angle flips between distinctive states of the individual spins composing a linear array. The effects of intramolecular correlations on small angle librations of a linear array of spins are given by Shore and Zwanzig [40]. In the case of a single dipolar spin i in a long chain, the first autocorrelation function is given as

$$M_1^{ii}(\tau) = \frac{\mu^2 E/kT}{2N} \exp\left\{ -D\int_0^\tau ds\, e^{-2JDs/kT} I_0(2JDs/kT) \right\} . \qquad (46)$$

Here, μ is the magnitude of the dipole moment, E is the externally applied electric field, N is the number of spins in the array, D is the rotational diffusion constant, J is the spin-spin coupling constant, and I_0 is the zeroth order modified Bessel function. Shore and Zwanzig showed that the normalized decay function $M_1^{ii}(\tau)/M_1^{ii}(0)$ may be approximated by three simple functions for the short, intermediate and long times as

$$\frac{M_1^{ii}(\tau)}{M_1^{ii}(0)} = \begin{cases} e^{-D\tau} & \text{for } JD\tau/kT \ll 1 \\ \exp[-(D\tau kT/\pi J)^{1/2}] & \text{for } 1 \ll JD\tau/kT \ll N^2 \\ e^{-D\tau/N} & \text{for } N^2 \ll JD\tau/kT . \end{cases} \qquad (47)$$

In Fig. 4, the three expressions of Eq. (47) yield, for arbitrary values of parameters, three straight lines with slopes of 1, 0.5, and 1 corresponding to the short, intermediate and the terminal relaxation times, respectively. This behavior is shown by the dotted line in the lower part of Fig. 4.

The function $M_1^{ii}(\tau)$ given by Eq. (42) indicates the degree of correlation between the conformations of spins i and k. In the DRIS formalism, the counterpart of Eq. (42) reads, for the case k = i,

$$M_1^{ii}(\tau) = \sum_\alpha \sum_\xi \{\phi_i\}_\alpha \{\phi_i\}_\xi C_\tau(\Omega_\alpha | \Omega_\xi) P_0(\Omega_\xi) \tag{48}$$

where $\{\phi_i\}_\alpha$ denotes the state of the i-th bond during the configuration Ω_α of the chain. Using the matrix multiplication scheme of Sect. 2.1.4, Eq. (48) may be rewritten as

$$M_1^{ii}(\tau) = \sum_{\sigma_i'} \sum_{\sigma_i} \sigma_i' \sigma_i p_i^*(\sigma_i', \tau; \sigma_i, 0) \tag{49}$$

where σ_i' and σ_i refer to the values characteristic of isomeric states of bond i at times τ and 0, respectively. The a priori probability $p_i^*(\sigma_i', \tau; \sigma_i, 0)$ of the joint state $(\sigma_i', \tau; \sigma_i, 0)$ is computed with Eq. (36) on the basis of the whole space of conformational transitions. For a polyethylene-like chain with three rotational states t, g^+ and g^- accessible to each bond, the summation of Eq. (49) includes nine terms corresponding to the joint states of the i-th bond for two distinct times. Calculations performed for such symmetric chains with pairwise interdependent bonds, subject to Scheme 2 shown in Sect. 2.1.3, lead to the conformational autocorrelation functions displayed by the upper boldface curve in Fig. 4. This curve is obtained for the middle bond of a chain of 30 units, using the parameters $E_\sigma = 2.1$, $E_\omega = 9.0$, $E_\psi = 0.0$, and $E_\eta = -4.2$ kJ/mol for the equilibrium energies, $E_{act} = 14.6$ kJ/mol for all of the transition rates in Scheme 2, and adopting the arbitrary value $\sigma_t = 1$ with the corresponding values $\sigma_{g^+} = \sigma_{g^-} = -(\sigma_t/2)p_i^*(t)/p_i^*(g^\pm)$. The horizontal and the vertical position of the curve is chosen arbitrarily. Three regions with respective slopes 0.98, 0.66 and 0.78 are again distinguishable, in parallel with the conformational autocorrelations resulting from Glauber's model. For independent bonds the DRIS model, using $E_\omega = E_\psi = E_\eta = 0.0$ with all other parameters being kept constant, gives a line (not shown in Fig. 4 for reasons of clarity) with a single slope of $\beta = 1.00$ for short, intermediate and long times. This verifies that a single exponential relaxation mode operates in the absence of neighbor dependence. These results invite attention to several features. First, both Glauber model and DRIS approach lead to a stretched exponential behavior in the intermediate time regime. Secondly, the exponent decreases with the strength of coupling between first neighbors along the chain. Thus, in spite of the large number of differences between the two models, conformational correlation functions exhibit the same qualitative behavior. Likewise, the one-dimensional model of Shore and Zwanzig, including coupling between spins, yields a stretched exponent of $\beta = 0.5$ at intermediate times. Irrespective of the details of the type and mechanism of relaxation processes, two common features in all three models are mainly responsible for these observations: the chain connectivity and the short-range intramolecular coupling. These emerge as the major factors underlying the stretched exponential behavior observed in a large number of experiments. Thus, stretched exponential behavior may be viewed as a consequence of local cooperativity in conformational relaxations. The exact value of the exponent depends on the specific structural and conformational properties of the chain, and is equal to unity only if near neighbor interdependence is absent.

2.2.1.2 Comparison with Analytical OACF Expressions

An early expression derived by Valeur, Jarry, Gény and Monnerie (VJGM) for the second orientational correlation function of bonds is [45, 46, 96]

$$M_2(\tau) = \exp\{(1/\tau_1 - 1/\tau_2)\tau\}\,\text{erfc}\,\{(\tau/\tau_1)^{1/2}\}\,. \tag{50}$$

Here erfc is the error function complement, and τ_1 and τ_2 are parameters related to the reciprocal of the jump rate for on-lattice rotameric transitions. This expression follows from the continuous limit of a master equation approach applying to three-bond motions on a tetrahedral lattice. The major deficiency of this expression is its infinite slope at $\tau = 0$, which is physically unrealistic. This feature arises from the mathematical approximation of the process of discrete rotameric transitions by a continuous diffusion equation. Likewise, Jones and Stockmayer (JS) [48], Bendler and Yaris (BY) [49] and Hall and Helfand (HH) [50] have given concise expressions for the first OACF. These are semiempirical expressions and do not distinguish between the first and second OACFs described above. Accordingly, the superscripts 1 and 2, as well as the bond indices of the OACFs, will be omitted in $M_{ii}^1(\tau)$ and $M_{ii}^1(\tau)$ in the following.

In the absence of overall tumbling and small amplitude torsional librations, the JS model leads to the following expression for the first OACF:

$$M(\tau) = \sum G_k \exp(-\omega\lambda_k\tau)\,. \tag{51}$$

Here λ_k and G_k are given by

$$\lambda_k = 4\sin^2[(2k-1)\pi/4s]$$

$$G_k = 1/s + (2/s)\left\{\sum_q \exp(-\gamma q)\cos[(2k-1)\pi q/2s]\right\} \tag{52}$$

where $\gamma = 2\ln 3$ and $s = (m+1)/2$. The two adjustable parameters in the theory are ω, the rate for the three bond jump, and m, the number of coupled units at both sides of a central mobile unit. The form of Eq. (51) for the autocorrelation function is identical to that predicted by the DRIS model (see Eq. (26)). However, the eigenrates λ_k of Eq. (51) become smaller with increasing number of coupled units around the central bond while those obtained by the DRIS model

[97] become larger. The latter are given by

$$\lambda(k, m) = m\lambda_3 + (k - m)\lambda_2 + (N - k)\lambda_1, \quad 0 \leqslant m \leqslant k \leqslant N. \tag{53}$$

Here, $\lambda(k, m)$ represents the set of the 3^N eigenrates corresponding to the array of N rotatable bonds with three states accessible to each bond. For chains composed of independent bonds obeying the kinetic Scheme 1 given in Sect. 2.1.3, $\lambda(k, m)$ is expressed as a combination of the three smallest eigenrates $\lambda_1, \lambda_2, \lambda_3$

$$\lambda_1 = 0$$

$$\lambda_2 = \lambda_3 = -(2r_1 + r_2) \tag{54}$$

corresponding to single bond rotameric transitions. Each of the $\lambda(k, m)$ eigenrates is $N(k, m) \equiv N!/m!(k - m)!(N - k)!$ fold degenerate. The decrease in relaxation rates with decreasing size of the kinetic unit in the DRIS model has been attributed to the smaller number of relaxation pathways accessible to a shorter segment. Such a decrease in mobility in the case of very short chain segments was first pointed out by Kuhn [22, 23], and referred to as the internal viscosity effect, as later elaborated by de Gennes [30] and Allegra [33]. This effect, however, is observed to be inverted, as the number of bonds in the kinetic unit gets larger [9].

Comparison of the results of the JS expression with those of the DRIS model shows that the latter allows for a more detailed analysis of the real chain in terms of torsional energy barriers and the three dimensional chemical structure. Furthermore, while the JS model originates from the VJGM model and accounts only for three-bond motions, the DRIS model shows the contribution of all transitions of the N bonds surrounding the bond of interest.

The approach of Bendler and Yaris, which is also based on the three-bond motions of the VJGM model, results in the following expression for the OACF:

$$M(\tau) = (\pi/\tau)^{1/2} (\tau_2^{-1/2} - \tau_1^{-1/2})\{erfc[(\tau/\tau_1)^{1/2}] - erfc[(\tau/\tau_2)^{1/2}]\} \tag{55}$$

where τ_1 and τ_2 are two parameters referred to as the short and long wavelength cutoffs. Comparison of the time decay of $M_1(\tau)$ resulting from the JS and BY models with the DRIS results indicates that the JS and the DRIS models are in relatively poor agreement, whereas the mean-square deviation between the OACF decay curves resulting from the DRIS formalism and those predicted by the BY model is minimal upon optimal choice of the two parameters τ_1 and τ_2 of Eq. (64) [7].

The HH model takes into account both correlated pair and isolated transitions, occurring at frequencies λ_1 and λ_0, respectively [50]. The pair transitions ensure the propagation of the motion along the chain while the isolated transitions are responsible for the damping. Thus, the HH model is more general than the JJ and the BY models. The first OACF proposed by the

HH model is

$$M_1(\tau) = \exp(-2\lambda_0\tau)\exp(-2\lambda_1\tau)I_0(2\lambda_1\tau) \tag{56}$$

where I_0 is the modified Bessel function. This expression was originally derived as a conformational correlation function [50] and therefore the mechanism of three dimensional motions leading to the decay of the first OACF is not built into it. However, its functional form has proven to fit well both the orientational and conformational decay curves. The decay of the first OACF obtained by the DRIS [7] model has shown best agreement with the results of the HH model, compared to other semi-empirical expressions. The parameters λ_0 and λ_1 which yield the best fit with the OACF curves of the DRIS formalism have been evaluated. It is interesting to note that resulting λ_0 value, which is attributed to the damping of the motion, is at least one order of magnitude smaller than λ_1 in agreement with previous curve fitting calculations carried out by Helfand et al. to reproduce the decay of the OACFs resulting from their BD simulations.

2.2.2 Comparison with Results of Brownian Dynamics (BD) Simulations

2.2.2.1 Conformational Correlation Functions

The computation of the a priori probability $p_i^*(\alpha, \tau; \xi, 0)$ of the time-delayed joint state $(\alpha, \tau; \xi, 0)$ of bond i has been described in Sect. 2.1.4. As mentioned above, $p_i^*(\alpha, \tau; \xi, 0)$ differs from the joint probability $p_i(\alpha, \tau; \xi, 0)$ of the i-th isolated bond deduced from the similarity transformation of $A^{(1)}$, inasmuch as the former accounts for the effect of chain connectivity through Eq. (36). An approximate method for the estimation of $p_i(\alpha, \tau; \xi, 0)$, including the near neighbor interdependence of bonds along the chain, is to consider the rate matrix $A^{(2)}_{i-1, i}$ operating on isolated bond pairs and use the corresponding joint probabilities $p_i(\beta\alpha, \tau; \eta\zeta, 0)$ in

$$p_i(\alpha, \tau; \zeta, 0) = \sum_\eta \sum_\beta p_i(\beta\alpha, \tau; \eta\zeta, 0) . \tag{57}$$

Here the summations are performed over all initial and final states of bond $i - 1$. Division of $p_i(\alpha, \tau; \zeta, 0)$ by the equilibrium probability $P_0(\zeta)$ of bond i, yields the conformational transition probability $C_i(\alpha, \tau | \zeta, 0)$ from state ζ to state α for the investigated bond i, when its conformational dynamics is assumed to be coupled to that of the preceding bond. The bond index in $C_i(\alpha, \tau | \zeta, 0)$ is conveniently omitted for chains with identical bonds, and the short-hand notation $C_\tau(\alpha | \zeta)$ is adopted for stationary processes.

Calculations of $C_\tau(\zeta | \zeta)$ for polyethylene have been performed according to the DRIS model [7] and the results have been compared with BD simulations performed by Weber and Helfand [98]. Conformational autocorrelation functions for the trans and gauche states based on the kinetic Scheme 2 of the DRIS formalism given above are shown in Fig. 5. The preexponential factor A_0 of the Kramers' rate expression is taken as $2.77 \times 10^{11}/s$, and the activation energies associated with specific transitions are determined on the basis of the two-dimensional contour maps resulting from molecular mechanics computations. For comparison, the results from BD simulations [98] are displayed by the dotted curves. It should be noted that the interdependencies of second neighbors along the backbone, such as the counterrotations of the bonds surrounding a given trans bond, which play a major role in local chain dynamics, are not considered in the DRIS calculations. Also, bond lengths and bond angles are taken to be fixed here while they are allowed to fluctuate in BD simulations. On the other hand, BD simulations neglect the pairwise interdependence of conformational kinetics, which renders the $g^+ g^-$ state, for example, highly improbable relative to $g^+ g^+$. This first neighbor interdependence is rigorously accounted for in the DRIS formalism. Despite these different approximations of the two models, the conformational autocorrelation functions of the two methods are in satisfactory agreement. This might be attributed to the fact that single bond conformational kinetics is a major factor controlling local chain dynamics, and perturbations due to effects such as short- or long-range bond interdependence are averaged out when attention is confined to the stochastics of a single bond.

A more detailed examination of conformational stochastics is possible by considering the transition probabilities of pairs of consecutive bonds. Second order transition probabilities $C_\tau(\zeta\alpha \mid \eta\beta)$ directly obtained from the transformation of the transition rate matrices $A^{(2)}$ for polyethylene are displayed in Figs. 6a, b. Curves in Fig. 6a correspond to the case $\{\zeta\alpha\} = \{\eta\alpha\}$. These are drawn for

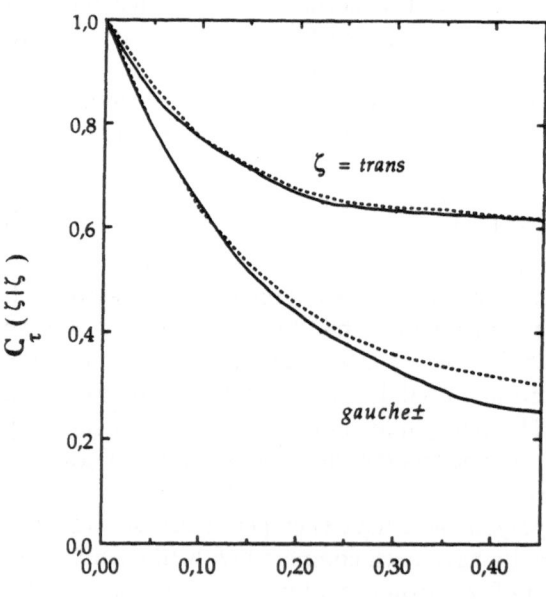

Fig. 5. Conformational autocorrelation functions $C_\tau(\zeta \mid \zeta)$ for internal bonds in polyethylene, calculated from DRIS, compared to results from BD simulations of Weber and Helfand, shown by *the dotted curves* [98]

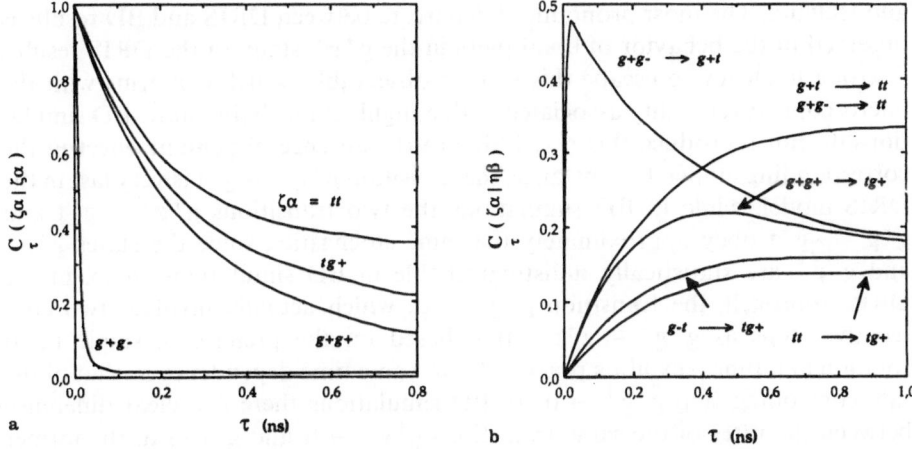

Fig. 6a, b. Second order transition probabilities: **a** $C_\tau(\zeta\alpha|\zeta\alpha)$ for internal bonds in polyethylene, evaluated by DRIS approach, **b** $C_\tau(\zeta\alpha|\eta\beta)$ for internal bonds in polyethylene, evaluated by DRIS approach

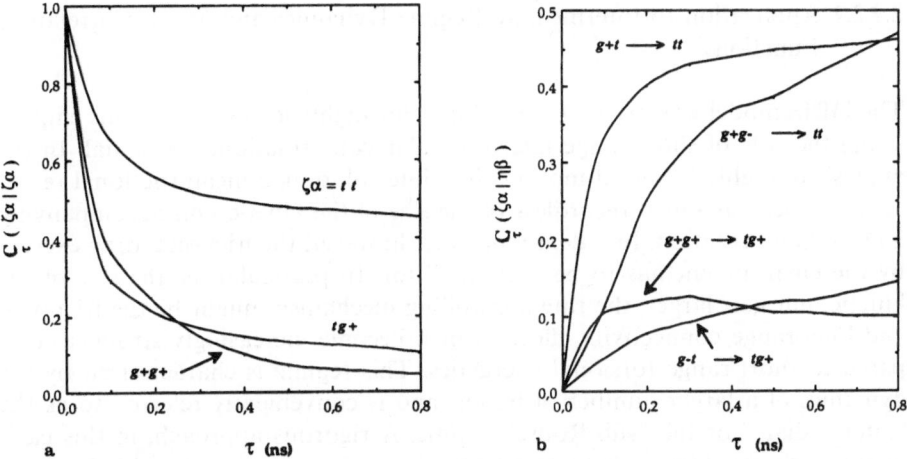

Fig. 7a, b. Second order transition probabilities: **a** $C_\tau(\zeta\alpha|\zeta\beta\alpha)$ for internal bonds in polyethylene, obtained from BD simulations, using independent bond rotational potentials. **b** $C_\tau(\zeta\alpha|\eta\beta)$ for internal bonds in polyethylene, obtained from BD simulations, using independent bond rotational potentials

a few isomeric pairs of interest, as indicated. Those in Fig. 6b represent the transition probabilities between distinct states. The initial and final states in each case are indicated by the labels. The mirror images of the indicated states obeying same stochastics are not explicitly shown. Figures 7a, b are the counterparts of Figs. 6a, b, respectively, which are obtained by BD simulations of total duration 2.5 ns, using the energy and geometry parameters of the work of Weber and Helfand. The most pronounced departure between DRIS and BD results is observed in the behavior of bond pairs in the g^+g^- state. In the DRIS results, a strong tendency to escape this state is observable, which conforms with the microscopic reversibility associated with a highly improbable state. BD simulations do not reproduce this result, due to the absence of pentane effect in the corresponding model. For instance, the transition $g^+g^- \rightarrow g^+t$ is very fast in the DRIS model, while in BD simulations the two transitions $g^+g^- \rightarrow g^+t$ and $g^+g^+ \rightarrow g^+t$ obey approximately the same stochastics, since the states g^+g^- and g^+g^+ are statistically indistinguishable in BD simulations. Also, in the DRIS approach, the transition $g^+g^- \rightarrow tt$, which actually involves two consecutive steps as $g^+g^- \rightarrow g^+t \rightarrow tt$ – based on the premise of single bond rotation at a time – exhibits practically the same time dependence as that of the rate controlling step $g^+g^+ \rightarrow tt$. In BD simulations there is a clear difference between the rates of the same transitions $g^+g^- \rightarrow tt$ and $g^+t \rightarrow tt$, the former being considerably slower. It is interesting to note that $g^-t \rightarrow tg^+$ appears as a relatively slow transition in both approaches, whereas $g^+t \rightarrow tt$ is invariably one of the most probable transitions.

2.2.2.2 Application to Intermediate Regime Dynamics and Mode Correlation Functions

The DRIS model was originally developed for highly localized motions emphasizing the role of short-range intermolecular conformational potential. In the analysis of highly localized motions in dilute solution, a mean frictional resistance has been assumed regardless of the size of the kinetic unit accompanying a given bond rotation. In a dense medium, however, the frictional drag exerted by the environment has to be accounted for. In particular, as the size of the kinetic unit gets larger, the rate-controlling mechanism might be the frictional and long-range connectivity effects, which become increasingly stronger compared to short-range torsional energetics. This regime is characterized by the dynamics of a larger number of bonds, and is conveniently referred to as the "intermediate" or the "sub-Rouse" regime. A rigorous approach, in this case, might be to consider each conformational transition, examine its implications insofar as the spatial coordinates of the atoms in the moving unit are concerned, and assign effective frictional coefficients on the basis of the particular distance/volume swept by all of the atoms set in motion. This approach, which has been adopted as an exploratory analysis [9], is prohibitively time-consuming for kinetic units of $\geqslant 6$ bonds, in view of the exceedingly large number of possible transitions. A computationally more efficient approximation has been introduced instead, based on the position-dependent effective frictional coefficient concept introduced by Jernigan [82], on the basis of the original work of Paul and Mazo [83].

The model adopted in the analysis of intermediate regime dynamics by Bahar et al. [10] considers a kinetic unit whose first bond defines a local reference frame OXYZ, as shown in Fig. 3. The portion of the kinetic unit between O and the vector **m**, rigidly attached to the n-th bond, is assumed to undergo in the configurational space all rotameric transitions, such as that shown by the curved arrow, while the second half is not explicitly considered in the analysis. The rate of a transition over bond i as seen by **m** is given by Eq. (31). $\zeta_{i,eff}$ is the effective friction coefficient expressed as [10]

$$\zeta_{i,\,eff} = 2\zeta^0 \sum_{p=i+1}^{n} \langle s_{ip}^2 \rangle \qquad (58)$$

where, ζ^0 is a constant of proportionality, s_{ip} is the separation of the p-th atom, and $i < p \leqslant n$, from the axis of rotation defined by the bond i which undergoes the rotameric transition. Angular brackets denote an average over all configurations accessible to the kinetic unit. From the moment of inertia considerations for a random coil, the summation in Eq. (58) may be expressed by the proportionality $(n - i)^{-2}$ and the expression for the effective rate of transition associated with the motion of **m** resulting from a transition over bond i reads

$$r_i \sim (n - i)^{-2} \exp(-E_{act}/RT) . \qquad (59)$$

The exponential part in Eq. (59) arises from the specific chain conformational energetics whereas the front term reflects the chain connectivity effect and frictional drag due to environment. The role of the latter becomes predominantly important in intermediate regime dynamics.

Bond orientational correlation functions and the associated correlation times for polyethylene have been evaluated by Bahar et al. [10]. The former are determined from the definitions given by Eqs. (8) and (9) where $\mathbf{m}_i$ in Eq. (9) is replaced by the bond vector $\mathbf{l}_i$. Cross-correlation times for orientations of bonds i and k are obtained as the area under the corresponding OACF decay curves according to Eq. (3). The analysis has also been extended to the domain of normal modes by adopting the transformation

$$\mathbf{q} = \mathbf{Q1} \qquad (60)$$

within the Gaussian model approximation. Here **1** is the column matrix of the bond vectors $\mathbf{1}_i$ and **q** is the column matrix of the normal mode vectors $\mathbf{q}_i$. **Q** is the symmetric n × n transformation matrix where Q_{ik} operates between the i-th bond vector and the k-th mode vector according to

$$Q_{ki} = \left[\frac{2}{n+1} \right]^{1/2} \sin \left[\frac{ik\pi}{n+1} \right] . \qquad (61)$$

The mode correlation functions $\langle \mathbf{q}_i(0) \cdot \mathbf{q}_j(\tau) \rangle$ may be expressed, using Eq. (60), as

$$\langle \mathbf{q}_i(0) \cdot \mathbf{q}_j(\tau) \rangle = \sum_{p=1}^{n} \sum_{q=1}^{n} [Q_{pi} Q_{qj} \langle \mathbf{1}_i(0) \cdot \mathbf{1}_j(\tau) \rangle] \qquad (62)$$

which, upon application of Eq. (3), yields the corresponding relaxation rates v_{ik} as

$$v_{ik}(\mathbf{q}) = \tilde{v}_{ik}(\mathbf{q}) \sum_{p=1}^{n} \sum_{q=1}^{n} Q_{pi} Q_{qk} [\langle \mathbf{1}_i \cdot \mathbf{1}_k \rangle - \langle \mathbf{1}_i \rangle \cdot \langle \mathbf{1}_k \rangle] \tag{63}$$

where

$$1/\tilde{v}_{ik}(\mathbf{q}) \equiv \sum_{p=1}^{n} \sum_{q=1}^{n} Q_{pi} Q_{qk} \int_{0}^{\infty} \{\langle \mathbf{1}_i(0) \cdot \mathbf{1}_k(\tau) \rangle - \langle \mathbf{1}_i \rangle \cdot \langle \mathbf{1}_k \rangle\} \, d\tau . \tag{64}$$

Results of calculations for the relaxation rates v_{nn} are shown by the empty circles in Fig. 8 as a function of $1/n$ [10]. Unlike the predictions of the Rouse model, a linear dependence between the relaxation rates and the size of the kinetic unit is observed. The filled circles are from the BD simulations of Fixman [99, 100] for a polyethylene-like chain. Dependence of v_{kk} on the mode number for a 16 bond polyethylene chain are shown by the empty circles in Fig. 9. The filled circles are from the BD simulations of Fixman [99, 100] for a similar chain. The curve is the best fitting curve through the empty circles. It is interesting to note that both the DRIS approach and Fixman's simulations yield a plateau value for the rate of most of the relaxational modes, while a few slowest modes exhibit distinct lower values.

2.2.2.3 Anisotropy of Motions

The decay rate of the OACFs depends critically on the direction of the associated vector, being higher along directions perpendicular to the chain axis. This

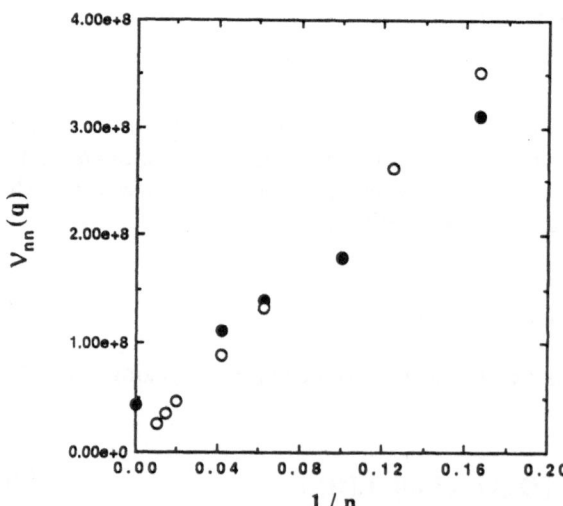

Fig. 8. Dependence of relaxation rates $v_{nn}(\mathbf{q})$ associated with the fastest mode on the size n of the kinetic unit. *Empty circles* are obtained with the DRIS model. *Filled circles* are from BD simulations [100]

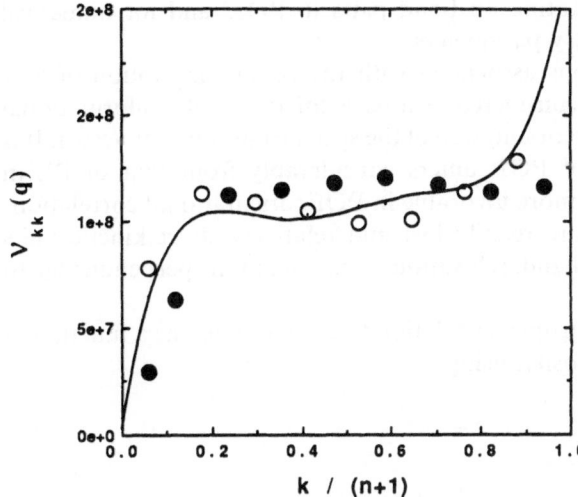

Fig. 9. Dependence of relaxation rates $v_{kk}(\mathbf{q})$ on mode number k. *Empty circles* are obtained with the DRIS model. *Filled circles* are from BD simulations [100]. The curve is the best fitting line through the *empty circles*

has been demonstrated for the polyethylene-like isolated chain by the BD studies of Weber and Helfand, and DRIS studies of Bahar et al. The anisotropy of local motion by these two techniques results from intramolecular effects. The directionality of mobility becomes even more significant in the melt, as shown by the recent MD simulations of Roe and collaborators. The anisotropy in this case results essentially from intermolecular effects, according to which a polymer chain may be considered to be confined to within a pipe. The correlation time of the OACF of a vector perpendicular to the chain axis is found to be about ten times larger than that of a vector along the chain contour in BD simulations of isolated chains, whereas MD studies of polymer melt yield a ratio of several hundreds.

2.2.3 Comparison with Experiments

2.2.3.1 Application to NMR Measurements in Dilute Polymer Solutions

Isotropic correlation times and spin lattice relaxation times measured by ^{13}C- and ^{1}H-NMR for polyoxyethylene (POE) solutions in a variety of solvents have been computed using the DRIS formalism for isolated polymer chains [8]. For this purpose, the conformational kinetics of POE has been analyzed and kinetic schemes of rotameric transitions have been estimated for the three distinct types of bond pairs (CO, OC), (OC, CC) and (CC, CO) on the backbone. The effective friction coefficient is deduced from the viscosity of the solvent, irrespective of the size of the kinetic unit, assuming environmental effects and chain connectivity constraints to be of secondary importance compared to torsional energy barriers. The reader is referred to [8] for explicit expressions of transition rate matrices operating on bond pairs in POE, and for numerical values of energy and geometry parameters.

The set of 6561 transitions associated with the rotameric motion of four consecutive bonds has been considered as a basis for the local conformational mobility of POE and for the reorientation of the specific internuclear vector. It is noted that the stochastics of POE differs considerably from that of PE in general. Gauche states being more favorable in POE, orientational correlations along the backbone bonds are readily lost and relatively short kinetic units operate. In PE, on the other hand, relaxation along directions perpendicular to the chain axis is faster.

Figure 10 displays the isotropic correlation times computed as a function of the reciprocal of solvent viscosity, using

$$\tau_{corr} = \frac{1}{1 - k_1} \sum_{j=2}^{3^N} \frac{k_j}{\lambda_j} \tag{65}$$

which readily follows from the integration of Eq. (26) of the DRIS formalism, according to Eq. (3). Here k_1 refers to the amplitude factor associated with the zero eigenvalue of the transition rate matrix, operating on the investigated kinetic segment. The filled circles represent results from the experiments of Lang et al. [101] at 30 °C, in a variety of solvents. Spin-lattice relaxation times

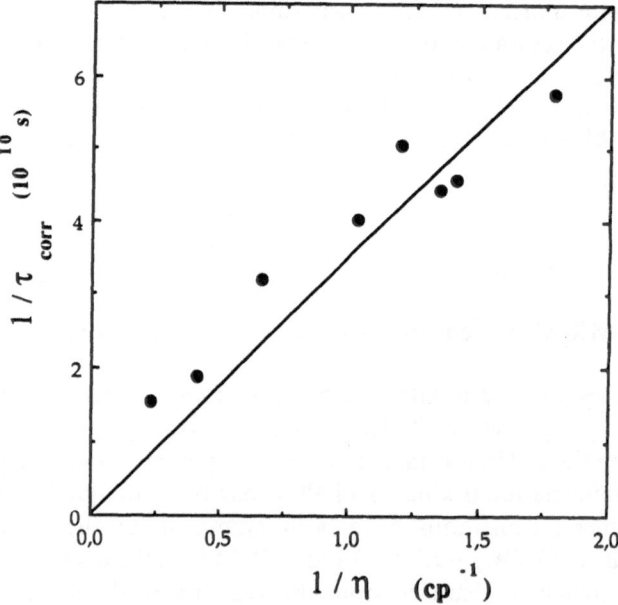

Fig. 10. Isotropic correlation times τ_{corr} as a function of reciprocal solvent viscosity for polyoxyethylene in dilute solution. *Circles* represent experimental data, *the line* is predicted by the DRIS model

T_{1H} from the ^{1}H-NMR experiments of Liu and Anderson [102], and Hermann and Weill [103] are also well reproduced by the DRIS calculations in which the same set of parameters is adopted. This behavior is illustrated in Fig. 11. The filled circles represent the experimental measurements of Liu and Anderson, the empty circles those of Hermann and Weill. It is important to point out that the theoretical approach here involves no adjustable parameter, but energy values readily following from molecular mechanics computations, and the front factor $A_0 = 2.8 \times 10^{11}\,s^{-1}/\eta$ (cp) which is in remarkable agreement with the value estimated by Friedrich et al. [104] from ESR experiments of dilute PEO solutions.

2.2.3.2 Temperature Dependence of Local Motions

The temperature dependence of correlation times for local motion in POE is displayed in Fig. 12. Points represent the experimental results by Lang et al. [101]. The solid line is obtained by the DRIS approach, using the same model and parameters as above. It is noted that perturbations arising from the temperature dependence of the viscosity are eliminated in this representation, inasmuch as the correlation times are normalized with respect to the solvent viscosity which is itself temperature dependent. The slope of the theoretical line

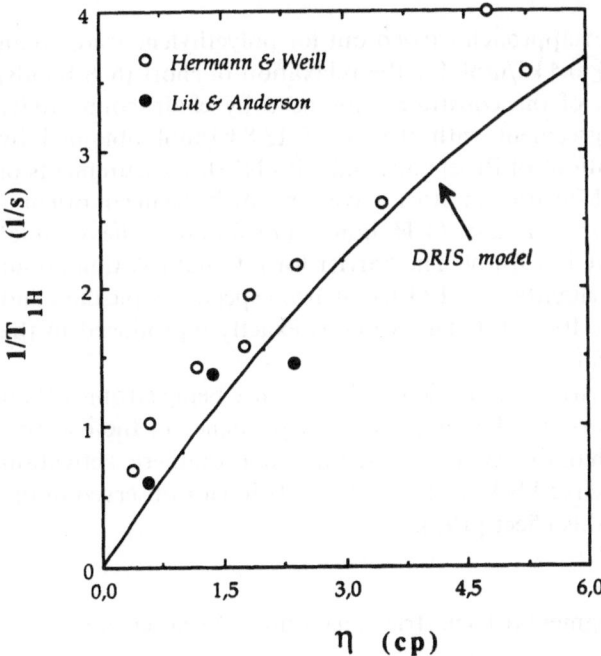

Fig. 11. Comparison of theory and NMR experiments for spin-lattice relaxation times T_{1H} for POE as a function of solvent viscosity

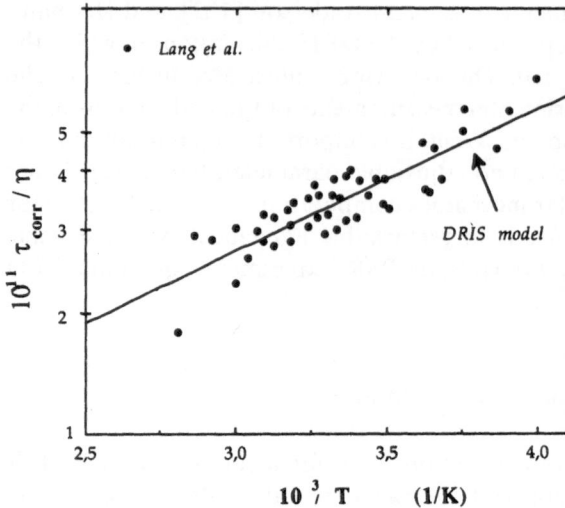

Fig. 12. Comparison of theory and ESR experiments for temperature dependence of correlation times associated with local motions of POE. An Arrhenius type temperature dependence with activation energy equal to 7.5 kJ/mol is obtained

yields an effective activation energy of 7.5 kJ/mol for local relaxation motions in poly(oxyethylene) in dilute solution, which is of the size of the energy barrier opposing a single isolated bond rotation. The good agreement between experiments and theory lends further support to the adoption of the DRIS model as a tool to investigate motions depending on short-range conformational energetics.

It is noted that a similar approach carried out for polyethylene leads to an activation energy of 13.6 ± 0.4 kJ/mol, for the relaxation of short (6–8 bonds) chain segments, regardless of the constraints imposed by chain connectivity [105]. This is in good agreement with the value 13.8 kJ/mol obtained by Brownian dynamics simulations of PE chains, and with NMR measurements of perfluoroalkane chains by Matsuo and Stockmayer, in which the mean orientational correlation times for C–F and C–H bonds are found to have equal activation energies of about 12 kJ/mol. The barrier for internal rotation about typical CH_2–CH_2 bonds is slightly over 12 kJ/mol, from spectroscopic data and conformational analysis [3, 106, 107]. This value is exactly reproduced in the present theoretical treatment.

In the bulk state, the environmental frictional resistance being stronger than intramolecular torsional barriers, the temperature dependence of the effective friction coefficient, rather than the Arrhenius type internal rotameric activation energies, dominates the observed behavior. The WLF behavior observed in the bulk state is attributed to this effect [108].

2.2.3.3 Interpretation of Segmental Dielectric Relaxation Measurements

The DRIS formalism has been recently used [84] for the theoretical interpretation of dielectric relaxation measurements of bulk *cis*-polyisoprene (*cis*-PIP) carried out by Boese and Kremer [109]. Experiments exhibit the presence of two distinct regimes of relaxational modes, normal and segmental, associated with the global and local motions of the chain, respectively. As first pointed out by Stockmayer, the reorientation of the component of the dipole moment vector perpendicular to the chain contour, gives a measure of the segmental motion of the chain, and is responsible for the relaxation spectrum observed at relatively high frequency range, whereas the component along the chain contour reflects the long-wavelength motion of the chain [21]. Here, we are interested in the segmental mode relaxation process, and in the implementation of the DRIS formalism to analyze the observed stretched exponential time decay of the dipole-dipole correlation function or the normalized response function.

The chemical structure and conformational characteristics of the chain have been considered in the analysis of local dynamics with the DRIS formalism. Statistical analysis of cis-PIP equilibrium structure and characteristics reveals that each of bond triplets separated by the double bond along the chain backbone, assume 12 rotameric states, characterized by strongly interdependent torsional angles, whereas successive triplets obey independent conformational energetics [110, 111]. The isomeric states of bond triplets are $s^{\pm} ts^{\pm}$, $s^{\pm} g^{\pm} s^{\pm}$, $s^{\pm} g^{\mp} s^{\pm}$, $s^{\mp} ts^{\pm}$, $s^{\mp} g^{\pm} s^{\pm}$, $s^{\mp} g^{\mp} s^{\pm}$, where $s^{\pm}$ refers to the skew$^{\pm}$ states with $\pm 60°$ rotations. Examination of energy contour maps constructed as a function of interdependent dihedral angles reveals the kinetic scheme of accessible rotameric transitions between those states, as well as the form of the transition rate matrix. The reader is referred to [84] for the particular form of $\mathbf{A}$ operating on repeat units in cis-PIP, and the geometry and energy parameters used in the DRIS model on the basis of short-range intramolecular energetics. Yet the dynamic interdependence of bonds is not limited to those belonging to the same monomeric unit, but extends to all bonds of a given kinetic unit, when the local chain relaxation occurs in a dense medium as delineated in the preceding section. This type of interdependence is accounted for through adoption of a position-dependent friction coefficient in the rate expression. Accordingly, the reorientation rate of a dipole moment vector appended to the n-th bond in the kinetic unit is slowed down by a factor of $(n - i)^2$, when the motion is initiated by the rotameric transition of the i-th $(i < n)$ bond. This is readily implemented in the model by taking the effective friction coefficient $\zeta_{i, eff}$ in Eq. (31) to be of the form $\zeta_{i, eff} \sim (n - i)^2$, as presented in Eqs. (58) and (59).

Inasmuch as the size of the kinetic unit is unknown, we chose to perform the DRIS calculations for chain segments of various size, ranging from 1 to 3 repeat units. The dipole-dipole correlation function $\Phi(\tau)$, given by Eq. (7) in normalized form, has been obtained as a function of time for each case. Calculations reveal the occurrence of two distinct mechanisms of relaxation, with respective stretched exponents close to unity and 0.5, characteristic of the above described short and intermediate time regimes [84]. The results obtained for a kinetic unit of three monomers are illustrated by the filled circles in Fig. 13. The best fitting lines through the results yield the exponents $\beta = 0.97$ and 0.34 as indicated. Moreover, a tendency to an enhancement of the exponent is distinguishable at

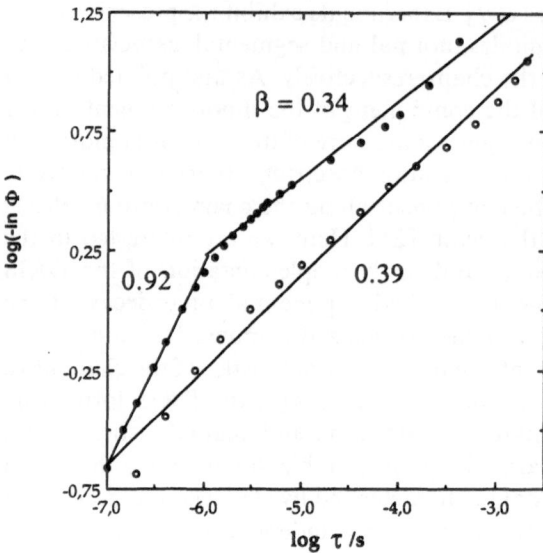

Fig. 13. Time dependence of dipole-dipole correlation function for segmental motion of polyisoprene in the bulk state. *Points* denote results of DRIS calculations, *filled circles* are obtained in the absence of intermolecular correlations. The two *straight lines* are best fits to short and long time regions. The *empty circles* show DRIS results in the presence of environmental fluctuations. Values of the KWW exponents are shown on each curve

long times. Repeating calculations for various size segments shows that, in general, the exponent β equates to 0.95 ± 0.02 at short times ($\tau \leqslant 10^{-6}$ s) whereas it is highly sensitive to the size of the kinetic unit at intermediate times, in the range $10^{-6} \leqslant \tau \leqslant 10^{-4}$ s. For kinetic units composed of 1, 2, and 3 coupled monomers, β decreases to the respective values of 0.62, 0.45, and 0.34, in this intermediate range.

The exponents obtained with kinetic units of 8–12 bonds (i.e., 2 or 3 monomeric units) approximate the experimentally observed behaviour (β = 0.39). However, experiments indicate the occurrence of a stretched exponential behavior throughout a wider range of time, covering several orders of magnitude. This suggests a larger extent of cooperativity, probably embodying not only successive units along the chain, but also units belonging to other chains in the close neighborhood of the relaxing kinetic unit. In fact, the adoption of a longer chain segment would not help, since the exponent is gradually lowered while the time range of stretched exponential behavior is not affected. Thus, the consideration of intermolecular restraints becomes mandatory. A two-state model has been adopted for the environment, as a first approximation, in parallel with previous work [108, 112–114]. MD simulations by Rigby and Roe also lend support to the occurrence of a bimodal distribution of free volume in bulk polymers. Accordingly, two states, slow and fast, with respective equilibrium probabilities p_{slow} and $p_{fast} = 1 - p_{slow}$, are assumed by the environment, thus rendering the total number of joint states accessible to a given repeat unit equal to 24. The rates of transitions between isomeric states are assumed to be affected by the state of the environment but the reverse is not true, i.e., the isomeric state of the chain sequence does not affect the transition rate of the environment from the slow to the fast state. The transition rate matrix $\tilde{A}$ under these conditions reads

$$\tilde{\mathbf{A}} = \begin{bmatrix} \mathbf{A} - r_{fast}\mathbf{E}_{12} & r_{slow}\mathbf{E}_{12} \\ r_{fast}\mathbf{E}_{12} & g\mathbf{A} - r_{slow}\mathbf{E}_{12} \end{bmatrix}. \tag{66}$$

Here $\mathbf{E}_{12}$ is the identity matrix of order 12, $\mathbf{A}$ is the transition rate matrix defined in Table IV of [84], and the coefficient $g < 1$ accounts for the decreased mobility of the chain in the slow medium, relative to the fast medium. r_{slow} refers to the rate of transition of the environment from the slow to the fast state. The reverse transition rate r_{fast} is fixed by the detailed balance principle as

$$r_{slow}/r_{fast} = (1 - p_{slow})/p_{slow} . \tag{67}$$

In the case of X repeat units collectively participating in the relaxation process, the operating transition rate matrix $\mathbf{A}^{(X)}$ may be found from the direct product of the individual transition matrices $\tilde{\mathbf{A}}$ with identity matrices of suitable order analogous to Eq. (29). Although this approach presents the advantage of yielding the dispersion of relaxational modes, it is clear that determination of the eigenvalues of $\mathbf{A}^{(X)}$ becomes prohibitively time-consuming for $X \geqslant 3$. The orientational correlation functions in this case are preferably computed using the matrix multiplication scheme described in Sect. 2.1.4. The empty circles in Fig. 13 display the result obtained [84] for a kinetic unit of 12 backbone bonds, in a two-state fluctuating environment. A smoother dependence on time is produced in this case, in which the two regimes with distinct exponents are hardly distinguishable, in contrast to relaxation in a homogeneous environment. The best fitting straight line through the DRIS results yields a slope equal to $\beta = 0.39$, in agreement with experiments. This analysis invites attention to the role of both intramolecular and intermolecular cooperativity in prescribing the mechanism of local motion.

3 Kinematics of Local Motions in Dense Media

3.1 General Motivation

The DRIS formalism is based on a relatively simple model, in which the contribution to local motion arising from the cooperative distortion of bond angles and lengths and of the fluctuations of bond torsional states about isomeric minima are ignored. Accordingly, the DRIS approach has two limitations: (i) only the changes in internal coordinates are treated without consideration of the large scale reorientation and translation of the chain, and (ii) the distortions of the various degrees of freedom of the chain are not considered; the compensating motions of the tails surrounding a given kinetic unit are not present. This latter question has been addressed by Skolnick and Helfand [69].

Among the various degrees of freedom of the chain contributing to local relaxation, fluctuations in dihedral angles appear to be the most important. This has also been confirmed by the BD and MD simulations. The problem of the coupling of several bonds has been addressed in the newly developed cooperative kinematics (CK) model. Here, the concerted fluctuations in bond dihedral angles are rigorously considered. In this approach, the emphasis is put on constraints imposed by the long-range chain connectivity and intermolecular effects, which becomes particularly strong in the bulk state or at temperatures approaching glass transition. Another question which awaits further elucidation is the size of the kinetic unit involved in local dynamics depending on the specific intermolecular and intramolecular conditions. The torsional rotations imparted by the random fluctuations of the environment are certainly localized along the chain, affecting in general a finite number of bonds in the neighborhood of the perturbation. The evidence for localization mechanism comes both from kinetic studies and Brownian dynamics simulations [41, 74, 98, 115] and from experiments [84, 116, 117]. The localization mechanism, although effective due to the inertia of the chain atoms, is particularly important in a restrictive environment where the energy cost of moving the chain atoms against environmental friction is relatively high. For an elucidation of the consequences of intermolecular and long-range intramolecular constraints, in the absence of internal barriers to rotameric transitions, a perfect tetrahedral model chain with freely rotating bonds is considered first. The perturbation of the kinematics in the presence of increasingly stronger internal barriers to rotational isomerization is presented next.

3.2 Model Chains with Freely-Rotating Bonds in Restrictive Environment

3.2.1 Model and Assumptions

The geometry of motion (kinematics) has been recently studied [11–13] for a model chain in which bond torsional motions are not opposed by any internal barriers. The latter is referred to as a freely-rotating chain (FRC) model. The chain is, however, assumed to be in a dense medium constraining the spatial rearrangements of atoms. A plausible behavior in the presence of such significant frictional resistance from the environment would be to minimize the displacements of the atoms following an external perturbation. The basic postulate of the proposed model is that, following any perturbation of an equilibrium configuration, the atoms rearrange cooperatively in space so as to minimize their overall square displacements. The formulation allows for the calculation of the changes in all of the degrees of freedom of the chain, internal and external, in response to a change in the dihedral angle of an internal bond [11, 12]. By the use of Lagrange multipliers, the formulation has been generalized to study the consequences of stretching the chain [13].

In the FRC model, a perfect tetrahedral structure of n backbone bonds, each of fixed length l, is considered. Bond angles are fixed and indicated as $\pi - \theta$, and only the torsional angles φ_i of bonds, $1 \leqslant i \leqslant n$, are assumed to vary. The chain is allowed to translate or rotate in space, and coordinates of chain atoms relative to a laboratory-fixed reference frame OXYZ are observed. The total number of degrees of freedom of the chain amounts to $n + 4$, $n - 2$ of them arising from the rotations of the internal backbone bonds $2 \leqslant i \leqslant n - 1$, and the remaining six associated with the absolute position and orientation of the chain. The internal motions of the chain are described in terms of bond-based coordinate systems following the notation by Flory [3]. Backbone atoms are labelled from 0 to n, as shown in Fig. 14. The absolute location of the chain in space is specified by the position vector $\mathbf{R}_0 = (X_0 \, Y_0 \, Z_0)^T$ of the zeroth atom in the frame OXYZ. The absolute orientation of the chain in space is prescribed by three Euler angles displayed in part (b) of Fig. 14, Φ showing the angle that the first bond makes with the Y-axis, ψ showing the angle that the projection of the first bond on the XZ plane makes with the Z-axis and χ representing the rotation of the first bond about its own axis. The latter angle is identified with the torsional rotation φ_1 about the first bond.

According to the basic postulate adopted in the study, succeeding an external perturbation, the atoms rearrange in space in a concerted fashion so as to preserve as much as possible the original position vectors. Physically, this postulate is a natural consequence of the energy minimization principle since, in the absence of energetic or enthalpic interactions, the energy change reduces to the work done by the system, which increases with the displacement of the atoms from their equilibrium positions. The chain is assumed to equilibrate prior to any conformational transition. The distortion of a given configuration is assumed to be opposed by harmonic potentials, forcing the backbone atoms to restore their original locations. Mathematically, this requirement is satisfied

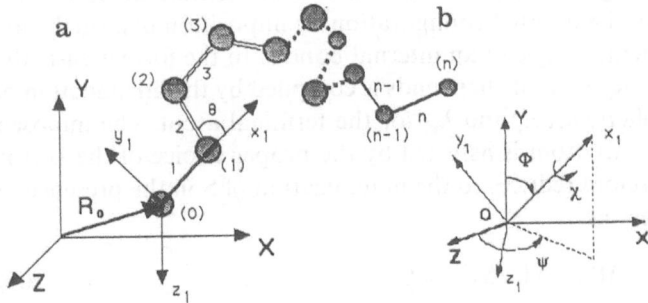

Fig. 14. a Schematic representation of the internal $(\varphi_i, 2 \leqslant i \leqslant n - 1)$ and external $(\mathbf{R}_0, \Phi, \psi \text{ and } \chi)$ coordinates of a chain of n bonds with respect to the laboratory-fixed frame OXYZ. **b** Absolute orientation of the chain, described by the two Euler angles Φ, ψ between the first bond and the axes of the frame OXYZ, and by the rotational angle $\chi = \varphi_1$ of the first bond

by minimizing the scalar function S of the incremental changes in position vectors,

$$S \equiv (n + 1)^{-1} \sum_{i=0}^{n} (\delta \mathbf{R}_i)^2 . \tag{68}$$

Here, $\delta \mathbf{R}_i$ is the differential displacement of the i-th atom resulting from the perturbation. For numerical purposes, $\delta \mathbf{R}_i$ will be replaced by $\Delta \mathbf{R}_i$ provided that the incremental changes in the position vectors are sufficiently small. The function S physically represents the mean-square displacement of atoms succeeding a perturbation of the original chain configuration. The present approach may be regarded as a zero temperature approach in which a coordinate is forced and the others respond in the absence of any stochastic force randomizing the coordinates. For instance, if one considers the Langevin equation of motion in the high-friction limit, which is adopted in most BD simulations, atoms tend to move in the direction of decreasing energy, as stipulated by the negative gradient of the potential, but also experience a random force which vanishes only in the limit of zero temperature. Such thermal fluctuations are absent here, and a unique solution corresponding to the lowest energy state accommodating an external perturbation is directly obtained.

3.2.2 Mathematical Formulation

The general method of approach for a systematic analysis of the kinematics of motion will be as follows. First, a chain which is sufficiently long to permit the internal reorganization of the atoms without substantial displacement of the tails, is considered. Bonds are originally assigned three types of equally probable torsional states, trans, gauche$^+$ and gauche$^-$, by a random number generator subroutine. The original configuration of the chain is then perturbed. Two different types of perturbations have been allowed for in previous work: (i) the deformation of chain length by application of a uniaxial tension at both ends, and (ii) the distortion of the original configuration by imposition of a small step change, $\Delta \varphi_s$, in the dihedral angle of an internal bond s. In the former case, the constraint of fixed displacement of chain ends is compiled by the introduction of three Lagrange multipliers, λ_x, λ_y and λ_z, for the terminal atom. The imposed displacement of the zeroth atom is asserted by the proper choice of the vector $\Delta \mathbf{R}_0$. The problem therefore reduces to the minimization of S in the presence of the Lagrange multipliers as

$$\partial[S - \lambda \cdot (\Delta \mathbf{R}_n - \Delta \mathbf{R}_{n,\text{ext}})]/\partial \Delta \varphi_m = 0 . \tag{69}$$

Here $\Delta \mathbf{R}_n - \Delta \mathbf{R}_{n,\text{ext}}$ represents the externally imposed displacement of the terminal atom of the chain, and the Lagrange multipliers are conveniently written as $\lambda = [\lambda_x \ \lambda_y \ \lambda_z]^T$. The solution of the above set of $(n - 1)$ homogeneous equations for m in the range $1 \leqslant m \leqslant n - 1$, together with the three identities in

$\Delta R_{n,ext} = \Delta R_n$, yields the $(n + 2)$ unknowns, $\Delta\varphi_m$, λ_x, λ_y and λ_z for the particular case of $\Delta\psi = \Delta\Phi = 0$. On the other hand, the second type of deformation, which will be presented here in some detail, may be readily handled without recourse to Lagrange multipliers. The imposition of incremenal changes, $\Delta\varphi_s$ to the torsional angle of a given bond s and the examination of the spatial rearrangement of chain atoms in response to these successive perturbations, would give insights as to the most probable mechanisms of local relaxational processes following the rotational transition of bond s. Thus, one variable among the $n + 4$ degrees of freedom is externally controlled and the induced changes in the other $n + 3$ variables are determined by the simultaneous solution of the set of equations

$$\partial S/\partial\Delta\varphi = 0 \tag{70}$$

$$\partial S/\partial\Delta X = 0 . \tag{71}$$

Here the incremental changes in the variables are organized, for convenience, in the form of two arrays

$$\Delta\boldsymbol{\varphi} \equiv \mathrm{col}(\Delta\varphi_1\,\Delta\varphi_2\,\ldots\,\Delta\varphi_{s-1}\,\Delta\varphi_{s+1}\,\ldots\,\Delta\varphi_{n-1}) \tag{72}$$

and

$$\Delta\mathbf{X} = \mathrm{col}(\Delta\psi\,\Delta\Phi\,\Delta X_0\,\Delta Y_0\,\Delta Z_0) . \tag{73}$$

Equation (70) is representative of $m = n - 2$ differentiations, each with respect to the dihedral angle $\Delta\varphi_m$ where $m \neq s$. Equation (71) accounts for the minimization of S with respect to the external degrees of freedom, except for χ, which is lumped into the set of dihedral angles.

The mathematical formulation comprises two major steps. First, a computationally convenient expression is written for the incremental changes in position vectors $\Delta\mathbf{R}_i$ as a linear combination of the incremental changes in the $n + 4$ variables controlling the kinematics of the chain. Second, the cumulative square displacements are minimized by application of Eqs. (70) and (71) to Eq. (68). A computationally suitable expression for $(\Delta\mathbf{R}_i)$, obtained in the first stage, reads [11, 12]

$$(\Delta\mathbf{R}_i)^2 =$$

$$\begin{bmatrix} \Delta\mathbf{r}_i \\ \Delta\psi \\ \Delta\Phi \\ \Delta\mathbf{R}_0 \end{bmatrix}^T \begin{bmatrix} 1 & \cdots & \cdots & \cdots \\ \mathbf{Dr}_i & \mathbf{Dr}_i\cdot\mathbf{Dr}_i & \cdots & \cdots \\ \mathbf{Br}_i & \mathbf{Br}_i.\mathbf{Dr}_i & \mathbf{Br}_i\cdot\mathbf{Br}_i & \cdots \\ \mathbf{T}(\psi)\mathbf{T}(\Phi) & \mathbf{T}(\psi)\mathbf{T}(\Phi)\mathbf{Dr}_i & \mathbf{T}(\psi)\mathbf{T}(\Phi)\mathbf{Br}_i & 1 \end{bmatrix} \begin{bmatrix} \Delta\mathbf{r}_i \\ \Delta\psi \\ \Delta\Phi \\ \Delta\mathbf{R}_0 \end{bmatrix} \tag{74}$$

where $\Delta\mathbf{r}_i$ refers to the position vectors with respect to the bond-based coordinate system appended to the first bond of the chain shown in Fig. 14, $\mathbf{T}(\psi)$

and $\mathbf{T}(\Phi)$ are the transformation matrices

$$\mathbf{T}(\psi) = \begin{bmatrix} \sin\psi & 0 & -\cos\psi \\ 0 & 1 & 0 \\ \cos\psi & 0 & \sin\psi \end{bmatrix} \tag{75}$$

and

$$\mathbf{T}(\Phi) = \begin{bmatrix} \sin\Phi & -\cos\Phi & 0 \\ \cos\Phi & \sin\Phi & 0 \\ 0 & 0 & 1 \end{bmatrix}. \tag{76}$$

$\mathbf{B}$ and $\mathbf{C}$ are defined as

$$\mathbf{B} \equiv \begin{bmatrix} 0 & 1 & 0 \\ -1 & 0 & 0 \\ 0 & 0 & 0 \end{bmatrix}; \quad \mathbf{C} \equiv \begin{bmatrix} 0 & 0 & 1 \\ 0 & 0 & 0 \\ -1 & 0 & 0 \end{bmatrix} \tag{77}$$

and

$$\mathbf{D} \equiv \mathbf{T}(\Phi)^{\mathrm{T}} \mathbf{C} \mathbf{T}(\Phi). \tag{78}$$

In the second stage of formulation, the differentiation of S following Eqs. (70) and (71) leads to

$$\begin{bmatrix} \mathbf{Q}_1 & \mathbf{Q}_2 \\ \mathbf{Q}_2^{\mathrm{T}} & \mathbf{Q}_4 \end{bmatrix} \begin{bmatrix} \Delta\varphi \\ \Delta\mathbf{X} \end{bmatrix} = \begin{bmatrix} \Delta\varphi_0 \\ \Delta\mathbf{X}^0 \end{bmatrix} \tag{79}$$

which is solved for $\Delta\varphi$ and $\Delta\mathbf{X}$ upon inversion of the leading matrix. Explicit expressions for the variables appearing in Eq. (79) are presented in the Appendix. The reader is referred to previous work, for the details of the derivation of Eq. (79). Thus, the optimal changes in the dihedral angles and in the absolute location and orientation of the chain segment accompanying the torsional motion of any internal bond, are computed from the inversion of Eq. (79), according to Eq. (A-16).

Inasmuch as the above theory is a first order approximation applicable to differential changes only, in either external or internal degrees of freedom, the incremental changes $\Delta\varphi_s$ in the rotational states have to be selected sufficiently small to avoid any nonlinear response. The occurrence of a nonlinear response, if any, may be monitored by observing the magnitude of the S function during a particular conformational rearrangement. An abrupt change in the latter is indicative of a major, large amplitude change in the spatial distribution of atoms, which is beyond the range of the applicability of the theory. Incremental changes, $\Delta\varphi_s$, of 0.05 radians have proven to be optimal inasmuch as the evolution of the dihedral angles of relevant bonds for $\Delta\varphi_s \leqslant 0.05$ radians exhibits closely the same pattern as that implied by $\Delta\varphi_s = 0.05$, and hence the adoption of smaller size steps is unnecessary. In order to extract information on the predominant mechanism of local relaxation phenomena in general, a suffficiently large number of Monte Carlo (MC) chains with a variety of initial configurations needs to be generated and the solution $\Delta\varphi$ and ΔX obtained in each case has to be analyzed.

3.2.3 Calculations

3.2.3.1 Cooperative Changes in Dihedral Angles

Calculations have been carried out for model FRCs of 25 bonds [12]. The torsional angle of the central bond is varied by 120° through a succession of 42 small steps and the induced changes in the conformations of the chain are recorded. The changes in the rotational state of the central bond were observed to be accommodated by the torsional motions of a few bonds (~ 8) in the neighborhood of the central rotating bond. The rotation of the central bond was observed to induce at least one counterrotation of substantial amplitude ($\sim 90°$) in either the first and second neighbor. A systematic analysis of the

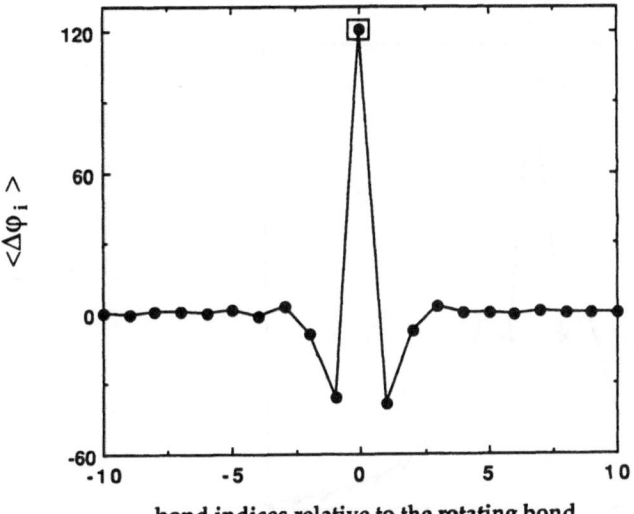

bond indices relative to the rotating bond

Fig. 15. Average change $\langle\Delta\varphi_i\rangle$ in bond dihedral angles predicted by the CK model in response to 120° rotation of the middle bond, presented as a function of location relative to the central bond as applied to FRC model. One thousand MC chains of various original configurations have been considered. The change in the dihedral angle of bond 13 is represented by a *square* around the *circle* to indicate that its rotation is externally imposed

response of the chain to the rotational transition of a given bond is carried out by solving Eq. (79) for a variety of original configurations, generated according to a Monte Carlo procedure assigning equal probabilities to the three isomeric states, t, g^+ and g^-. In Fig. 15 the average change $\langle \Delta\varphi_i \rangle$ in the dihedral angle of the i-th bond in response to 120° rotation of the central bond is shown for 1000 MC chains. The rotations of the first neighbors on both sides of the central rotating bond do not average out to zero but remain approximately equal to $- 40°$. This indicates the strong tendency of bonds to undergo counterrotations in response to the rotational isomerization of their first neighbors. On the other hand, when the absolute magnitudes of the changes in dihedral angles are considered, irrespective of their sense, the response of second neighbors turns out to be stronger than that of the first neighbors. This feature is illustrated by the solid curve in Fig. 16. The respective $\langle | \Delta\varphi_i | \rangle$ values for the first and second neighbors are equal to 45° and 51°. It should be pointed out that these two values represent about half of the actual rotational displacements observed in the majority of MC chains. This is due to the fact that a large amplitude rotation occurs, in general, on one side of the middle bond only, while the other side is weakly perturbed. As a consequence, the average $\langle | \Delta\varphi_i | \rangle$ for a give bond on one side is diminished by a factor of two, approximately. It should be added that calculations repeated for MC chains of n = 39 bonds lead, almost indistinguishably, to the same distribution curves as those displayed in Figs. 15 and 16

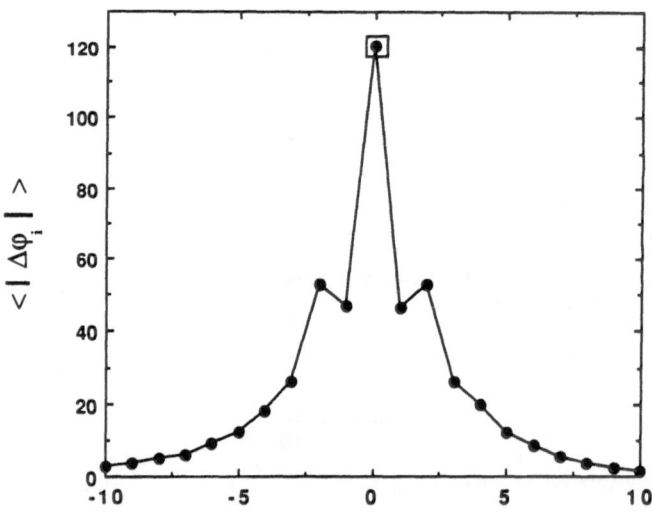

bond indices relative to the rotating bond

Fig. 16. Dependence of $\langle | \Delta\varphi_i | \rangle$ on location i relative to the middle bond, in response to 120° rotation of the middle bond, evaluated from 1000 MC chains of various original configurations. The change in the dihedral angle of bond 13 is represented by a *square* around the *circle* to indicate that its rotation is externally imposed

for the average changes in dihedral angles in response to the rotation of the middle bond. Thus, these curves are representative of the kinematics of freely-rotating chains in restrictive environment, irrespective of the size of the molecule.

3.2.3.2 Correlations Between Conformational Transitions

Calculations based on the CK theory show that a five bond trans sequence in the middle of the chain tends invariably to undergo the transition $ttttt \rightarrow g^- tg^+ g^-$, when a rotation of $120°$ is externally imposed on the central bond [12]. This observation supports the emphasis on type II motions observed in the BD simulations of polyethylene-like chains performed by Helfand and collaborators [98]. Similarly, the transition $g^{\pm} tt \leftrightarrow ttg^{\pm}$ resulting from gauche migration along the chain is systematically obtained as a natural consequence of the urge for minimizing cumulative displacements, which is in agreement with the BD observations. The two transitions (i) $ttt \leftrightarrow g^{\pm} tg^{\mp}$ and (ii) $g^{\mp} tt \leftrightarrow ttg^{\mp}$ account for more than 80% of the cooperative transitions observed in Brownian dynamics simulations [98]. Cooperative transitions, amounting to 29% of all observed transitions, refer therein to two successive rotational jumps undergone by bonds i and i + 2 within a time interval shorter than a cutoff value. Other transitions are also listed, belonging to that category, though their probability of occurrence is substantially lower. An example is the transition $ttt \leftrightarrow g^{\pm} tg^{\pm}$ in which two gauche bonds of the same sign are either created or annihilated. The application of the present mathematical formalism to a variety of MC chains, with central sequences originally in the ttt state, leads systematically to final configurations close to $g^{\pm} tg^{\mp}$ and not to $g^{\pm} tg^{\pm}$. Indeed, the final state $g^{\pm} tg^{\pm}$ is only obtained for some particularly favourable configurations of the neighboring bonds. It is also deduced in the calculations, on the basis of the mean squared displacements of atoms, that gauche migration is more probable compared to gauche pair production/annihilation, which is also in qualitative agreement with BD observations.

3.2.3.3 Spatial Reorientation and Transition in Response to Rotational Jumps

Orientational correlations have been studied between bonds by considering the final reorientation of all bonds in response to a given rotational transition, as predicted by the CK theory. In particular, the absolute orientation of the rotating bond itself is found to change in space, as its torsional state is changed. An average axial reorientation of about $35°$ is obtained for middle bonds of 1000 MC chains, from the solution of CK formulation. Thus the rotameric transition of a given bond is accommodated, among various readjustments, by the spatial reorientation of the bond itself, which overrules the previously proposed on-lattice descriptions of local motions in polymers. Examination of the angular deviation between the original and final orientations of all bonds indicates that first neighbors to a rotating bond are certainly most affected and rotate in space by about 50° on average, whereas angular displacements of other bonds decrease with increasing separation from the bond undergoing a rotational jump. The distribution of angular dispalcements for closest neighbors surrounding a rotating bond is displayed in Fig. 17. It is interesting to note that this distribution reveals the occurrence of a large number of small-amplitude distortions associated with rotational isomeric transitions, which conceals the experimental observation [118–120] of small angular displacements with the discrete process of rotational jumps. In fact, NMR experiments detect the spatial reorientation of transition moment vectors rigidly moving with the backbone bonds. In this respect, the motion of the first neighbor as a result of the rotation of a given bond, might be of interest. For a more detailed elucidation of spatial reorientations induced by rotational jumps, the angle α between the initial and final orientations of a first neighbor to a rotating bond has been considered. In the absence of any cooperative internal and external rearrangements, an isomeric jump of 120° by a given bond in a tetrahedral chain is expected to induce a spatial reorientation of $\alpha = 109.5°$ on its first neighbor, which is eventually evenly distributed over the neighbors on both sides. On the other hand, contrary to the expectations for on-lattice transitions, a wide variety of angular displacements, changing in the range $0° \leqslant \alpha \leqslant 95°$, was obtained, by solving Eq. (79) for 1600 MC of various original configurations. The distribution

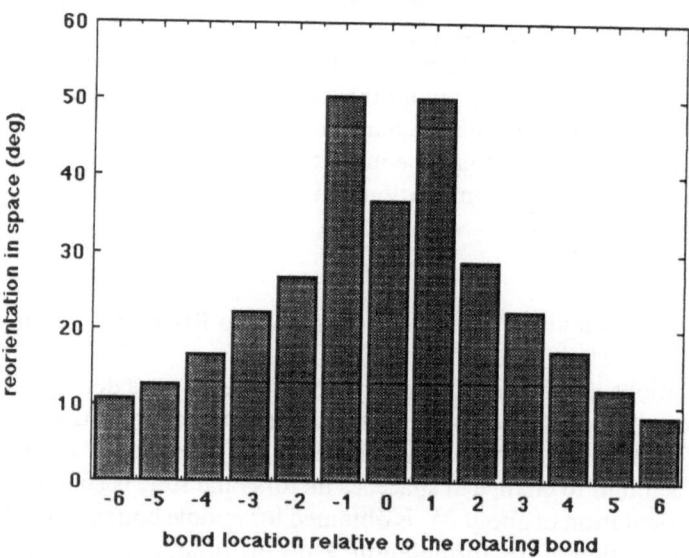

Fig. 17. Mean angular displacements of backbone bonds in response to the rotameric transition of the central bond. The reorientation of each bond is given as a function of its location relative to the rotating bond

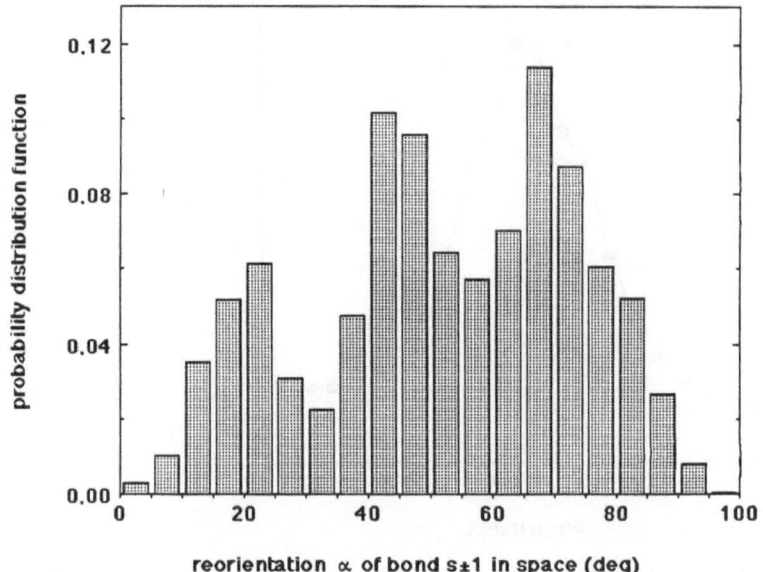

Fig. 18. Probability distribution of angular displacement α of backbone bonds in response to the rotameric transition of their first neighbors along the chain

exhibits three peaks centered at about 20°, 45°, and 70°, approximately, as displayed in Fig. 18. This broad range of angular displacements might contribute to the experimentally observed small amplitude distortions which are generally identified as librational motions. Thus, the experimental observation of a wide range of small amplitude reorientations does not necessarily imply the absence of jumps between rotational isomeric states. Although a complete 120° rotation is undergone by a given bond in a restrictive environment, it is reflected upon its neighbors as relatively small amplitude distortions, only.

The average displacement $\langle \Delta R_i \rangle$ of chain atoms accompanying the rotation of the central bond is shown by the solid line in Fig. 19 as a function of atom index. It is noted that the two atoms belonging to the rotating bond undergo the largest displacement in space, and the displacements gradually decrease with separation from the rotating bond. Finite displacements remaining at the tails indicate the overall translation and reorientation of the chain in space. Calculations performed for longer chains indicate that the displacement of the tails rapidly approaches zero as the length of the chain increases.

3.3 Contribution of Short-Range Conformational Energetics

The FRC model represents a relatively flexible chain, in which bond torsional motions are not hindered by internal conformational barriers. Bond length and

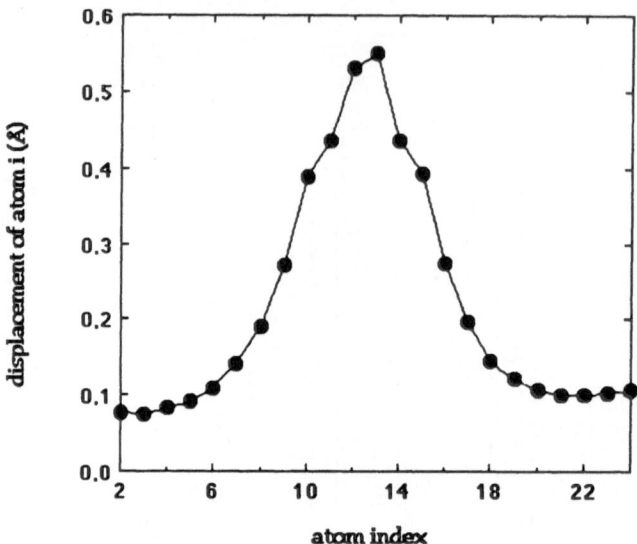

Fig. 19. Mean spatial dispalcements $\langle \Delta \mathbf{R}_i \rangle$ of backbone atoms i in resposne to the rotameric transition of the central bond, in freely-rotating chains of 25 bonds. The length of each bond is taken as 1 Å. The displacement of each atom is given as a function of its atom index along the chain

bending angle restrictions are therein the only internal factors geometrically constraining the motion. Real chains are however subject to bond rotational potentials favouring particular torsional states. A simple approximation is to assume bond rotational potentials to be independent of each other. A fifth order polynomial expression of the form

$$V_\varphi(\varphi_i) = k_\varphi \sum_{m'=0}^{5} a_{m'} \cos^{m'}\varphi_i, \quad i = 2, \dots, n-1 \tag{80}$$

suitably fits tristate potentials, as previously used for polyethylene-like chains. The coefficients $a_0, \dots, a_5$ are assigned values depending on the particular shape of the torsional potential. It is convenient to factor out in this expression a proportionality constant k_φ controlling the absolute magnitude of torsional potential, and examine its contribution to the kinematics of motion. The overall conformational potential experienced by the chain becomes

$$V = \sum_{i=2}^{n-1} V_\varphi(\varphi_i) \tag{81}$$

in the absence of short-range torsional coupling between bonds.

The energy associated with a given motion includes two contributions, the intermolecular effect opposing the displacement of atoms in space, and the intramolecular potential favouring torsions towards lower energy states. These may be viewed as the work and heat contributions to motion, in the terminology of classical thermodynamics, and sum up to yield the differential change in internal energy

$$dE = \delta S + \delta V \tag{82}$$

involved in a given conformational change. Minimization of E with respect to all degrees of freedom implies, in analogy, to Eqs. (70) and (71),

$$\partial E/\partial \Delta \varphi = 0 , \tag{83}$$

$$\partial E/\partial \Delta X = 0 . \tag{84}$$

Equation (83) may be rewritten as

$$\frac{\partial E}{\partial \Delta \varphi_m} = \frac{\partial S}{\partial \Delta \varphi_m} + \frac{\partial V}{\partial \Delta \varphi_m} = 0 \tag{85}$$

which upon expansion of V in Taylor series as

$$V = V_0 + \sum_i \frac{\partial V}{\partial \varphi_i} \Delta \varphi_i + \frac{1}{2} \sum_i \sum_j \frac{\partial^2 V}{\partial \varphi_i \partial \varphi_j} \Delta \varphi_i \Delta \varphi_j + \cdots \tag{86}$$

yields

$$\frac{\partial S}{\partial \Delta \varphi_m} + \frac{\partial V}{\partial \varphi_m} + \frac{\partial^2 V}{\partial \varphi_m^2} \Delta \varphi_m = 0 . \tag{87}$$

The set of Eqs. (83) and (84) may be rearranged by mathematical manipulations similar to those previously described to obtain the same matrix equation equivalent to the one in Appendix 3, i.e.,

$$\begin{bmatrix} \Delta \varphi \\ \Delta X \end{bmatrix} = \begin{bmatrix} Q_{1,E} & Q_2 \\ Q_2^T & Q_4 \end{bmatrix}^{-1} \begin{bmatrix} \Delta \varphi_E^0 \\ \Delta X^0 \end{bmatrix} . \tag{88}$$

Here all variables retain their previous definitions given in Appendix 3, with the exception of $Q_{1,E}$ and $\Delta \varphi_E^0$ which become

$$Q_{1,E} = Q_1 + \text{diag}(\partial^2 V/\partial \varphi_m^2) \tag{89}$$

and

$$\Delta \varphi_E^0 = \Delta \varphi^0 + \text{col}(\partial V/\partial \varphi_1 \ \partial V/\partial \varphi_2 \ldots \partial V/\partial \varphi_{s-1}$$
$$\partial V/\partial \varphi_{s+1} \ldots \partial V/\partial \varphi_{n-1})^T . \tag{90}$$

Exploratory calculations performed in this direction indicate that with increasing strength of intramolecular resistance to torsional motion, which is monitored by varying k_φ in Eq. (80), the kinematics of motion undergoes the following changes. (1) The mean amplitudes of counterrotations of immediate neighbors to the rotating bond become smaller. (2) The absolute changes in dihedral angles exhibit the same qualitative dependence on bond location relative to rotating bond, as the one displayed in Fig. 16, i.e., the response of the second neighbor remains the strongest even in the presence of torsional potentials. However, the absolute changes in dihedral angles are diminished. (3) Mean displacements of all atoms along the backbone increase steadily with increasing torsional barriers, though the shape of the distribution curve displayed in Fig. 19 is preserved. (4) The distribution function shown in Fig. 18 for the spatial reorientation of first neighbors is smoothed out; its trimodal character is lost.

4 Conclusion and Discussion

The rotational isomeric state model is a well established technique for deriving the static properties of polymers from the chemical structure. The dynamic rotational isomeric state approach is its dynamic counterpart. It is the first step for gaining insights as to the role of the intrinsic chemical structure on observed properties. It may be applied to the prediction of the intrinsic dynamics of polymers, i.e., those occurring in the absence of external effects, and to the examination of the relative relaxation rates of different units along a given chain. This will pertain to both homopolymers composed of different types of backbone atoms, and to copolymers built of different monomeric units.

In the DRIS approach, jumps between rotameric states constitute the only operating process for local relaxation. Both the constraints imposed by the covalently bonded portions of the chains beyond nearest neighbors, and environmental effects are approximated by a mean-field formalism only, in which bond rotations leading to reorientation of larger size kinetic segments experience higher resistance. In spite of these approximations, the DRIS approach has proven to be useful in explaining several experimental observations, which are also confirmed by BD simulations, such as (i) the difference in the local relaxation rates of polymers depending on their chemical structure, as exemplified by the relative behavior of polyethylene and polyoxyethylene chains, (ii) the stretched exponential time decay of correlation functions at intermediate times becoming more apparent with the degree of cooperativity of units along the chain, (iii) the activation energies for local motions of the order of the rotational barrier to single bond isomerization, (iv) the anisotropic character of segmental motions favoring motions perpendicular to chain contour rather than those along the backbone, (v) the predominant influence of the specific chemical structure of the chain and solvent viscosity on local relaxation processes in dilute solution. Yet in spite of the rigorous mathematical foundations of the stochastic model and its usefulness for gaining an understanding of the basic features of local motions, the theory involves some approximations and limitations, which are suitably complemented by the cooperative kinematics model.

The precise cooperative relaxation mechanism comprising the compensating motions implied by chain connectivity and the correlated rotameric transitions effectively localizing the motion are rigorously accounted for by the CK model. The model provides information on the precise geometry of motion, on the size of the chain segment engaged in cooperative relaxation, and on the amplitude and range of compensating motions accompanying a given bond rotation. Correlated transitions of the type $g^{\pm} tt \leftrightarrow ttg^{\pm}$ and $ttt \leftrightarrow g^{\pm} tg^{\mp}$, which amount to about 20% of all transitions observed in BD simulations of polyethylene-like chains, are readily predicted with the deterministic solution of CK theory applied to a perfect tetrahedral FRC model, which invites attention to the predominant role of chain geometry in inducing a specific relaxation mechanism. Also, the relative response of bond dihedral angles to rotameric transitions occurring in their close neighborhood, displayed in Fig. 16 for example, is in very good agreement with BD simulations, and is quite encouraging for future progress. Indeed, with its newly developed extension to incorporate real chain torsional energetics, the kinematic approach is easy to use for any type of chemical structure. It only requires correct geometry and energy parameters, which can now be obtained in a routine way from molecular mechanics software. Compared to MD and BD simulations, CK calculations are faster by several orders of magnitude. Also results obtained in simulations at the expense of long computational time and extensive data analysis are obtained from the unique solution of the set of equation based on energy minimization.

Acknowledgement. This work was supported by NATO grant no. CRG 910422.

5 Appendices

5.1 Correspondence Between Master and Fokker-Planck Equations

The Fokker-Planck equation may be obtained as an approximation to the master equation, as first derived by Planck [78]. Starting from Eq. (18), which, for a continuous random variable process with probability distribution $P(y)$, may be written in integral form instead of summations, as

$$\partial P(y)/\partial \tau = \int A(y|y - r) P(y - r)dr - P(y)\int A(y - r|y)dr . \tag{A-1}$$

Using the Taylor expansions of $P(y - r)$ and $A(y|y - r)$ about $P(y)$ and $A(y + r|y)$, respectively, up to second order in the continuous random variable y [188],

$$P(y - r) = P(y) - r \cdot \nabla_r P(y) + \tfrac{1}{2} rr : \nabla_r \nabla_r P(y) \tag{A-2}$$

$$A(y|y - r) = A(y + r|y) - r \cdot \nabla_r A(y + r|y) + \tfrac{1}{2} rr : \nabla_r \nabla_r A(y + r|y) \tag{A-3}$$

one may readily obtain the expression

$$\partial P(y)/\partial \tau = - \int \{r \cdot \nabla_r [P(y)A(y + r|y)]\}dr$$

$$+ \tfrac{1}{2}\int \{rr : \nabla_r \nabla_r [P(y)A(y + r|y)]\} dr \tag{A-4}$$

for the time evolution of the probability distribution function P(y). This expression relies on the approximation that the transition rate varies slowly with y. Also, the identity

$$\int A(y + r|y)\,dr = \int A(y - r|y)\,dr = a_0(y) \tag{A-5}$$

has been used in Eq. (A-4) for eliminating the last term of Eq. (A-1) with the leading term resulting from the Taylor expansion. Using the definitions

$$\langle r \rangle \equiv \int r\,A(y + r|y)\,dr; \quad \langle rr \rangle \equiv \int rr\,A(y + r|y)\,dr \tag{A-6}$$

for the jump moments per unit time, Eq. (A-4) may be rewritten in the Fokker-Planck form as

$$\partial P(y)/\partial \tau = - \nabla_r \cdot [\langle r \rangle P(y)] + \tfrac{1}{2}\nabla_r \nabla_r : [\langle rr \rangle P(y)] \ . \tag{A-7}$$

The Fokker-Planck equation, also referred to as the Smoluchowski equation or the generalized diffusion equation, neglects the moments of order larger than 2. Higher order terms appears in the Kramers-Moyal expansion [78].

5.2 Stochastic Weight Matrices and Computation of Correlation Functions

Using the definition

$$v_i(\alpha\beta; \xi\eta) = p_i(\alpha\beta, \tau; \xi\eta, 0)/p_i(\alpha, \tau; \xi, 0) \tag{A-8}$$

for the stochastic weight associated with the isomeric transition $\{\xi, \eta\} \to \{\alpha, \beta\}$ undergone by the pair of bonds $(i - 1, i)$ at a given time τ, the stochastic weight matrices for Markov chains read [85, 86]

$$\mathbf{V}_2(\tau) = \begin{bmatrix} p(\alpha; \alpha) & & & \\ & p(\beta; \beta) & & \\ & & \cdots & \\ & & & p(v; v) \end{bmatrix} \tag{A-9}$$

for bond 2, and

$$\mathbf{V}_i(\tau) = \begin{bmatrix} v_i(\alpha\alpha; \tau) & v_i(\alpha\beta; \tau) & \cdots & \\ v_i(\alpha\beta; \tau) & v_i(\beta\beta; \tau) & & \\ \cdots & \cdots & & \\ \cdots & \cdots & \cdots & \\ \cdots & \cdots & \cdots & \\ v_i(\alpha v; \tau) & v_i(v\beta; \tau) & v_i(vv; \tau) \end{bmatrix} \tag{A-10}$$

for the pair of bonds $(i - 1, i)$ in the range $3 \le i \le N - 1$. Here $\mathbf{v}_i(\alpha\alpha; \tau)$ is the $v^2 \times v^2$ matrix defined as

$$
\mathbf{v}_i(\alpha\alpha; \tau) =
\begin{bmatrix}
\mathbf{v}_i(\alpha\alpha; \alpha\alpha) & \mathbf{v}_i(\alpha\alpha; \alpha\beta) & \cdots & \cdots \\
\mathbf{v}_i(\alpha\alpha; \beta\alpha) & \mathbf{v}_i(\alpha\alpha; \beta\beta) & \cdots & \cdots \\
\mathbf{v}_i(\alpha\alpha; \gamma\alpha) & \mathbf{v}_i(\alpha\alpha; \gamma\beta) & \cdots & \cdots \\
\cdots & \cdots & \cdots & \cdots \\
\mathbf{v}_i(\alpha\alpha; v\alpha) & \mathbf{v}_i(\alpha\alpha; v\beta) & \cdots & \mathbf{v}_i(\alpha\alpha; vv)
\end{bmatrix} .
\tag{A-11}
$$

Thus $\mathbf{v}_i(\alpha\alpha; \tau)$ is the matrix of the weights associated with the transition to a given final state $\alpha\beta$ for the pair of bonds $(i - 1, i)$. We note that $\mathbf{v}_i(\alpha\beta; \xi\eta)$ is time-dependent but the time argument is omitted here for brevity.

The stochastic weight matrices are used in the computation of correlation functions as follows. The orientational correlation function between two vectors $\mathbf{m}$ and $\mathbf{n}$ affixed to the i-th and j-th bond-based frame is written in terms of the frame transformation operators $\mathbf{T}\{\Omega_\alpha\}$ and $\mathbf{T}\{\Omega_\beta\}$ using Eq. (37). $\mathbf{T}\{\Omega_\alpha\}$ and $\mathbf{T}\{\Omega_\beta\}$ are given by the serial multiplication of the transformation matrices $\mathbf{T}_k(\tau)$ associated with the passage from local frame $k + 1$ to frame k according to

$$
\mathbf{T}\{\Omega_\alpha\} = \prod_{k=1}^{i-1} \mathbf{T}_k(0) ,
$$

and

$$
\mathbf{T}\{\Omega_\beta\} = \prod_{k=1}^{j-1} \mathbf{T}_k(\tau) .
\tag{A-12}
$$

The arguments account for the implicit dependence of $\mathbf{T}_k(\tau)$ on time, through the rotating torsional angle ϕ_k. The ensemble average of Eq. (37) over the $v^{j-2} \times v^{j-2}$ possible conformational transitions leads to Eq. (38) in which the matrix $\mathbf{D}(\tau)$ is defined as

$$
\mathbf{D}(\tau) \equiv \mathbf{F} \left\langle \prod_{k=2}^{j-1} [\mathbf{T}_k(0) \otimes \mathbf{T}_k(\tau)] \right\rangle .
\tag{A-13}
$$

Here $\mathbf{T}_k(0)$ is set equal to $\mathbf{E}$, for $k > 1$, and the product $\left\langle \prod_{k=2}^{j-1} [\mathbf{T}_k(0) \otimes \mathbf{T}_k(\tau)] \right\rangle$ is computed with the matrix multiplication scheme

$$
\left\langle \prod_{k=2}^{j-1} [\mathbf{T}_k(0) \otimes \mathbf{T}_k(\tau)] \right\rangle =
$$

$$
Z_j(\tau)^{-1} (\mathbf{J}^T \otimes \mathbf{E}_{v^2}) \prod_{k=2}^{j-1} [(\mathbf{V}_k(\tau) \otimes \mathbf{E}_{v^2}) \| \mathbf{T}_k \otimes \mathbf{T}_k \|] (\mathbf{J} \otimes \mathbf{E}_{v^2})
\tag{A-14}
$$

where $\mathbf{E}_{v^2}$ is the identity matrix of order v^2 and $\|\mathbf{T}_k \otimes \mathbf{T}_k\|$ is the diagonal supermatrix of the form

$$
\|\mathbf{T}_k \otimes \mathbf{T}_k\| = \begin{bmatrix} \mathbf{T}_\alpha \otimes \mathbf{T}_\alpha & & & \\ & \mathbf{T}_\alpha \otimes \mathbf{T}_\beta & & \\ & & \cdots & \\ & & & \mathbf{T}_v \otimes \mathbf{T}_v \end{bmatrix}_k . \tag{A-15}
$$

$\mathbf{T}_\alpha, \mathbf{T}_\beta, \ldots, \mathbf{T}_v$ in Eq. (A-15) are the forms assumed by the transformation matrix $\mathbf{T}_k$ when bond k takes on the respective rotational isomeric states $\alpha, \beta, \ldots, v$.

5.3 Matrix Formulation of the Cooperative Dynamics Model

In the cooperative kinematics theory applied to the FRC model, the optimal changes in the variables $\Delta\varphi$ and $\Delta\mathbf{K}$ are found from

$$
\begin{bmatrix} \Delta\varphi \\ \Delta\mathbf{X} \end{bmatrix} = \begin{bmatrix} \mathbf{Q}_1 & \mathbf{Q}_2 \\ \mathbf{Q}_2^T & \mathbf{Q}_4 \end{bmatrix}^{-1} \begin{bmatrix} \Delta\varphi^0 \\ \Delta\mathbf{X}^0 \end{bmatrix}, \tag{A-16}
$$

which readily follows from Eq. (79). Here $\mathbf{Q}_1$ is the symmetric matrix of order $n - 2$,

$$
\mathbf{Q}_1 \equiv \begin{bmatrix} u_{11} & u_{12} \cdots u_{1,s+1} & u_{1,s+1} \cdots u_{1,n-2} & u_{1,n-1} \\ u_{21} & u_{22} \cdots u_{2,s-1} & u_{2,s+1} \cdots u_{2,n-2} & u_{2,n-1} \\ & \cdots & \cdots & \\ u_{n-1,1} & u_{n-1,2} \cdots & & u_{n-1,n-1} \end{bmatrix} \tag{A-17}
$$

$\mathbf{Q}_2$ is defined as

$$
\mathbf{Q}_2 \equiv \begin{bmatrix} p_1 & w_1 & v_1^T \\ p_2 & w_2 & v_2^T \\ \cdots & & \\ p_{s-1} & w_{s-1} & v_{s-1}^T \\ p_{s+1} & w_{s+1} & v_{s+1}^T \\ \cdots & & \\ p_{n-1} & w_{n-1} & v_{n-1}^T \end{bmatrix} \tag{A-18}
$$

and $\mathbf{Q}_4$ is the 5×5 symmetric matrix

$$
\mathbf{Q}_4 \equiv \begin{bmatrix} \sum_{i=0}^{n} \mathbf{Dr}_i \cdot \mathbf{Dr}_i & \cdots & \cdots \\ \sum_{i=0}^{n} \mathbf{Dr}_i \cdot \mathbf{Br}_i & \sum_{i=0}^{n} \mathbf{Br}_i \cdot \mathbf{Br}_i & \cdots \\ \mathbf{T}(\psi)\mathbf{T}(\Phi) \sum_{i=0}^{n} \mathbf{Dr}_i & \mathbf{T}(\psi)\mathbf{T}(\Phi) \sum_{i=0}^{n} \mathbf{Br}_i & (n+1)\mathbf{E} \end{bmatrix} \tag{A-18}
$$

with

$$
p_m \equiv \sum_{i=m+1}^{n} (\mathbf{a}_{im} \cdot \mathbf{Dr}_i) \tag{A-20}
$$

$$
w_m \equiv \sum_{i=m+1}^{n} (\mathbf{a}_{im} \cdot \mathbf{Br}_i) \tag{A-21}
$$

$$
\mathbf{v}_m \equiv \mathbf{T}(\psi)\mathbf{T}(\Phi) \sum_{i=m+1}^{n} \mathbf{a}_{im} \tag{A-22}
$$

$$
\Delta\varphi^0 \equiv -\operatorname{col}[u_{1s} \; u_{2s} \ldots u_{s-1,s} \; u_{s-1,s} \ldots u_{n-2,s} \; u_{n-1,s}]\Delta\varphi_s \tag{A-23}
$$

$$
\Delta\mathbf{X}^0 = -\operatorname{col}[p_s \; w_s \; \mathbf{v}_s^T]\Delta\varphi_s \tag{A-24}
$$

$$
u_{mj} = \sum_{i=k+1}^{n} \mathbf{a}_{im} \cdot \mathbf{a}_{ij} \quad \text{with } k = m \text{ if } m > j \text{ and } k = j \text{ if } j > m \tag{A-25}
$$

$$
\mathbf{a}_{ij} \equiv \left[\prod_{k=0}^{j-1} \mathbf{T}_k\right] \mathbf{A} \, \mathbf{T}_j [\mathbf{E} \; 0] \left[\prod_{k=j+1}^{i-1} \begin{bmatrix} \mathbf{T}_j & 1 \\ 0 & 1 \end{bmatrix}\right]\begin{bmatrix} 1 \\ 1 \end{bmatrix}. \tag{A-26}
$$

$\mathbf{T}_0$ is taken as the identity matrix of order 3.

7 References

1. Rouse PEJ (1953) J Chem Phys 21: 1272
2. Zimm BH (1956) J Chem Phys 24: 269
3. Flory PJ (1969) Statistical mechanics of chain molecules. Interscience, New York
4. Jernigan RL (1972) In: Karasz FE (ed) Dielectric properties of polymers. Plenum, New York, p 99
5. Bahar I, Erman B (1987) Macromolecules 20: 1368

6. Bahar I, Erman B (1988) J Chem Phys 88: 1228
7. Bahar I, Erman B, Monnerie L (1989) Macromolecules 22: 431
8. Bahar I, Erman B, Monnerie L (1989) Macromolecules 22: 2396
9. Bahar I, Erman B, Monnerie L (1990) Macromolecules 23: 1174
10. Bahar I, Erman B, Monnerie L (1991) Macromolecules 24: 3621
11. Bahar I, Erman B, Monnerie L (1992) Macromolecules 25: 6309
12. Bahar I, Erman B, Monnerie L (1992) Macromolecules 25: 6315
13. Erman B, Bahar I (1992) In: Noda I, Rubingh DN (eds) Polymer solutions, blends, and interfaces. Elsevier Science Publ. B.V., p 197
14. Ferrarini A, Moro G, Nordio PL (1990) Liquid Crystals 8: 593
15. Monnerie L, Vivoy JL (1986) In: Winnik MA (ed) Photophysical and photochemical tools in polymer science. p 193
16. Heatley F (1979) Progress in NMR spectroscopy. Pergamon, London, vol. 13
17. Heatley F (1986) Academic Press, London, vol. 17
18. Spiess HW (1985) Adv in Polym Sci 66: 23
19. Ediger MD (1991) Annu Rev Phys Chem 42: 225
20. Williams G (1972) Chemical Reviews 72: 55
21. Stockmayer WH (1967) Pure Appl Chem 15: 247
22. Kuhn W, Kuhn H (1945) Helv Chim Acta 28: 1533
23. Kuhn W, Kuhn H (1946) Helv Chim Acta 29: 609 830
24. Cerf, R (1957) J Polym Sci 23: 125.
25. Peterlin A (1967) J Polym Sci A-2 5: 179
26. Iwata K (1971) J Chem Phys 54: 12
27. Cerf R (1977) J Phys (Paris) 38: 357
28. MacInnes DA (1977) J Polym Sci, Polym Phys Ed 15: 465
29. Bazua ER, Williams MC (1973) J Chem Phys 59: 2858
30. deGennes PG (1979) Scaling concepts in polymer physics. Cornell University Press, Ithaca and London
31. Allegra G (1974) J Chem Phys 61: 4910
32. Allegra G (1975) J Chem Phys 63: 599
33. Allegra G (1978) J Chem Phys 68: 3600
34. Allegra G, Ganazzoli F (1981) J Chem Phys 74: 1310
35. Allegra G, Ganazzoli F (1981) Macromolecules 14: 1110
36. Allegra G, Ganazzoli F (1982) J Chem Phys 76: 6354
37. Allegra G, Higgins JS, Ganazzoli F, Lucchelli E, Bruckner S (1984) Macromolecules 17: 1253
38. Ganazzoli F, Allegra G, Higgins JS, Roots J, Brückner S, Lucchelli E (1985) Macromolecules 18: 435
39. Glauber RJ (1963) J Math Phys 4: 294
40. Shore JE, Zwanzig R (1975) J Chem Phys 63: 5445
41. Helfand EJ (1971) J Chem Phys 54: 4651
42. Shatzki TF (1962) J Polym Sci 57: 496
43. Boyer RF (1963) Rubber Chem Technol 34: 1303
44. Monnerie L, Gény F (1970) J Polym Sci Pt C 30: 93
45. Valeur B, Jarry JP, Gény F, Monnerie L (1975) J Polym Sci, Polym Phys Ed 13: 667
46. Valeur B, Monnerie L, Jarry JP (1975) J Polym Sci, Polym Phys Ed 13: 675
47. Dubois-Violette E, Gény F, Monnerie L, Parodi O (1969) J Chim Phys (Paris) 66: 1865
48. Jones AA, Stockmayer WH (1977) J Polym Sci, Polym Phys Ed 15: 847
49 Bendler JT, Yaris R (1978) Macromolecules 11: 650
50. Hall CK, Helfand E (1982) J Chem Phys 77: 3275
51. Orwoll RA, Stockmayer WH (1969) Adv Chem Phys 15: 305
52. Bozdemir S (1981) Phys Status Solidi B 13: 459
53. Bozdemir S (1981) Phys Status Solidi B 104: 37
54. Skinner JL (1983) J Chem Phys 79: 1955
55. Moro G, Nordio PL (1985) Mol Phys 56: 255
56. Moro G, Nordio PL (1986) Mol Phys 57: 947
57. Moro G (1987) Chem Phys 118: 167, 181
58. Moro G, Ferrarini A, Polimeno A, Nordio PL (1989) In: Dorfmüller T (ed) Reactive and flexible molecules in liquids. Kluwer Academic, Dordrecht, p 107
59. Moro G (1991) J Chem Phys 94: 8577
60. Ferrarini A, Moro G, Nordio PL (1988) Mol Phys 63: 225

61. Ferrarini A, Nordio PL, Moro RH, Crepeau RH, Freed JH (1989) J Chem Phys 91: 5307
62. Coletta F, Moro G, Nordio PL (1987) Mol Phys 61: 1259
63. Coletta F, Ferrarini A, Nordio PL (1988) Chem Phys 123: 397
64. Kramers A (1940) Physica 7: 284
65. Mashimo S (1981) J Polym Sci, Polym Phys Ed 19: 213
66. Iwata K (1973) J Chem Phys 58: 4184
67. Boyd RH, Breitling SM (1974) Macromolecules 7: 855
68. Blomberg C (1979) Chemical Physics 37: 219
69. Skolnick J, Helfand E (1980) J Chem Phys 72: 5489
70. Roe RJ (ed) (1991) Computer simulation of polymers. Prentice Hall, Englewood Cliffs, New Jersey, p 404
71. Zuniga I, Bahar I, Dodge R, Mattice WL (1991) J Chem Phys 95: 5348
72. Bahar I, Neuburger N, Mattice WL (1992) Macromolecules 25: 4619
73. Haliloglu T, Bahar I, Erman B (1992) J Chem Phys 97: 4428
74. Adolf DB, Ediger MD (1991) Macromolecules 24: 5834
75. Adolf DB, Ediger MD (1992) Macromolecules 25: 1074
76. Ediger MD, Adolf DB, in this volume
77. Roe RJ, in this volume
78. Van Kampen NG (1990) Stochastic processes in physics and chemistry. Elsevier Science Publishers, Amsterdam, North-Holland, p 419
79. Oppenheim I, Shuler KE, Weiss GH (1967) Adv Mol Relax Processes 1: 13
80. Gardiner CW (1990) Handbook of Stochastic Methods for Physics, Chemistry and Natural Sciences, 2nd edn. Springer-Verlag, London, p 442
81. Volkenstein MV (1963) Configurational Statistics of Polymeric Chains. Wiley-Interscience, New York
82. Jernigan R, Szu SC (1976) J Polym Sci, Polym Symp 54: 271
83. Paul E, Mazo RM (1971) Macromolecules 4: 424
84. Bahar I, Erman B, Kremer F, Fischer EW (1992) Macromolecules 25: 816
85. Bahar I (1989) J Chem Phys 91: 6525
86. Bahar I, Mattice WL (1990) Macromolecules 23: 2719
87. Kramers HA, Wannier GH (1941) Phys Rev 60: 252, 263
88. Newell GF, Montroll FW (1953) Rev Mod Phys 25: 353
89. Lifson S (1957) J Chem Phys 26: 727
90. Lifson S (1959) J Chem Phys 30: 964
91. Nagai J (1959) J Chem Phys 31: 1169
92. Birshtein TM, Ptitsyn OB (1959) J Tech Phys (USSR) 29: 1048
93. Williams G, Watts DC (1970) Trans Faraday Soc 66: 80
94. Williams G (1979) Adv Polym Sci 33: 59
95. Ngai KL, Mashimo S, Fytas G (1988) Macromolecules 21: 3030
96. Viovy JL, Monnerie L, Brochon J (1983) Macromolecules 16: 1845
97. Kloczkowski A, Mark JE, Bahar I, Erman B (1990) J Chem Phys 92: 4513
98. Weber TA, Helfand E (1983) J Phys Chem 87: 2881
99. Fixman M (1978) J Chem Phys 69: 1527
100. Fixman M (1978) J Chem Phys 69: 1538
101. Lang MC, Laupretre F, Noel C, Monnerie L (1979) J Chem Soc Faraday Trans 2 75: 349
102. Liu KJ, Anderson JE (1970) Macromolecules 3: 163
103. Hermann G, Weill G (1975) Macromolecules 8: 171
104. Friedrich C, Laupretre F, Noel C, Monnerie L (1980) Macromolecules 13: 1625
105. Bahar, I, Erman B (1987) Macromolecules 20: 2310
106. Herzberg G (1945) Infrared and Raman spectra; Van Nostrand-Reinhold, New Yor
107. Eliel EL, Allinger NL, Angval SJ, Morrison GA (1967) Conformational Analysis. Interscience, New York
108. Bahar I, Erman B, Monnerie L (1990) Macromolecules 23: 3805
109. Boese D, Kremer F (1990) Macromolecules 23: 829
110. Mark JE (1966) J Am Chem Soc 88: 4354
111. Abe Y, Flory PJ (1971) Macromolecules 4: 219
112. Anderson JE (1970) J Chem Phys 52: 2821
113. Ullman R (1968) J Chem Phys 49: 831
114. Ullman R (1967) J Chem Phys 47: 4879
115. Helfand E (1984) Science 226: 647

116. Viovy JL, Frank CW, Monnerie L (1985) Macromolecules 18: 2606
117. Jones AA (1989) In: Klafter J, Drake JM (eds) Molecular dynamics in restricted geometry, Wiley-Interscience, New York, p 247
118. Wefing S, Kauffman S, Spiess HW (1988) J Chem Phys 89: 1234
119. Westermark B, Spiess HW (1988) Makromol Chem 189: 2367
120. Schaefer D, Spiess HW, Suter UW, Fleming WW (1990) Macromolecules 23: 3431

Received: June 1993

Dynamics of Small Molecules in Bulk Polymers

A.A. Gusev[1], F. Müller-Plathe[2], W.F. van Gunsteren[2], and U.W. Suter[1]
[1] Institute of Polymers, [2] Laboratory of Physical Chemistry, ETH-Zentrum, CH-8092 Zürich, Switzerland

The mechanisms operative in diffusion and solubility of light gases in amorphous, dense polymer systems are discussed. Two types of methods used, Molecular Dymanics simulation and an approach based on Transition State Theory, are described. The methods prove to be complementary and to yield identical results in the areas where both can be independently applied.

Advances in Polymer Science, Vol. 116
© Springer-Verlag Berlin Heidelberg 1994

1 Introduction

The transport of low-molar-mass substances through polymers is a topic of broad interest. The technological relevance of such behavior has become evident in recent years through the rapidly growing demand for polymers with specified gas-transport properties. Examples are selective membranes for separation technology, e.g., gas separation membranes (employed in the fastest growing among gas separation methods) [1], or barrier membranes, i.e., highly impermeable yet flexible films. Despite this manifest increase in technological demand, however, there has been, until very recently, a considerable lack of fundamental understanding of the mechanisms underlying the solubility and diffusion of even the smallest molecules in macromolecular substances.

It is well known that the transport coefficients of low-molar-mass substances in polymers are strongly dependent on a number of factors, prominent among them the chemical structure of the macromolecules. Nevertheless, the models applied for interpretation of transport of small molecules through macromolecular systems have usually been of a rather non-specific nature. The polymer has commonly been viewed as a material continuum, sometimes as a "two-phase" structure as in the "dual-mode" model [2–12] where the medium contains cavities that act as Langmuir troughs. The processes operative in polymeric materials have often been interpreted on the basis of the concept of "free volume" [13–18], again in a structurally unspecific manner (note that the term "free volume" in the context of gas diffusion through solids usually means "volume not occupied by the polymer atoms" and should not be confused with the term of the same name used in dynamic theories and employed for the interpretation of, e.g., the glass transition or the viscosity of liquids). Molecular models also have been proposed and even early theories were based on a detailed consideration of the molecular structure [19], but the task proved difficult for theoretical approaches and the researchers were left with the options either to introduce unphysical simplifying assumptions or retain many parameters that could not be determined a priori [3, 18, 19–24]. Simulation as a method for the estimation of the transport coefficients of gases in polymers has only lately become a widely used approach, although early examples of such endeavors exist [25].

Here, we are concerned with amorphous polymers. Hence, amorphous structures have to be generated, by some means, that are low in energy, have an appropriate distribution of torsional angles, a physically acceptable distribution of unoccupied volume, and so on. In order to avoid surface effects, it is customary to employ periodic continuation conditions: The primary simulation box is surrounded by periodic images of itself and when an atom exits the box through one face of the box it reenters it through the opposite face. In order to keep the effect of chain ends minimal, we usually employ a single polymer chain per simulation box. For the chain the periodic continuation conditions imply, that it is "folded" into the box (if it leaves the box on one face, it reenters it immediately on the opposite face). The system is therefore an "amorphous crystal" in which a single polymer chain interacts with its own periodic images. It should be noted that during our simulations (in Molecular Dynamics [MD] runs the time interval covered is up to a few ns) the polymer only undergoes local change. The global habit of the polymer chain stays as it started out. One can, therefore, not hope for the polymer to "relax" to another configuration during the simulation and a few starting structures have to properly represent the entirety of the polymer sample. Here, we employ the modified rotational isomeric state method [26] in order to generate the iniatial configurations. Other methods, such as the reptation Monte Carlo [27] or MD with soft-core potentials [28] have also been used.

In this review we will focus on recent atomistic modeling of gas transport through polymeric materials, glassy solids as well as rubbery liquids, described in atomistic detail. We aim to reveal the probable mechanism of gas mobility and diffusion in polymers as brought forth through the combined use of Molecular Dynamics (MD) and a Transition-State Approach (TSA).

2 Solubility and Diffusion

Experimental methods for the determination of the transport coefficients of low-molar-mass substances in polymers have been well established for a long time (for an overview see Ref. 29). Complications arise due to the often observed dependence of solubility and diffusivity on solute concentration in the polymer [8, 29, 30]; however, for the lightest gases the diffusivity is often independent of concentration in the pressure range of experimental interest, and for the purpose of determination of the fundamental mechanisms of solubility and diffusivity pursued here, and for comparison with computer simulations, the limit of low concentration is sufficient. There, the transport coefficients are independent of penetrant concentration. The solubility is given by

$$c = Sp \quad \text{for } p \to 0 \tag{1}$$

where c is the solute concentration, p the pressure of the gas phase consisting of the low-molar-mass substance, and S the solubility coefficient. The steady state permeation rate (flux) J (per unit area) through a membrane is

$$J = D \nabla c = DS \nabla p \tag{2}$$

where ∇c is the concentration gradient and ∇p the corresponding pressure gradient across the membrane, and D the diffusion coefficient. The permeability coefficient P is defined by

$$P = DS \tag{3}$$

S can be obtained by direct measurement of the increase in the sample mass in a sorption experiment, but this is rather imprecise when S is small, as it tends to be for the lightest gases. A relatively simple and commonly used technique to determine P and D simultaneously, and hence S, is through a set-up in which the polymer membrane in question separates a chamber with the sample gas at controlled pressure from a chamber initially set to a lower pressure or with vacuum; permeation of the gas increases the pressure in the second chamber which is monitored with time and the resulting data are used to extract the required information. Specifically, the analysis is based on a simple one-dimensional differential equation, $\delta c/\delta t = D \, \delta^2 c/\delta x^2$, with the following initial and boundary conditions (the membrane thickness is l) in the membrane: $c(t=0)=c_0$, $c(x=0)=c_1$, $c(x=l)=c_2$. [31, 32]. When $c_1 \gg c_2$, as is usually the case, the product DS is obtained through the flux at steady state [Eq. (2): $DS = Jl/p$] and the diffusion coefficient D through linear backward extrapolation of the integrated steady state flux, since then

$$\int_0^t J dt \to \frac{Dc}{l}\left(t - \frac{l^2}{6D}\right), \quad t \to \infty \tag{4}$$

and, hence, the "lag time" θ, i.e., the intercept of the straight line fitted to the steady state integrated flux vs t, is

$$\theta = l^2/6D. \tag{5}$$

Care must be taken that a true steady state is indeed reached, a situation sometimes difficult to attain with glassy polymers [8].

A consideration relevant for later comparison of the results of simulations to experimental data concerns the experimental accuracy. The transport coefficients are notoriously difficult to measure and the steady state flux method outlined above is not very precise; large values for the transport coefficients are easier to determine and reported values are often of respectable precision, but even for high values (e.g., $D \sim 10^{-5}$ cm^2/s) the accuracy is typically not better than 10–20%, while it decreases for barrier membranes (e.g., $D \sim 10^{-9}$ cm^2/s and smaller) to that of an order-of-magnitude estimate. Hence, comparison with experiments will be made taking account of these variabilities.

The gas diffusion coefficient D may be extracted from a simulation in various ways. In a Molecular Dynamics simulation, if the hydrodynamic limit is reached (i.e., the simulation time is long enough), D may be calculated either from the evolution of the penetrants' instantaneous velocities by means of the Green–Kubo formalism or, alternatively, with the aid of the Einstein relation from their spatial positions. For TSA, there is only the Einstein-route, via the mean-square displacement $\langle r^2 \rangle = \langle |\mathbf{r}(t) - \mathbf{r}(0)|^2 \rangle$ which is plotted against time, best in a log-log plot. The portion of the curve that represents $\langle r^2 \rangle \propto t$ is used to fit a least-squares line, the slope of which (in a $\langle r^2 \rangle$ vs t plot) or the position of which (in a log $\langle r^2 \rangle$ vs log (t) plot) yields the diffusion coefficient:

$$D = \frac{\langle r^2 \rangle}{6t}, \quad t \to \infty \tag{6}$$

where t is the time and the angle brackets ⟨...⟩ denote an ensemble average realized as an average over many simulation attempts (in MD it is common to use several penetrants and to take the average over multiple time-origins).

For MD, there also is a reason for preferring the mean-square displacement method to the velocity autocorrelation function (i.e., the Green–Kubo formalism): the velocity autocorrelation function tends to be "noisy" at large t and it is often necessary to make assumptions about the analytical form of its long time tail [33].

3 Methods for Studying Small Molecule Diffusion in Polymers

3.1 Deterministic Dynamics (Molecular Dynamics)

The most straightforward way of studying the motion of individual atoms or molecules is molecular dynamics (MD) because time is explicitly present in its formulation. The MD method has been well described in the literature [33–35] so that a short overview is sufficient here.

In its simplest form, the MD method for a system of N atoms in the Cartesian coordinates $r_\alpha (\alpha = 1, N)$ works as follows. Given a potential energy function $\Phi = \Phi(r_\alpha)$ that depends on the positions of the atoms, Newton's equations of motion

$$\frac{d^2 r_\alpha}{dt^2} = -\frac{1}{m_\alpha} \frac{\delta \Phi}{\delta r_\alpha} \tag{7}$$

are then solved, subject to suitable boundary conditions and some initial conditions $r_\alpha = r_\alpha(0)$ and $v_\alpha = v_\alpha(0)$, where v_α are the velocities of the atoms.

Examining the above statement in detail we find that the potential energy function Φ describes the Born–Oppenheimer surface for the motion of the nuclei. It would be desirable to obtain this surface quantum-chemically, i.e, by solving the electronic Schrödinger equation of the system for every configuration of nuclei. Using ab initio or semiempirical quantum-chemical methods, this is possible at present for systems of a few atoms and simulations of up to several hundred time steps. Using, density functional methods, larger systems have been treated for longer times [36]. However, treating polymeric systems of several hundreds or thousands of atoms for several million time steps, as is required for the problem of diffusion, is still impossible with present-day computational resources. One therefore uses empirical analytical representations of Φ, so-called forcefields. A typical force-field for a polymeric system might look as follows:

$$\Phi = \sum_{\text{bonds}} \frac{k_r}{2}(r - r_0)^2 + \sum_{\text{angles}} \frac{k_\theta}{2}(\theta - \theta_0)^2 + \sum_{\text{torsions}} k_\phi[1 + s\cos(n\phi)]$$

$$+ \sum_{i>j} \left(\frac{A_{ij}}{r_{ij}^{12}} - \frac{B_{ij}}{r_{ij}^{6}} + \frac{q_i q_j}{r_{ij}} \right) \tag{8}$$

where the first two sums with the stiffness constants k_r and k_θ describe the bond-length and bond-angle deformations, the third sum with parameters k_ϕ, s, and n gives the intrinsic torsional potential energy functions, the constants A_{ij} and B_{ij} are the parameters of the non-bonded van-der-Waals interaction, and the q_i are the partial atomic charges.

There is a wide variety of functional forms for the force-field and various forms are used for its terms (bond angles harmonic in $\cos \theta$ rather than in θ, exponential terms instead of the r^{-12} repulsion expression, etc.). Some force-fields have cross-terms. It is also common to use rigid bond contraints instead of flexible harmonic bonds (as is done in the work described here) [37, 38]. Bond constraints are, strictly speaking, not a part of the forcefield, but correspond to solving the equations of motion in some lower-dimensional space. The reasons for their use are both technical (one can use longer time steps if the high-frequency bond vibrations are "frozen") and physical (only the vibrational ground states are populated at room temperature so the bond lengths are nearly constant).

In order to solve Newton's equations of motion they are discretised in time. Various schemes exist, but for the purpose of this discussion, it makes little difference which one is used. We employ the leap-frog scheme [33]. Newton's equations of motion conserve the total energy of the system and lead to a micro-canonical statistical-mechanical ensemble. In practice, one is usually more interested in ensembles in which the temperature (canonical ensemble) or the temperature and the pressure (isothermal-isobaric ensemble) are conserved. For this, Newton's equations of motion have to be slightly modified to couple the system temperature T (or pressure p) to a temperature (or pressure) bath of temperature T_0 (or pressure p_0). There are several such constant-temperature and constant-pressure schemes [33, 39]. We use the loose-coupling algorithm [40] which implements a first-order thermostat and manostat.

$$\frac{dT}{dt} = -\frac{T - T_0}{\tau_T}$$
$$\frac{dp}{dt} = -\frac{p - p_0}{\tau_p} \tag{9}$$

where τ_T and τ_p are the coupling times.

The MD simulation is started with a dense amorphous polymer micro-structure (see above) with appropriate initial conditions, i.e., the initial position and velocity of every atom must be specified. For the polymer, the initial coordinates are created by the modified RIS method [26], while the velocities are randomly picked from a Maxwell–Boltzmann distribution at the desired temperature. The diffusants are started off in some cavities in the polymer, found by randomly trying to insert them into the structure until a place is found where their energy is below a certain threshold. The straightforward technique is now to follow the displacement of the penetrants during the MD simulation and to apply the mean-square displacement method discussed above.

3.2 Transition-State Approach

The early theoretical treatments of gas dynamics in solid polymers were based on Transition-State theory. It was deduced [19, 20] fifty years ago that small gas molecules move through dense polymers in a series of activated "hops" between holes in the polymeric matrix. Surmising that a small molecule dissolved in the polymer must behave as a three-dimensional oscillator trapped by the surrounding chains, a vibrational frequency v_0 connected with the acquisition of the activation energy was evaluated [21] from solubility data. For light gases dissolved in dense polymers at 300 K, the value of v_0 was found to be nearly independent of the polymer and was about $10^{12} s^{-1}$. This frequency range indicated that the solute's jumps could be treated as an elementary process, thus justifying the use of Transition-State Theory [41] for evaluating the rates of the solute's hops. The problem was to account for the coupling between the solute mobility and the polymer (i.e., the matrix) dynamics. It was shown [20] that elastic displacement of the polymer atoms could considerably facilitate the solute's hops by causing a large reduction in the energy barriers encountered on diffusion. The separation of polymeric chains by thermal agitation was widely regarded [3, 20] as the process controlling the transition rates and, consequently, heuristic molecular-level models [22–24] were created in order to relate the solute dynamics to small-amplitude torsional motion of polymer chains. Nevertheless, the limitations to this phenomenological approach became obvious: guessing the structure of amorphous polymers in bulk and the number of polymer segments involved in the activation process, one was confronted with the need to specify a number of adjustable parameters without precise physical meaning.

Fortunately, results of MD studies [27, 28, 42–49] demonstrate (see also Sect. 4 below) that the MD trajectories of small gases in atomistic microstructures of dense polymers are consistent with those expected from the hopping mechanism for gas motion and the structural relaxation of the polymer chains does not seem to contribute appreciably to the rate constants of the solute hops, thus supporting the adequacy of the hopping mechanism for the motion of a light solute in dense polymers and lending credibility to the simple modeling of the transport processes.

For the work reviewed here, two relevant mechanistic features pertaining to gas motion in dense polymers have been conjectured:

- small molecules move in dense polymers in a series of activated hops;
- the solute dynamics is coupled to the elastic motion of dense polymers, but, to a first approximation, is independent from the structural relaxation of the polymeric matrix.

The second assumption has far-reaching consequences: if the solute dynamics is not coupled to the structural relaxation of the polymer, the problem becomes much easier – instead of solving a formidable dynamic multi-body problem one describes the behavior and properties of the solute with a time-independent single-particle distribution function $\rho(\mathbf{r})$, thus reducing the problem to that of an ideal gas subjected to an external field stationary in time.

3.2.1 Describing the Thermal Motion of the Polymer Atoms

Thermal motion causes the polymer matrix to move in its configuration space. At short times, the vibrational modes of motion dominate: vibrations of chemical bonds and bond angles, small-amplitude rotational motions of side groups, or wiggling of torsion angles are some of the relevant examples of mobility in this time domain. We term this vibrational behavior "elastic motion". At longer times, the system tends to perform structural relaxation. Torsional transitions in the main chain or in side groups are examples of such processes. To specify an upper bound for times at which the system at hand can be treated as essentially executing elastic motion only, one should compare the time scales of correlation functions describing elastic and structural relaxation processes. In practice, this comparison can be conducted on the basis of an appropriate MD trajectory or by means of analyzing suitable experimental data (e.g., X-ray or neutron scattering, dielectric or NMR relaxation data, etc.).

Elastic motion implies that the atoms of the matrix fluctuate about their equilibrium positions $\langle \mathbf{x}_\alpha \rangle$ (where $\alpha \in \{1, \ldots, N\}$) and we introduce here the normalized probability density function $W(\Delta_1, \ldots, \Delta_N)$, with the deviations $\Delta_\alpha = \mathbf{x}_\alpha - \langle \mathbf{x}_\alpha \rangle$, to describe the elastic fluctuations. We shall use below $\{\Delta\}$ as shorthand for $\Delta_1, \ldots, \Delta_N$. Imagine now a small dissolved molecule residing in the polymer matrix and suppose that it is legitimate to neglect the correlations between the structural relaxation in the matrix and the dynamics of the dissolved molecule. In this case the solute distribution function $\rho(\mathbf{r})$ can be written as [50]

$$\rho(\mathbf{r}) = \int d\{\Delta\}\, W(\{\Delta\})\, e^{-U(\mathbf{r}, \{\mathbf{x}\})/kT} \tag{10}$$

where $U(\mathbf{r}, \{\mathbf{x}\})$ is the potential energy of interaction between the dissolved molecule at point $\mathbf{r}$ and the host atoms at the positions $\{\mathbf{x}\}$ (i.e., at $\mathbf{x}_\alpha$, $\alpha \in \{1, \ldots, N\}$), and integration is over all possible values of the deviations $\Delta_\alpha = \mathbf{x}_\alpha - \langle \mathbf{x}_\alpha \rangle$ of the host atoms.

We follow Debye's approach [51], suggested many years ago for analyzing the thermal vibrations in solids, and approximate $W(\{\Delta\})$ by the assumption of isotropic elastic motion in the following form [50]

$$W(\{\Delta\}) = C \exp\left(-\sum_\alpha \frac{\Delta_\alpha^2}{2\langle \Delta_\alpha^2 \rangle} \right) \tag{11}$$

where the normalization constant C is obtained from the equation

$$\int d\{\Delta\}\, W(\{\Delta\}) = 1 \tag{12}$$

and $\langle\Delta_\alpha^2\rangle$ stands for the mean-square deviation of host atom α from its average position. A similar parameter is employed in the analysis of X-rays diffraction patterns, the so-called "Debye–Waller factor" [52, 53].

Of course, the elastic motion of the chain molecules is anisotropic and a refinement might be to consider a $3N \times 3N$ smearing tensor which describes the elastic fluctuations of the atomic positions. Nevertheless, we tend to think that the simple approximation of Eq. (11) is enough in order to study the fundamentals involved in the transport of gas molecules through solid polymers.

3.2.2 A Guest Molecule in a Matrix with Isotropic Elastic Motion

In the pair approximation, the potential energy $U(\mathbf{r}) = U(\mathbf{r},\{\mathbf{x}\})$ of the interaction between the dissolved molecule at point $\mathbf{r}$ and the host atoms at the positions $\{\mathbf{x}\}$ is,

$$U(\mathbf{r}) = \sum_\alpha U_\alpha(\mathbf{r} - \mathbf{x}_\alpha) \tag{13}$$

where the term $U_\alpha(\mathbf{r} - \mathbf{x}_\alpha)$ depends only on the separation between the dissolved molecule and host atom α. With Eqs. (11) and (13) we immediately have from Eq. (10)

$$\rho(\mathbf{r}) \propto \prod_\alpha \int d\Delta_\alpha \exp\left[-\frac{\Delta_\alpha^2}{2\langle\Delta_\alpha^2\rangle} - \frac{U_\alpha(\mathbf{r} - \mathbf{x}_\alpha)}{kT} \right] \tag{14}$$

where the integration in $d\Delta_\alpha$ extends over all possible deviations Δ_α of host atom α. Since, after integrating, the terms on the right-hand side of Eq. (14) depend only on the separation between the guest at point $\mathbf{r}$ and the equilibrium positions $\langle\mathbf{x}_\alpha\rangle$ of the host atoms, we have,

$$\rho(\mathbf{r}) = \prod_\alpha \rho_\alpha(\mathbf{r} - \langle\mathbf{x}_\alpha\rangle) \tag{15}$$

where $\rho_\alpha(\mathbf{r} - \langle\mathbf{x}_\alpha\rangle)$ stands for the probability density function describing the interacting pair of host atom α and the dissolved molecule, considered separate from the influence of all other host atoms.

The isotropic approximation to elastic motion makes the treatment remarkably simple by converting the evaluation of the solute distribution function $\rho(\mathbf{r})$ to a trivial multiplication of independent terms describing the pair interaction between the solute molecule and the polymer atoms localized at their average positions $\langle\mathbf{x}_\alpha\rangle$. A set of mean positions $\langle\mathbf{x}_\alpha\rangle$ describes the structure of the polymer matrix, while a set of $\langle \Delta_\alpha^2 \rangle$ gives the effective stiffness ξ_α of the "springs" by which the host atoms can be thought to be bound to their mean positions $\langle \mathbf{x}_\alpha \rangle$:

$$\xi_\alpha = kT/\langle \Delta_\alpha^2 \rangle \tag{16}$$

The presence of the dissolved molecule influences the thermal vibrations of the host atoms and causes an elastic displacement in the matrix, strongest for the nearest host atoms. This intrinsic elasticity of the host matrix is formally accounted for by Eq. (14) and leads to a change in the repulsive part of the guest-host interaction that is nearly harmonic and depends on the effective stiffness ξ_α of the springs rather than on the functional form of the guest-host interaction [50].

The isotropic elastic motion of the polymer matrix is specified by the spectrum of the $\langle \Delta_\alpha^2 \rangle$. In the homogeneous approximation we neglect the differences between the amplitudes of elastic motion of different polymer atoms and describe the elastic motion of all polymer atoms with an identical smearing factor $\langle \Delta^2 \rangle$. One would expect $\langle \Delta^2 \rangle^{1/2}$ to assume a value similar to the Debye–Waller factors [53] obtained from the scattering of X-rays or neutrons with dense polymers, i.e., about 0.2–0.4 Å at ambient conditions. Of course, the local elasticity of the polymer matrix may deviate somehow from the mean elasticity introduced by the isotropic homogeneous smearing factor $\langle \Delta^2 \rangle$ and a refinement might be advisable to account for the spectrum of $\langle \Delta_\alpha^2 \rangle$.

3.2.3 What Can Be Learned by Assuming the Polymer Matrix Is Rigid

In the limiting cases of very small molecules or of very low temperatures one can surmise [54] that the rigid (static) approach to the polymer matrix might be useful as a first approximation to estimate the solute distribution function $\rho(\mathbf{r})$. In this situation, $\rho(\mathbf{r})$ is given by the limiting case of Eq. (14) [or Eq. (10)]:

$$\rho(\mathbf{r}) \propto e^{-U(\mathbf{r})/kT} \tag{17}$$

where the potential energy $U(\mathbf{r})$ describes the interaction between a structureless solute molecule at position $\mathbf{r}$ and the atoms of the "frozen" polymer matrix.

The potential energy $U(\mathbf{r})$ of the solute determines the spectrum of the normal mode frequencies ν_α that characterize the vibrations of solute molecules in the vicinity of the local minima of $U(\mathbf{r})$. To obtain this spectrum, at each local minimum of $U(\mathbf{r})$ the components of the Hessian matrix at the local minimum can be computed and the spectrum of ν_α can be determined from its eigenvalues. It has been found that for different small molecules dissolved in the rigid-microstructures of both glassy and rubbery polymers at 300 K, the spectrum of ν_α has its maximum at about $10^{12}\,\text{s}^{-1}$ (see Fig. 1), in impressive agreement with the above-mentioned deduction by Barrer[21] from solubility data, more than 40 years ago. It has also been established that the condition $h\nu_\alpha < 3 \cdot kT$ holds at

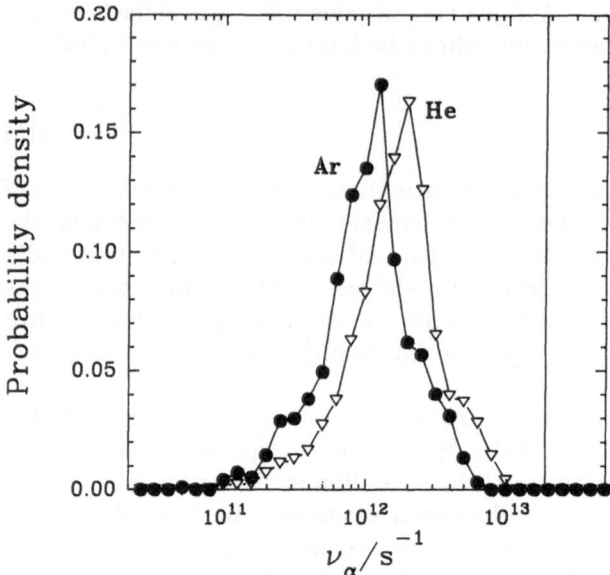

Fig. 1. The distribution functions for the frequencies γ_α of the normal mode vibrations of a solute molecule in local maxima of $\rho(\mathbf{r})$ (i.e., local minima of the Helmholtz energy of residence of the solute molecule). Data were collected from 5×10^2 sites for He and from 1×10^2 sites for Ar from two micro-structures of PC. The vertical line indicates the threshold $3 \cdot kT/h$, significantly below which all the frequencies must lie if the quasi-classical approximation is appropriate

300 K for the vibrational frequencies v_α of light gases at the sites of minimum potential energy in dense polymers. Hence, the quasi-classical approximation [55] is appropriate for studying the behavior and properties of small molecules in dense polymers and we shall use this approximation below.

3.2.4 Solubility (The Spatial Fermi Gas and the Dual-Mode Sorption Model)

Suppose that the solute distribution function $\rho(\mathbf{r})$ is known. The real trajectories of solute molecules that yield this distribution function are very complex, but an important feature of this movement can be elucidated explicitly: the solute molecules should spend a large fraction of time in oscillations in the vicinity of the local maxima of the distribution function $\rho(\mathbf{r})$. Any two adjacent local maxima, i and j, of this distribution function are separated by a common crest surface Ω_{ij} on which the following condition holds:

$$\left.\frac{\delta\rho}{\delta\mathbf{n}}\right|_{\Omega_{ij}} = 0 \tag{18}$$

where $\delta/\delta\mathbf{n}$ denotes differentiation along the normal $\mathbf{n}$ to surface Ω_{ij}. A set of surfaces $\{\Omega_{ij}\}$ surrounding a particular maximum i delimits that part of space which belongs to maximum i and we define site i as that part of space.

If we assume that not more than one solute molecule can occupy a site at one time, the solute molecules will obey statistics of the Fermi–Dirac type; an ensemble of solute particles obeying this principle is termed a "spatial Fermi gas" [54]. The fundamental thermodynamic relations, such as the equilibrium distribution function, the free energy, the pressure dependence of the concentration of solute in contact with the ideal-gas phase, and the equation of state have been established in closed form for a spatial Fermi gas of identical as well as different particles [54]. Some of the results shall be reviewed here.

At equilibrium between the spatial Fermi gas and the ideal gas of the same molecules the mean number $\langle n_j \rangle_{eq}$ of solute molecules at site j is [54]

$$\langle n_j \rangle_{eq} = \frac{1}{\dfrac{kT}{pZ_j} + 1} \tag{19}$$

where T is the temperature, p is the ideal-gas pressure, and Z_j is the configurational partition function of a solute molecule localized at site j,

$$Z_j = \int_{V_j} \rho(\mathbf{r}) dV \tag{20}$$

Here, V_j stands for the volume of site j. The solute concentration c is then simply the sum over the equilibrium occupation numbers $\langle n_j \rangle_{eq}$ per unit volume,

$$c = \sum_{j=1}^{K} \frac{b_j p}{1 + b_j p} \tag{21}$$

where K denotes the total number of sites, per unit volume, and the constant b_j characterizes the sorption ability of site j.

Equation (21) is a description of the sorption isotherm. It provides the total dissolved concentration c at any pressure p. Its usefulness lies in the fact that for any temperature T the local classical configuration partition function Z_j can be evaluated for each site j, once the distribution function $\rho(\mathbf{r})$ is known for a solute molecule as a function of its position $\mathbf{r}$ in the polymer matrix.

For very small pressures·p we neglect the term $b_j p$ in the denominator of Eq. (21) and find that the solute concentration c is proportional to the ideal gas pressure p.

$$c = p \sum_{j=1}^{K} b_j = Sp \tag{22}$$

where Henry's constant S is obtained by integration over the entire volume V of the polymer structure.

$$S = \frac{1}{kTV} \int_V \rho(\mathbf{r}) dV \qquad (23)$$

At elevated pressures, the dependence of the solute concentration on the ideal gas pressure can be represented as a sum of linear terms, $b_j p$ (for sites j where $b_j p \ll 1$), and "Langmuir-like" terms, $b_j p /(1 + b_j p)$ (for sites j where the term $b_j p$ is at least comparable to unity). If a system consists only of "linear" sites and a set of identical "Langmuir" sites, the well-known "*dual-mode-sorption*" model [2] is obtained [54],

$$C = k_D p + C_H \frac{b_H p}{1 + b_H p} \qquad (24)$$

where k_D, C_H, and b_H are constants. This form is widely used [11] as a convenient functional form for fitting experimental data.

It is often implied that the dual-mode-sorption model has a physical basis in two distinct mechanisms of gas solubility in dense polymers. The first one is assumed to be associated with a "liquid-like" solubility, while the second one is due to gas solubility in some "preexisting holes" in a polymer structure. To check the microscopic basis of this approach, one can analyze the distribution of b_j values, obtained through numerical evaluation of the solute's distribution function in atomistic micro-structures of dense polymers; if the dual-mode-sorption model is meaningful for this case, then the value b_H should stand out among all others in the "spectrum" of b_j values.

The spectrum of b_j values has been evaluated from the distribution function $\rho(\mathbf{r})$ of methane molecules in rigid micro-structures of bisphenol-A-polycarbonate at 300 K (for details, see Ref. [54]). It was found that the parameter b_H obtained by fitting the experimental sorption isotherm with Eq. (24) agrees to within an order of magnitude with that segment of the spectrum of b_j that provides the major contribution to methane sorption, thus demonstrating that the dual-mode-sorption model is indeed physically reasonable for the description of gas sorption in glassy polymers (although its results should not be taken to indicate the actual existence of Langmuir-troughs as cavities in the polymer matrix).

3.2.5 Dynamics of a Solute Molecule

Following Transition-State Theory [41], we write [56] the rate constant R_{ij} of solute transition from a site i to adjacent site j as:

$$R_{ij} = \kappa \frac{kT}{h} \frac{Q_{ij}}{Q_i} \qquad (25)$$

where k is the Boltzmann constant, h the Planck constant, Q_i and Q_{ij} are the partition functions of a solute molecule in site i and at the activated state for a transition from site i to site j, respectively, and κ is a transmission factor taken to be $\kappa \approx 1/2$ [50].

The quasi-classical partition function of a solute particle of mass m, located at site i, can be written as:

$$Q_i = \left[\frac{2\pi mkT}{h^2}\right]^{3/2} \int_{V_i} \rho(\mathbf{r}) dV \tag{26}$$

where V_i is the volume of site i. The partition function of the activated (transition) state between sites i and j is

$$Q_{ij} = \frac{2\pi mkT}{h^2} \int_{\Omega_{ij}} \rho(\mathbf{r}) ds \tag{27}$$

with ds denoting the surface element. Substituting Eqns. (26) and (27) into Eqn. (25), we find [56, 57]:

$$R_{ij} = \sqrt{\frac{kT}{8\pi m}} \int_{\Omega_{ij}} \rho(\mathbf{r}) ds \bigg/ \int_{V_i} \rho(\mathbf{r}) dV = \sqrt{\frac{kT}{8\pi m}} \frac{Z_{ij}}{Z_i} \tag{28}$$

where Z_i and Z_{ij} are the configurational parts of the corresponding partition functions.

Equation (28) gives the (absolute) rate constants for site-to-site transitions in the quasi-classical approximation. Evaluation of the R_{ij} requires integration of the distribution function $\rho(\mathbf{r})$ over the volume of site i and crest surface Ω_{ij} separating sites i and j.

3.2.6 Motion on the Network of Sites

Let us focus on a solute molecule at site i. The probability w_{ij} of its transition from site i to a particular adjacent site j is proportional to the rate constant R_{ij} for that transition

$$w_{ij} = \tau_i R_{ij} \quad \text{with} \quad \sum_n w_{in} = 1 \tag{29}$$

where the summation is made over all adjacent sites n (i.e., over all sites having common crest surfaces with site i). The normalization constant τ_i is the mean residence-time of the solute at site i,

$$\tau_i \equiv 1 \bigg/ \sum_n R_{in} \tag{30}$$

where n again denotes all adjacent sites.

The sites form a spatial network [56] on which the gas molecules can "jump" through the dense structure. This network comprises the spatial connectivity of the sites and the residence times τ_i of the solute there.

3.2.7 Polyatomic Molecules as Solutes

All considerations outlined above have focused on solutes consisting of a single interaction site. However, most solute molecules in practice are polyatomic and it is mandatory to assess the effect of this added complexity. We focus here on diatomic molecules as an example.

In the gas phase, the energy level of a diatomic molecule is approximately [55] the sum of the internal vibrational energy of the molecule, the rotational energy of the molecule as a whole, and the translational energy of its center of mass. At ambient conditions, the intervals in the vibrational terms are always large compared with the thermal energy kT (k is the Boltzmann constant and T is the temperature), so that the vibrational degree-of-freedom is almost always in its ground state. As an illustration, we quote the characteristic vibrational temperature [55], the "vibrational quantum" $T_{vib} = h\nu/k$ (h is the Planck constant) for some gases, H_2: 6100 K, NO: 2700 K, O_2: 2200 K. This stiff vibrational degree of freedom is probably never affected by the dissolution process and, at ambient conditions, its effect can be ignored.

The characteristic rotational temperature, the "rotational quantum" $T_{rot} = h^2/8\pi^2 kI$ (I is the moment of inertia) of some diatomic molecules [55] is, H_2: 85 K, O_2: 2.1 K, NO: 2.4 K, Cl_2: 0.4 K. As a consequence, the rotation of the diatomic molecules is classical at ambient conditions and each rotational degree of freedom has the energy 1/2 kT at equilibrium. That immediately yields an estimate of the rotational correlation time τ_{rot}, in the absence of an external field [58]:

$$\frac{1}{\tau_{rot}} = \frac{4\pi k}{h}\sqrt{T \cdot T_{rot}} \tag{31}$$

For the four molecules just mentioned this yields at T = 300 K the following values for τ_{rot}, H_2: 0.02 ps, O_2: 0.15 ps, NO: 0.14 ps, Cl_2: 0.35 ps.

A solid matrix supplies an external field in which the potential energy $U(r, \theta, \phi)$ of the diatomic molecule depends on both the position r of its center of mass and its spatial orientation given by two angles θ and ϕ. The translational and rotational degrees-of-freedom may be coupled by means of the external field and it has to be determined to which extent the organic polymer matrices are capable of influencing the rotation of diatomic molecules. Results of a Molecular Dynamics study of O_2 dynamics in poly(isobutylene) (PIB) at 300 K indicate [59] that the rotation of the O_2 molecules is well separated from their translational motion and the rotational correlation time $\tau_{rot} \approx 0.1$ ps derived from the Molecular Dynamics trajectories [59] agrees well with the value of 0.15 ps deduced above; one can conclude that the PIB matrix does not affect the coupling between translation and rotation in this case. Consequently it seems appropriate to treat the translational and rotational degrees of freedom of diatomic molecules dissolved in dense polymers as separable and to write [58]

$$\rho(\mathbf{r}) \propto \int d\theta \, d\phi \, d\{\Delta\} \, W(\{\Delta\}) \exp\left\{ -\frac{U(\mathbf{r},\{\Delta\},\theta,\phi)}{kT} \right\} \tag{32}$$

where the distribution function $\rho(\mathbf{r})$ depends only on the position $\mathbf{r}$ of the center of mass of the molecule, as for a single-site particle.

3.2.8 How to Actually Carry Out a TSA Simulation

Evaluating the configurational partition function. For the purpose of numerically evaluating the solute distribution function $\rho(\mathbf{r})$ according to Eq. (14), given a smearing factor $\langle \Delta^2 \rangle$, we construct an orthogonal, equispaced lattice with a grid interval $d \approx 0.2$ Å inside the atomistic micro-structure of interest and compute $\rho(\mathbf{r})$ (or the Helmholtz energy for the solute's residence) at each lattice point. The "sites" of residence) (see Sect. 3.2) and the crest surfaces between them are then determined [54] by a steepest-descent gradient method: a steepest-descent path is started from every grid point; the paths terminates, by necessity, in one of the local maxima of $\rho(\mathbf{r})$, thus providing a unique assignment of the grid points to the set of sites. The volume integrals Z_i of Eq. (20) are then simply the sums of the contributions from the appropriate grid points.

The surface integrals Z_{ij}, see Eq. (28), can now also be determined: the crest surface Ω_{ij} must be located between adjacent grid points with allocation to sites i and j and one would expect (from the definition of the crest surface Ω_{ij} [see Eq. (18)]) that the distribution function $\rho(\mathbf{r})$ be practically constant along the surface normal, and we use the grid points allocated to the (initial) site i to evaluate the surface integral Z_{ij}. The appropriate value of the surface element Δs for numerical summation is found as follows: If a grid point, allocated to site i, has only one of its six nearest grid neighbors allocated to site j, one uses $\Delta s = d^2$; the case of two neighbors with allocation to site j requires $\Delta s = 2^{1/2} d^2$; from empirical trials it has been determined that it is suitable to take $\Delta s = 1.41 \, d^2$ when three neighbors are members of site j, $\Delta s = 2.0 d^2$ when four neighbors belonged to site j, and $\Delta s = 0$ for the remaining cases. The algorithm was tested with simple figures (sphere, various polyhedra, etc.) and yields an accuracy of better than 7% in area estimation.

Micro-reversibility requires that there be no difference in the values of Z_{ij} and Z_{ji} and we use $(Z_{ij} + Z_{ji})/2$ for evaluating the solute partition function on surface Ω_{ij} (this prevents effects of errors from numerical integration).

The Monte-Carlo procedure. To study the dynamics of the solute molecules on the network of sites, a simple Monto Carlo procedure [56, 57] is employed:

– At equilibrium, the probability of finding a solute molecule at site j is proportional to Z_j and we use this fact to start a solute's random walk: we randomly select a number from the interval $[0, \Sigma Z_j]$ to specify the initial site for a particular solute trajectory;

– Suppose that at time t a dissolved molecule is at site i. Equation (29) allows use of a random number from a uniform distribution in the interval (0, 1) to select one of the adjacent sites to which the dissolved molecule jumps. The jump is associated with a time increment $\Delta t = \tau_i$ (simulation time is updated). Repeating the procedure we obtain the stochastic trajectory of the dissolved molecule;

– The trajectories of about 10^3 independent solute molecules are then averaged;

– One could also account for the distribution of life times at site i by using $\Delta t = -\tau_i \log \xi$, where ξ is a random number uniformly distributed in (0, 1), for the time update. However, nothing that $(\log \xi) = -1$, we shall simply use $\Delta t = \tau_i$, implying that it is the result of time pre-averaging over the distribution of life times at site i.

The smearing factor $\langle \Delta^2 \rangle$. To start the TSA simulation one needs to specify the smearing factor $\langle \Delta^2 \rangle$. We can surmise that, at ambient conditions, the value of $\langle \Delta^2 \rangle$ should be some fractions of an Å and should not be very sensitive to the chemical details of the polymer matrix. On the other hand, given the above approximations involved in describing the elastic motion of polymer atoms, a "universal" method for evaluating $\langle \Delta^2 \rangle$ can hardly be expected; the researcher should rather rely on intuition and experience in order to specify a suitable value of $\langle \Delta^2 \rangle$ for the problem at hand. Two possibilities are reviewed in the following.

Short-scale MD trajectories of the polymer matrices in the absence of solute molecules were used to evaluate the smearing factor $\langle \Delta^2 \rangle$ [50]. In the microstructures of both rubbery and glassy polymers studied [50], the smearing factor $\langle \Delta^2 \rangle$ was found not to level off to a constant value over the time-scale of simulation (up to ca. 10^{-8} s); hence, the elastic motion is not stationary and one must specify the time interval Δt on which the smearing factor should be evaluated. Jumps of the dissolved molecule between adjacent local maxima of $\rho(\mathbf{r})$ displace it by several Å and, very probably, completely change its nearest neighbors, i.e., those with which it has the most important interactions (the short-range repulsive ones). It is, therefore, reasonable to use the most frequent residence times τ_i to specify the averaging interval Δt; however, the τ_i are distributed and cannot necessarily be represented by a single value. In equilibrium, the probability density of finding the dissolved molecule at sites with a given τ is proportional to $\tau \cdot \rho(\tau) = \rho(\log \tau)$ and intuitively one can identify [50] Δt with the maximum of $\rho(\log \tau)$. Again, no physical reason demands precisely this choice and we regard it only as an-order-of-magnitude estimate.

Another possible way to evaluate the smearing factor $\langle \Delta^2 \rangle$ is to match the short-time region of the $\langle r^2 \rangle$ vs t curves obtained from TSA with those from MD calculations. As we shall see below, the broad domain of the anomalous diffusion [48, 50, 56], where $\langle r^2 \rangle \propto t^n (n < 1)$, pertains to the dynamics of small solutes in dense polymers. The value of n is sensitive to the smearing factor $\langle \Delta^2 \rangle$ employed, thus making possible to evaluate $\langle \Delta^2 \rangle$ from mapping the results of MD and TSA in the mutually accessible time-domain (i.e., 10^{-12}–10^{-8} s).

4 Results and Insights from Molecular Dynamics

4.1 Diffusion Proceeds by Hopping Between Voids

To date, penetrant diffusion through polymers has been studied by MD only in amorphous polymers. This is mainly due to the commonly held belief that only very few penetrants can enter the tightly packed crystallites and that those which do, get trapped at defect locations and hardly move afterwards. Therefore, diffusivities in the crystallites should be orders of magnitude smaller than those in amorphous regions and diffusion in a semi-crystalline material should be dominated by its amorphous part, even if it is small. In addition, if diffusion is excessively slow, one is unlikely to observe it on the time scale accessible to MD (i.e., ns).

The simplest way of studying diffusion of an individual penetrant is to inspect its path through space. In Fig. 2, we show a typical trace of an oxygen molecule moving through amorphous polyisobutylene (PIB) [48, 60]. We can clearly discern two types of motion: 1) For some periods of time the penetrant stays in certain small regions of space and during such a quasi-stationary period, it explores the small regions thoroughly, but it does not move beyond the confines of the area it resides in; 2) the quasi-stationary periods are interrupted by quick leaps from one region into another one that is close by. As a first result we note that diffusion proceeds by hopping of the penetrant from one attractive site to the next. Similar motion patterns have been found in all MD studies of the diffusion of small penetrants in amorphous polymers. It is interesting to note that a related mechanism is responsible for the superionic conductivity of sodium in β''-alumina at higher temperatures (≈ 1200 K), as has recently been found by MD simulations [61]: in between the two-dimensional alumina layers, Na^+ ions can perform hops between lattice sites.

The nature of the attractive sites in polymers can also be investigated. One can, for instance, change the representation to gain further insight. In Fig. 3 the displacement of three penetrant molecules (H_2, O_2, and CH_4) in amorphous polypropylene from their initial positions is shown, as a function of the simulation time [60]. The hops can clearly be seen for all three penetrants. However, there are also differences in the motion patterns that mainly relate to the size of the penetrants. Firstly, the larger the penetrant the more pronounced are the step-like hops in the graph. For CH_4, these are quite articulate, whereas for the small H_2 they are blurred to some extent. One may wonder if penetrants smaller than H_2 would have displacement curves completely without steps, indicative of a diffusion mechanism without discrete jump events (as is commonly assumed for liquid-like diffusion). Secondly, the behaviour during the quasi-stationary periods is also related to the penetrant's size. The larger the penetrant, the smaller are its positional fluctuations at the attractive site. Since the gases studied have no strong specific interaction with the polymer, one may surmise already at this stage that the attractive sites in the polymer are merely cavities large enough to hold a penetrant molecule. For a given cavity, a small penetrant

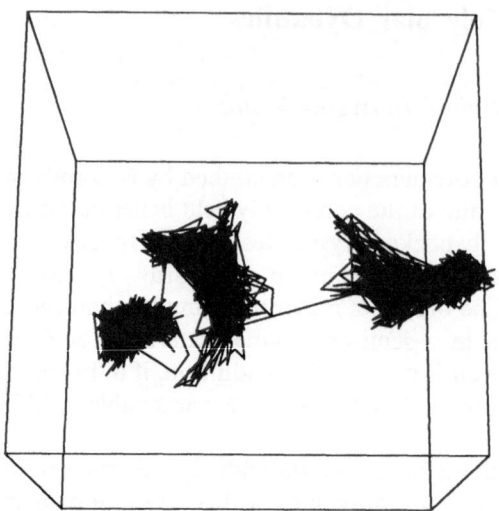

Fig. 2. A typical trace of an oxygen molecule through amorphous polyisobutylene (PIB) at 300 K, obtained by a MD run

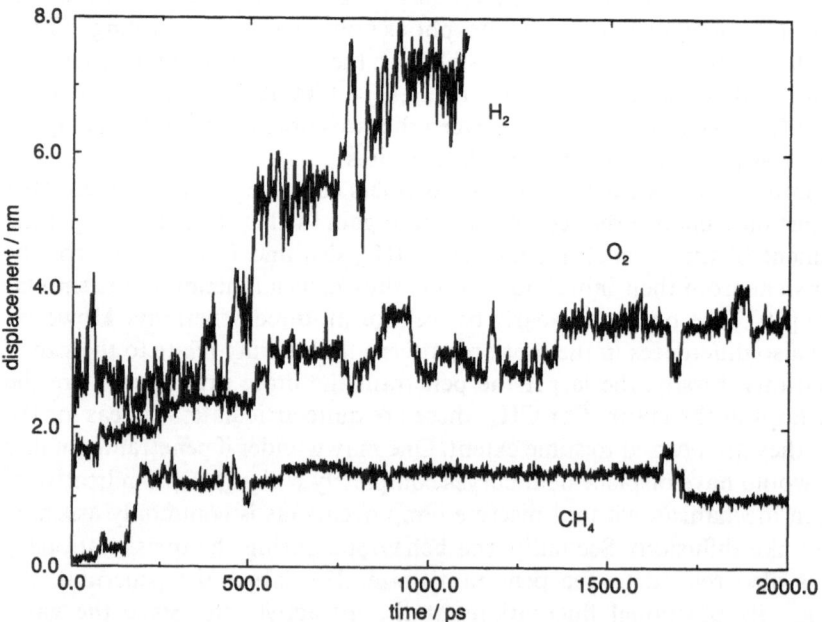

Fig. 3. Displacements from the original position vs time of hydrogen, oxygen, and methane in amorphous polypropylene at 300 K, determined by MD simulation. Note the offsets for H_2 and O_2

is able to access more of its volume than a larger penetrant and this explains the qualitative difference in the positional fluctuations. Thirdly, even in this qualitative picture one sees that the mobility of a penetrant is related to its size. One may suspect the diffusion coefficients to be ordered as $H_2 > O_2 > CH_4$, an order indeed found. The increased mobility is brought about by an increase in the jump frequency, whereas the jump distance changes only moderately. It is the jump rate that is related strongest to the penetrant's size.

4.2 Penetrants in Cavities

Of the two modes of motion of a penetrant, i.e., local motion in a cavity and jump between cavities, the former is more frequently observed; however, it does not contribute to diffusion and is therefore less interesting in the context of this article. Hence, we provide only a short summary about the behaviour of penetrants during their quasi-stationary periods. More detail can be found elsewhere [59].

Mobility in this region is dominated by short-time motion, typically < 2 ps. After that time, all correlation of molecular motion is lost due to frequent collisions with the cavity walls. The center-of-mass velocity autocorrelation function of the penetrant exhibits typical liquid-like behavior with a negative region due to velocity reversal when the penetrant hits the cavity wall [59]. This picture has recently been confirmed by Pant and Boyd [62] who monitored reversals in the penetrant's travelling direction when it hits the cavity walls. The details of the velocity autocorrelation function are not very sensitive to the force-field parameters used. On the other hand, the orientational correlation function of diatomic penetrants showed residuals of a gas-like behavior. Reorientation of the molecular axis does not have the signature of rotational diffusion, but rather shows some amount of free rotation with rotational correlation times of the order of a few tenths of a picosecond, although dependent in value on the Lennard-Jones radii of the penetrant's atoms.

It can be noted that, in principle, the rotational correlation functions can be measured by various molecular spectroscopies (e.g., IR, Raman, NMR) if suitable penetrants are chosen, but we are not aware of such measurements. Experimental correlation functions, together with atomistic modelling of the polymer-penetrant system, could yield important information about size and, possibly, shape distributions of the cavities inside the polymer, as probed by small molecules.

4.3 The Jump Event

During a jump event, a penetrant moves from a cavity to a neighboring cavity in a very short time compared to the residence time in the cavities. The question arises, what atomic motions are involved in such an event, and, in particular, how the polymer matrix participates in the event.

We have repeatedly tried to relate jump events to sudden conformational changes in the polymer. Major torsional angle changes or ring flips would be candidates, for example. However, we have never been able to find any such correlation. It appears rather that many degrees of freedom of the polymer are involved in a distributed and vague fashion. Interesting insight can be derived from displaying the volume accessible to the penetrant (rather than individual polymer atoms) [63]; a two-dimensional sketch of the process is shown in Fig. 4a–e where a matrix consisting of polymer with embedded cavities is displayed. These cavities exist whether there is a penetrant or not. They can fluctuate in size and shape to some degree, but they do not move, at least not on typical MD time scales. They are normally not connected. However, by means of fluctuations occurring naturally in the polymer, passages between the cavities open for very short times. If a penetrant is present in the vicinity and it happens to have the right vectorial velocity, it can take advantage of such a channel and slip into a neighbouring cavity before the channel closes again. This behavior has also been found by others [28, 44], for different polymers and for a variety of small gas molecules and it is probably safe to assume that it is the primary mechanism for the motion of small penetrants in dense polymers.

From molecular graphics it might appear that the rate-determining step is the opening of the channel and that the penetrant migrates through it without much effort, once the channel is formed. This is not quite correct: in Fig. 5 we see a profile of the kinetic energy of H_2 moving through polyisobutylene, averaged over many jump events [64]. For the preparation of this figure, we have searched the trajectory of the H_2 molecules for all instances in which the center-of-mass of an H_2 moved farther than 0.7 nm within 1 ps, and have taken this as the indication of a jump event. Although the energy profiles of individual jumps vary strongly among each other (the instantaneous kinetic energy of individual molecules is subject to large fluctuations) we can pin down the general trend by averaging over all jumps. We find that during a time slice of about 2 ps around the jump the total kinetic energy of H_2 is significantly increased over the simulation temperature of 300 K. This indicates that even after channel formation there exists some energy barrier which has to be overcome by the penetrant. This is logical since the channels are relatively narrow on average. When one looks at the individual components of the kinetic energy, one finds that the excess energy is in the translational degrees of freedom. Rotationally, the jumping diatomic molecules are at the simulation temperature.

A question frequently brought up is, whether there is a sign of the penetrant "forcing" its way through the polymer. For a small molecule the only possible way of forcing the opening of a channel would be by impacting on a cavity wall with high velocity. Given the harmonic nature of the motion and the disparity in the (effective) masses, it is very likely that the penetrant will bounce back. Indeed, we have never observed anything like a "forced opening of a channel", except in non-equilibrium MD (see below).

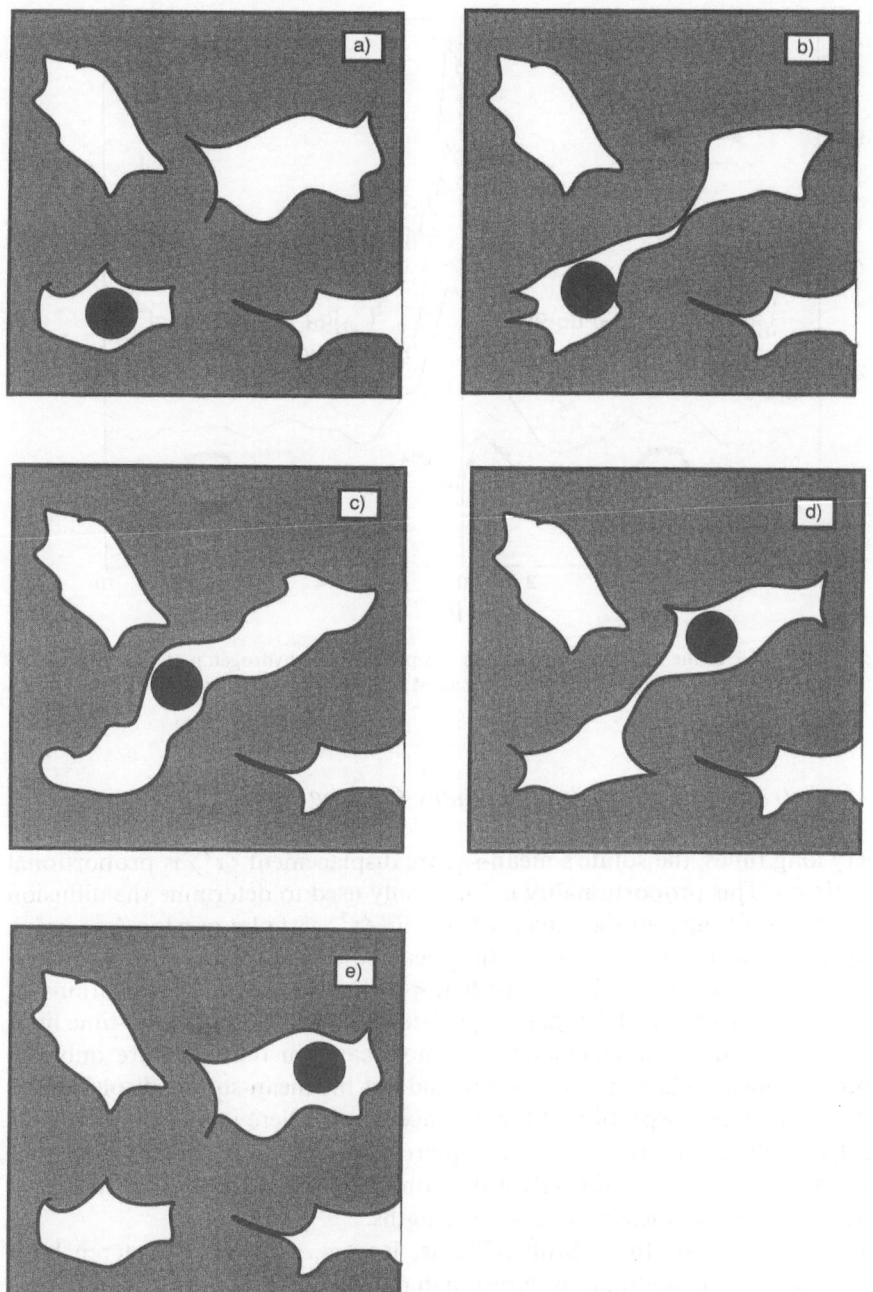

Fig. 4. A simple sketch in two dimensions of a hopping event in a polymer matrix with embedded fluctuating "cavities". This is the picture typically found in MD simulations

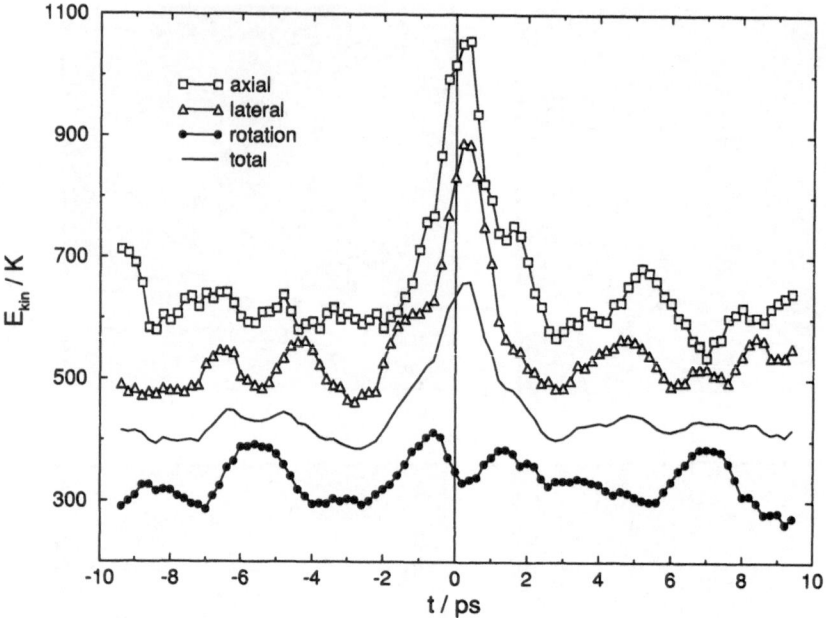

Fig. 5. Time profile of the kinetic energy and its components of a hydrogen molecule as it moves through a polyisobutylene matrix, extracted from an MD run and averaged over many jump events

4.4 Penetrant Diffusion in Polymers Can be Anomalous

At very long times, the solute's mean-square displacement $\langle r^2 \rangle$ is proportional to the time t. This proportionality is commonly used to determine the diffusion coefficients by fitting a straight line, either in an $\langle r^2 \rangle$ vs t plot or a log $\langle r^2 \rangle$ vs log t graph. At short times, however, the mean-square displacement may obey a different power law, $\langle r^2 \rangle \propto t^n$ with $n < 1$. In most of the calculations of diffusion coefficients by MD it has tacitly been assumed that the long-time limit has been reached in the calculation. In most cases in the literature only the diffusion coefficients have been reported and not the mean-square displacement curves; upon closer inspection of the displacements, where reported [27, 28, 60, 65, 66], one often notes that the mean-square displacement is a wobbly line that can be interpreted as straight only if the nonlinearity is attributed to statistical uncertainties for insufficient simulation lengths.

In a more recent study, Müller-Plathe, Rogers, and van Gunsteren have pointed out a case of anomalous diffusion in polyisobutylene near room temperature [48], in harmony with the findings by TSA for gas motion in dense polymers [56]. For He in PIB, anomalous behavior could be clearly shown, and the transition to normal diffusion at around 0.1 ns could be captured. The log-log plot that shows this crossover, is reproduced in Fig. 6. For the much slower diffusing O_2, the mean-square displacement data were not accurate enough to determine unambiguously if the curve represented diffusive behavior

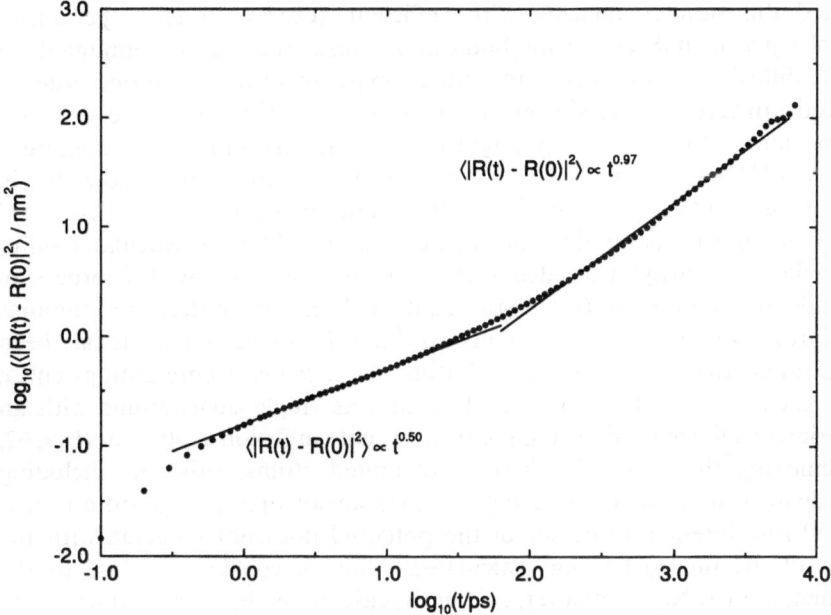

Fig. 6. Double-logarithmic plot of the displacement of Helium atom in amorphous polyisobutylene vs time, simulated by MD at 300 K

or if it followed a different power law. However, in the light of current knowledge it is likely that the Einstein diffusion regime had not yet been reached. It is interesting to note that the anomalous diffusion is clearly visible also for gas molecules diffusing through zeolites [67].

These results also imply that diffusion coefficients determined from the paths shorter than a few nm (the length-scale of the cross-over to diffusive behavior) would probably not be representative for the macroscopic effect. Hence, calculated diffusion coefficients have to be taken with caution unless the Einstein behavior has clearly been established.

The presence of anomalous diffusion can be explained by the structure of the polymer matrix, which – at the length-scale of a few hopping distances – restricts the penetrant motion to the effect that – on this scale – the penetrant's paths cannot be truly random. The polymer environment, at the same time, causes a separation of time-scales consistent with the hopping mechanism (short-time in-cavity motion vs long-time diffusive motion). This, in turn, is another cause for anomalous diffusion.

4.5 Calculation of Diffusion Coefficients

If simulation methods are going to be useful tools in the area of penetrant diffusion in polymers they have to be able to reproduce experimentally measured diffusion coefficients with sufficient accuracy. Given the uncertainty of experiment (Sect. 2) a method can be considered accurate enough if, for absolute diffusion coefficients, it reproduces experiment to within one order of magnitude. In certain cases, for instance when only relative diffusion coefficients are important, a lower accuracy might suffice, since errors may compensate.

While MD studies of penetrants in polymers were rather successful in revealing the qualitative aspects of the diffusion mechanism, many of them failed to reproduce experimental diffusion coefficients. In early work, calculated diffusion coefficients always exceeded measured ones, usually by 1–2 orders of magnitude. In a number of recent studies, it has been shown that shortcomings of the force-fields have played a considerable role. In particular, it has been demonstrated that isotropic united-atom force-fields (representing entire groups, such as methylene or methyl groups, as single quasi-atoms with an appropriately adjusted radius) tend to overestimate diffusion coefficients [49, 62, 65]. Removing the artificial sphericity of united atoms, either by including explicit hydrogen atoms or by using so-called anisotropic united atom potentials [68] (the interaction center of the potential does not coincide with the position of the nucleus) brings calculated diffusion coefficients close to the experimental ones. Note, however, that in the case of low-barrier polymers, such as polydimethylsiloxane or atactic polypropylene, united-atom models often seem to work well. A selection of diffusion coefficients calculated by MD that show reasonable agreement with experiment, is given in Table 1.

4.6 Temperature and Density Dependence of Diffusion Coefficients

The dependence of the diffusion coefficient on system parameters such as temperature [42, 65, 66] and density [43] has been a topic of the very first MD simulations of penetrant diffusion in polymers. Studies of the temperature dependence of D are aimed at determining the parameters of models such as those described by the Arrhenius or Williams–Landel–Ferry (WLF) equations. Initially, for CH_4 in polyethylene, an Arrhenius behavior had been found with apparent activation energies of 6.4–10.8 kJ/mol [42], 10.5 kJ/mol [66], or 11.9 kJ/mol [65]. However, in the temperature range of 140 to 180°C, the experimental activation energies are 32–40 kJ/mol [70]. Furthermore, if a wider temperature range is considered, the behavior of CH_4 and of many other gases in polyethylene is found to obey a WLF equation rather than a Arrhenius equation. In view of this, Pant and Boyd [62, 65], have recently studied the temperature dependence of D for CH_2 in polyethylene and polyisobutylene using anisotropic united atom potentials. For polyethylene they establish WLF behavior in agreement with experiment. For PIB, their Arrhenius plot also shows excellent agreement with experiment [70].

The dependence of D on the density (or equivalently the pressure) has usually been analyzed in terms of the so-called free-volume v_f. The free volume

Table 1. Diffusion coefficients from Molecular Dynamics simulations and from experiment, at 300 K unless indicated otherwise

Polymer	Penetrant	Diffusion coefficient (in units of 10^6 cm^2/s)		Reference
		calculated	experimental[a]	
atactic polypropylene	H_2	44	5.7	60
	O_2	4.0	≈ 1.5[b]	60
	CH_4	0.48	≈ 0.6[b]	60
polydimethyl-siloxane	He	18	10	28
	CH_4	2.1	2.0	28
polyethylene	CH_4	0.5	0.3–0.6	65
	CH_4	1	—	62
polyisobutylene	He	30	5.93	64
	H_2	9.2	1.52	64
	O_2	0.047–0.17	0.081	64
	CH_4	0.4 (350 K)	1.7 (375 K)[c]	62

[a] From Ref. [69]
[b] In various natural and synthetic rubbers with gas diffusion coefficients very similar to those in atactic polypropylene.
[c] Ref. [70]

fraction, loosely identified with "the volume not occupied by polymer atoms" divided by the total volume, seems to contain some physical significance, despite all the arbitrariness in its definition, since D indeed increases with available free volume. The relationship can often be fitted by the simple exponential form known from free-volume theories [14, 42–44, 62].

$$D = A\,e^{-B/v_f} \qquad (33)$$

Comparisons with experiment have not yet been carried out.

A related observation can be found in Table 1: The higher of the two values for the diffusion coefficient for O_2 in PIB corresponds to a constant-volume simulation at the experimental density of PIB, the lower to a constant-pressure simulation at a slightly higher (3%) density. The authors believed that the difference between the two simulation results stems largely from the increase in density when the constant-volume constraint is relaxed.

4.7 Influence of the Force-field

4.7.1 Penetrant Parameters

The interaction between the penetrant and the matrix is dominated by parameters that describe the solute: the nonbonded interaction is typically given by the coefficients ε and σ of a Lennard–Jones potential energy function

$$U_{\alpha\beta}(r) = 4\varepsilon_{\alpha\beta}\left[\left[\frac{\sigma_{\alpha\beta}}{r}\right]^6 - \left[\frac{\sigma_{\alpha\beta}}{r}\right]^{12}\right] \tag{34}$$

where the subscripts α and β denote the types of site involved in the interaction (we do not know of results on penetrants with Coulombic interactions). The effect of changing ε is discussed below in some detail. The effect of increasing the penetrant's size (σ) should parallel that of an increase in the polymer density or, in vague terms, that of a decrease in the free volume. Experimentally, it has long been known that for a given polymer the diffusion coefficient decreases with the "diameter" of the penetrant [71–73]. Similar relationships among the diffusion coefficients of different penetrant species have been found computationally (H_2, O_2, CH_4 in polypropylene) [60]). Sonnenburg et al. [43] have systematically varied the diameter of a fictitious penetrant and have found a relationship resembling an exponential form. Müller-Plathe et al. [49] have tested two commonly used models for O_2: the one with smaller σ (0.295 vs 0.309 nm) had a larger ε (0.51 vs 0.36 kJ/mol) which gives rise to counteracting effects (because of its smaller size, the first O_2 model should diffuse faster, but because of its stronger interaction with the matrix, it should diffuse slower). Even though the relative difference in ε was much larger than that in σ, the size argument won; the "slimmer" O_2 diffused faster. Radii of interaction sites have to be chosen more carefully than interaction energies.

4.7.2 Polymer Parameters

The polymer-penetrant interaction also depends on the nonbonded parameters within the polymer matrix. The deficiencies of isotropic united-atom models have already been mentioned. They have been explained [49] by less efficient packing of united-atom chains. Even if the united-atom model provides the correct density or free volume, it generates a different distribution of volume not occupied than either the anisotropic united-atom model or an all-atom description. The latter pack more efficiently and leave smaller interstitial cavities. For a detailed discussion, see Ref. [49].

Correlations of the diffusion coefficient with the mobility of the polymer atoms, as given by the intramolecular force-field, have also been attempted. Takeuchi and Okazaki [42] have studied the effect of neglecting the torsional potential in a polyethylene chain, i.e., of allowing rotation around backbone bonds without intramolecular barriers. This increased the diffusion coefficient of CH_4 by only 60% at 300 K. Pant and Boyd [62], however, found that raising the torsional barrier of polyethylene by 50% brought the diffusion of CH_4 virtually to a standstill at 400 K. It was argued that the different observations originate in a change in the diffusion mechanism: At lower T, the hopping mechanism seemed to prevail with diffusion rates determined by the opening of channels effected by cooperative small displacements of polymer atoms (for which the torsional barrier is only one of many components). At higher T, diffusion becomes more liquid-like and the mobility of individual polymer atoms becomes more important. This, in turn, seems to be strongly affected by the torsional potential energy functions. Takeuchi, Roe, and Mark [74] also have studied the effect of changing the backbone bond angles in polyethylene. They found a marked decrease of D as they increased the equilibrium value from 100° to 150°. This was rationalized by the effects of a "stretching out" of the chains that allows different chains to approach each other more closely and leaves less room between them for the penetrant.

4.8 Where Can Molecular Dynamics Be Usefully Applied?

The range of polymers and penetrants for which the MD method can be usefully applied in the calculation of diffusion coefficients (assuming that technical problems such as force-fields, anomalous diffusion, etc., have been sorted out) can be estimated as follows. Firstly, there are general considerations; one has to be able to simulate a system large enough to sample the configurational statistics of the polymer sufficiently; to obtain a reasonable cross-section of polyethylene configurations one may need a few hundred repeat units or a few hundred to a few thousand atoms. One might need many more repeat units if, for instance, the polymer is stiff. Also, if the monomers are large, many more atoms are involved. Secondly, there are factors arising from the mobility of the penetrant itself: At equilibrium and assuming hopping motion in an isotropic matrix, the diffusion coefficient can be written as

$$D = \frac{\langle \lambda \rangle^2}{6 \langle \tau \rangle} \tag{35}$$

where $\langle \lambda \rangle$ is the average jump distance and $\langle \tau \rangle$ is the average residence time between jumps. To be sufficiently precise one needs to observe, say, 10 jump events for a single penetrant (this is probably the bare minimum). Assuming a typical mean jump distance of 0.5 nm one finds that in a 1 ns simulation one will encounter 10 jumps on average, if the diffusion coefficient is 4×10^{-6} cm^2/s. Nowadays simulation lengths are often of the order of 10 ns, which brings diffusion coefficients of 4×10^{-7} cm^2/s within reach. Improvement in the precision of the estimate requires longer simulation times (the precision increases with the square root of the simulation time). An improvement is possible if one can use several penetrants at the same time, and thereby improve sampling. This is usually possible if the diffusion coefficient is not very sensitive to the penetrant concentration, as is often the case in rubbery polymers where the penetrant concentration is low. This discussion indicates that MD can be usefully employed, if the penetrant is a small gas molecule or the temperature is high and the polymer presents a reasonably low barrier. At room temperature, MD is therefore useful to predict absolute diffusion constants for gases through siloxanes, natural and synthetic rubbers, polyalkanes, etc. However, the study of barrier materials such as polyvinylchloride or polyvinylidenechloride with diffusion coefficients of 10^{-9}–10^{-12} cm^2/s at room temperature is certainly out of the question for several generations of supercomputers to come. Hence, the role of MD will probably be in areas where high throughput is required, such as separation membranes.

4.9 Can MD Methods Be Extended to Slower Diffusion Processes?

There have been two attempts to enhance sampling of diffusive motion in simulation by artificially increasing the diffusion rate of the gas molecules in amorphous polymers. The first attempt involved scaling the polymer-penetrant interaction by a factor ξ $(0 < \xi \le 1)$ and involving an Arrenius-type law to recover the diffusion coefficient at full interaction [46]. It could be demonstrated for CH$_4$ in PE that ξ can be as small as 0.1–0.2 for the scaling relation to hold with a corresponding increase in the diffusion coefficient by a factor of 4–5. Further decrease of ξ leads to new ("unphysical") pathways becoming available to the penetrant and the quantitative scaling relation is no longer applicable.

The second attempt used non-equilibrium molecular dynamics (NEMD) [64]. Here, an artificial force F acts on the penetrants and the diffusion coefficient is calculated assuming a linear response

$$D = \frac{kT}{F} \langle v \rangle \tag{36}$$

where $\langle v \rangle$ is the average drift velocity of the penetrants in the direction of F. NEMD, however, proved to be of insufficient usefulness in the simulation of O$_2$ in PIB. The main reason was that, in order to observe a significant speed-up of diffusion, one had to apply a force so large that the linear response relation [see Eq. (36)] no longer held. The reason for this is that too large a force not only increases the jump probability for a penetrant across a barrier that would also be surmounted without the force, albeit at a lesser frequency, but that pathways normally not accessible to the diffusant become available.

5 Results from TSA

5.1 Three Modes of Solute Motion

Figure 7 depicts [75] the mean square displacement $\langle r^2 \rangle$ of Helium molecules in elastically responding micro-structures of glassy bisphenol-A-polycarbonate (below, glassy PC) vs time, generated [76] with a force-field developed for

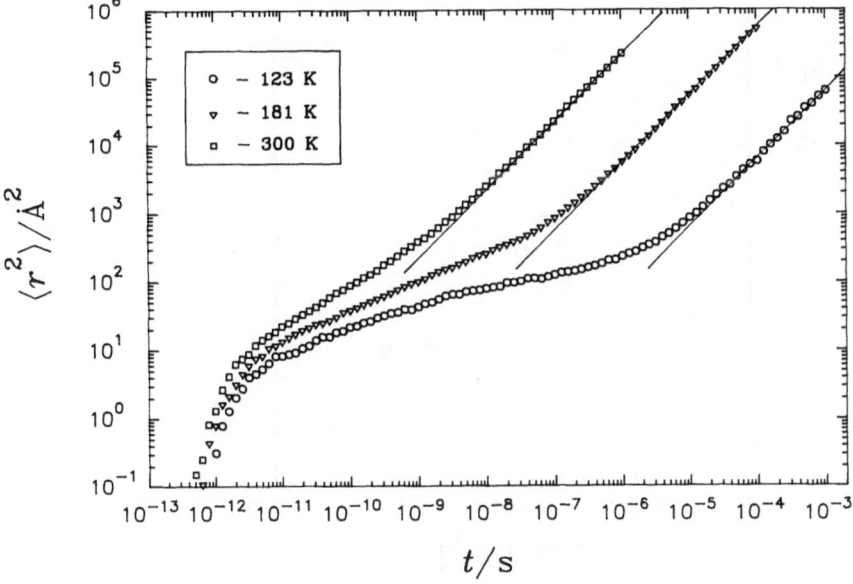

Fig. 7. Mean-square displacement $\langle r^2 \rangle$ of Helium atoms in glassy PC. The *straight lines* represent the equation $\langle r^2 \rangle = 6Dt$, where D is the diffusion coefficient. *Symbols* stand for means over 1500 independent solute trajectories

modeling dense polymers. The smearing factor $\langle \Delta^2 \rangle^{1/2}$ for elastic motion in the PC matrices was obtained from mapping computed and experimental diffusion coefficients and was 0.43 Å at 300 K; the smearing factor at 181 K and 123 K was then computed through $\langle \Delta^2 \rangle \propto T$ [see Eq. (16)].

Figure 8 shows that experimental and computed diffusion coefficients are in satisfactory agreement over the entire temperature range studied of about 200 K. Henry's constant S of Helium sorption in the PC micro-structures were evaluated through Eq. (23) and the equation $P = S \cdot D$ was employed to evaluate the permeability coefficients P, see Fig. 9.

The diffusion coefficients D and permeability coefficient P are two independent transport coefficients and the agreement achieved between computed and experimental data indicates [75] that we are dealing with a physically appropriate model of solute motion.

The three modes [56] of Helium motion in glassy PC are evident from Fig. 8:

- at short times one observes *local mobility*;
- the solute molecules follow then an *anomalous diffusion* behavior with $\langle r^2 \rangle \propto t^n$ (n < 1);
- at long times the Helium motion is described by the *diffusive* form $\langle r^2 \rangle \propto t$.

The distribution function $\rho(\mathbf{r})$: Visual examination of the surfaces $\rho(\mathbf{r}) = 1$ in microstructures of dense polymers reveals its cavernous structure [56, 57].

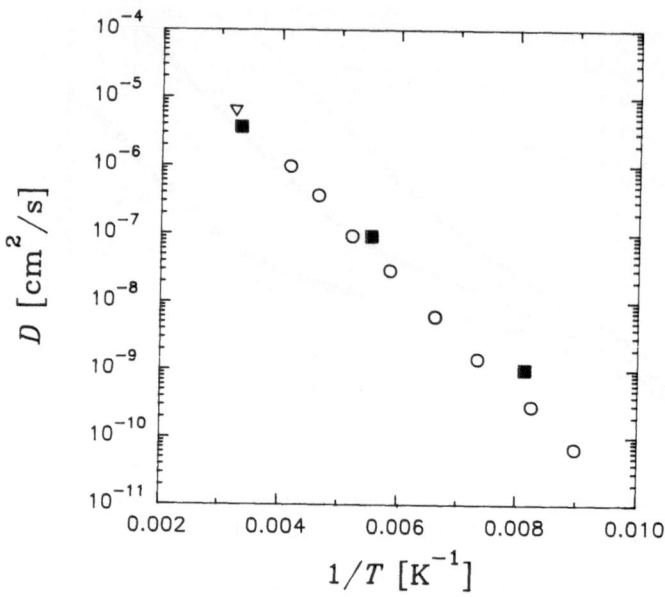

Fig. 8. Diffusivity of Helium in Bisphenol-A-polycarbonate. *Empty symbols* are experimental diffusion coefficients D, *filled squares* are computed

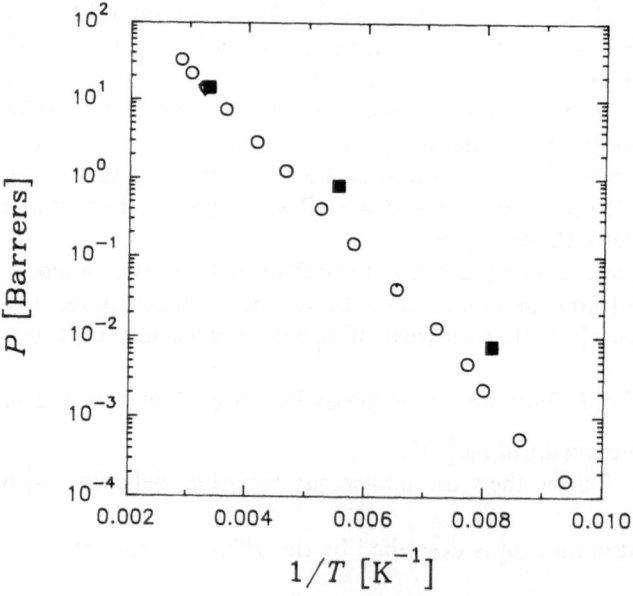

Fig. 9. Permeability of Helium in Bisphenol-A-polycarbonate. *Empty symbols* are experimental permeability coefficients P, *filled squares* are computed

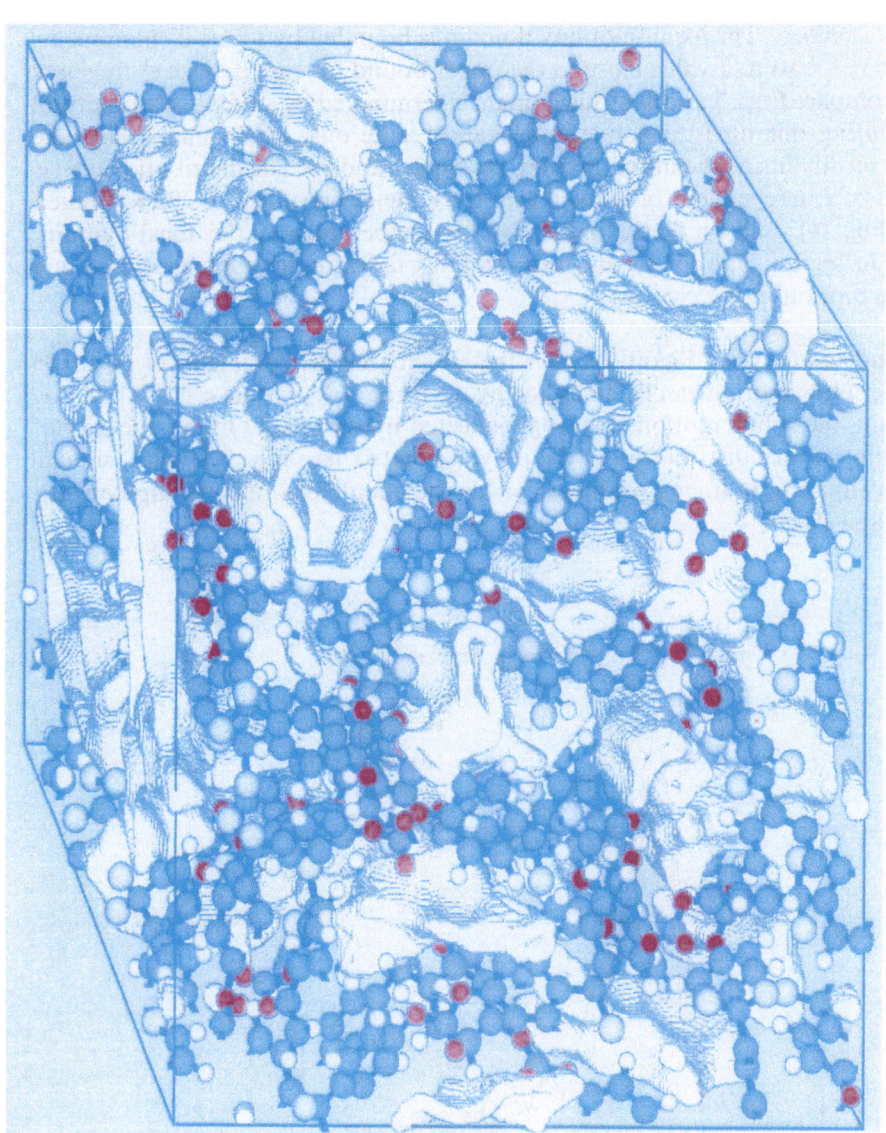

Fig. 10. Visualization of the surface $\rho(\mathbf{r}) = 1$ of a Helium molecule as function of location in one of the PC micro-structures investigated. The surface shown contains loci of $\rho(\mathbf{r}) > 1$. The size of the polymer atoms is set to $\delta/4$ for convenience

240   A.A. Gusev et al.

Figure 10 presents this surface for Helium molecules in a micro-structure of glassy PC at 300 K. Typically, the cavities are 5–10 Å in size and contain few local maxima of the distribution function ρ(r) connected by small barriers (less than kT). The cavities are separated by some bottle-neck channels of 5–10 Å in length and 1–2 Å in width that prevent easy passage from one cavity to another.

Local mobility: The local -mobility domain is bounded by solute displacements of $\langle r^2 \rangle^{1/2} \approx 10$ Å, a value close to the upper bound for the "cave size" in glassy PC, compare Figs. 7 and 10. This domain is terminated by a time-scale of several ps. During this time the solute molecules execute only a few jumps that take place mostly inside the initial cavities, see Figs. 7 and 10. The distribution of τ_i is practically independent of temperature over the entire range of 200 K studied (see Fig. 11), and similarly insensitive to temperature is the local-mobility domain, see Fig. 7. All in all, the local-mobility mode of the solute motion is due to the motion inside cavities.

Anomalous diffusion: Figure 7 shows that at squared displacements of 10 to 10^3 Å^2 the solute motion can be described by the form $\langle r^2 \rangle \propto t^n$ (n < 1) [56]. In this domain, the solute motion is strongly affected by the spatial inhomogeneity of the empty-space distribution in the polymer matrix (see Fig. 10) which, similar to the situation in fractal problems, yields the anomalous relationship between $\langle r^2 \rangle$ and t^n.

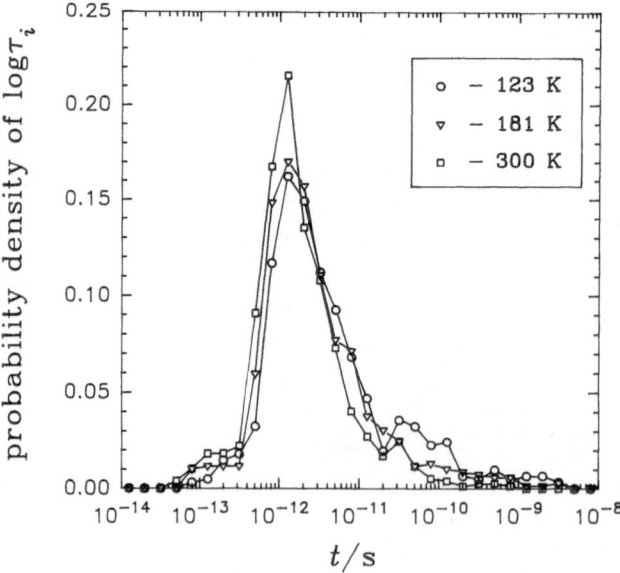

Fig. 11. The distribution of the mean-residence time τ_i of Helium molecules in local maxima of the distribution function ρ(r)

Diffusive mode: At long times the solute motion is diffusive and the Einstein equation $\langle r^2 \rangle = 6\,Dt$ holds. The border between anomalous and diffusive modes of motion corresponds to displacements ≈ 20–30 Å; this is close to the size of the micro-structures employed, and it is not yet clear to what extent finite size effects might be operative.

Corollary: It is surprising that times of the order of ns to µs and longer are required for the solute motion in dense polymers to become diffusive. This is at variance with the case of simple liquids, where the diffusive mode of motion can be observed already at times shorter than ps. To properly evaluate the diffusion coefficient of solute molecules in dense polymers, one needs to consider a time interval on which the solute motion does follow diffusive behavior. Otherwise, misleading results could be obtained.

For illustration, suppose that we are limited in our study to times of maximum 1 ns, an interval typical in many investigations. From Fig. 8 we have $\langle r^2 \rangle \approx 100$ Å^2 at 181 K and $\langle r^2 \rangle \approx 50$ Å^2 at 123 K and, by use of the Einstein relation $\langle r^2 \rangle = 6\,Dt$, we obtain $D \approx 2 \times 10^{-6}$ cm^2/s at 181 K and $D \approx 1 \times 10^{-6}$ cm^2/s at 123 K. At 181 K this estimate is within an order of magnitude from that obtained from the diffusive domain, namely 1×10^{-7} cm^2/s, but at 123 K the correct (diffusive) value is 1×10^{-9} cm^2/s, three orders of magnitude lower than the "naive" estimate.

5.2 The Scope of the Elastic Mechanism of Gas Motion in Dense Polymers

We have surmised (see Sect 3.2) that the structural relaxation in dense polymers is too slow to notably contribute to the transport of small molecules. This premise is intuitively acceptable for glassy polymers, but we apply it equally to gas motion in both glassy and rubbery polymers [77].

Glassy and rubbery polymers: We have generated four cubic micro-structures of glassy atactic polyvinylchloride (glassy PVC, $T_g \approx 350°C$) and liquid polydimethylsiloxane (rubbery polydimethylsiloxane, $T_g \approx 120$ K) of 30 Å edge length, using a general purpose force-field (cvff [78]). The transport of inert gases through these micro-structures has been studied at about room temperature by means of the approach outlined above. The smearing factors for elastic motion of the matrices, 0.25 to 0.5 Å, have been obtained by mapping computed and experimental permeability coefficients P.

Tables 2 and 3 show that in all cases computed diffusion coefficients D and Henry's constant S agree with experimental data [71, 73] to within an order of magnitude. The computed Henry's constants S do not scatter much among the individual micro-structures, whereas there is a considerable scattering of the diffusion coefficients D for Ar and larger molecules. The computed Henry's constants S are not sensitive to the smearing factor $\langle \Delta^2 \rangle$ employed: The values

Table 2. Transport of inert gases in 30 Å micro-structures of PVC at 318 K. Henry's constants S are in cm³ (STP)/cm³·atm, diffusion coefficients D in cm²/s, and permeability coefficients P in cm³ (STP)/cm·s·atm[a]

		He	Ne	Ar	Kr
$\langle \Delta^2 \rangle^{1/2}$ [Å]		0.25	0.33	0.43	0.50
S	for structure 1	0.054	0.075	0.25	0.62
	for structure 2	0.045	0.061	0.19	0.43
	for structure 3	0.045	0.063	0.20	0.48
	for structure 4	0.040	0.053	0.16	0.36
	$\langle S \rangle_{comp}$	0.046	0.063	0.20	0.47
	S_{exp}	0.007	0.015	0.06	0.19
$10^6 \cdot$ D	for structure 1	4.0	0.6	0.007	0.0004
	for structure 2	0.3	0.07	0.008	0.0006
	for structure 3	1.4	0.09	<0.00001	<0.00001
	for structure 4	1.2	0.1	<0.00001	<0.00001
$10^6 \cdot \langle D \rangle_{comp}$		1.7	0.2	0.004	0.0003
$10^6 \cdot D_{exp}$		4	0.4	0.005	0.001
$10^8 \cdot \langle P \rangle_{comp}$		8	1	0.08	0.008
$10^8 \cdot P_{exp}$		3	0.6	0.03	0.02

[a] From Ref. [77] experimental values from Refs. [71–73]

Table 3. Transport of inert gases in 30 Å micro-structures of polydimethylsiloxane at 273 K. Henry's constants S are in cm³ (STP)/cm³·atm, diffusion coefficients D in cm²/s, and permeability coefficients P in cm³ (STP)/cm·s·atm[a]

		He	Ne	Ar	Kr	Xe
$\langle \Delta^2 \rangle^{1/2}$ [Å]		0.25	0.31	0.35	0.40	0.42
S	for structure 1	0.16	0.26	1.7	5.2	39
	for structure 2	0.18	0.29	1.9	5.7	40
	for structure 3	0.16	0.26	1.5	4.1	28
	for structure 4	0.20	0.32	2.0	6.2	39
	$\langle S \rangle_{comp}$	0.17	0.28	1.8	5.3	39
	S_{exp}	0.043	0.09	0.34	0.98	4.0
$10^6 \cdot$ D	for structure 1	37	12	5	3	0.6
	for structure 2	25	6	0.1	0.02	<0.001
	for structure 3	34	9	0.5	0.1	<0.001
	for structure 4	69	26	6	3	1
$10^6 \cdot \langle D \rangle_{comp}$		41	13	3	1.5	0.4
$10^6 \cdot D_{exp}$		41	16	12	8	4
$10^6 \cdot \langle P \rangle_{comp}$		7	4	5	7.5	16
$10^6 \cdot P_{exp}$		2	1.5	4	7.5	19

[a] From Ref. [77] experimental values from Refs. [71–73]

presented in Tables 2 and 3 are only a factor of 2 smaller than the rigid-matrix values. In contrast, the diffusion coefficients are very sensitive to the smearing factor used: Rigid-matrix diffusion coefficients of Ar and larger inert gases are several orders of magnitude less than those from experimental data.

One and the same approach has been taken for modeling the gas transport through both glassy PVC and rubbery polydimethylsiloxane. In both situations, the transport coefficients, obtained with physically acceptable values of the smearing factor $\langle \Delta^2 \rangle$, agree well with experimental data. Hence, even far above the glass transition region (in polydimethylsiloxane $T_g \approx 150$ K, the simulations were carried out at 273 K) the rate of structural relaxation is still too slow to be important for the motion of small solute molecules through dense polymers, indicating that the mechanism of motion of small molecules in both glassy nd rubbery polymers is essentially the same.

Diatomic molecules: The distribution function of the center of mass of Oxygen molecules in micro-structures of glassy PVC of ca. 40 Å has been evaluated [58] assuming separability between the translational and rotational degrees of freedom of the O_2 molecules [see Eq. (32)]. Two different diatomic force fields for the O_2 molecule have been used. In both cases, the same three modes of solute motion have been found and computed transport coefficients were consistent with experimental data. Nevertheless, we feel that the use of the explicit diatomic approach is impractical today: In view of the inaccuracy which intrinsically pertains to the force-field approach, one can always conveniently "optimize" the parameters of a united-atom force-field to properly reproduce experimental transport coefficients of the diatomic molecules in dense polymers.

The smearing factor $\langle \Delta^2 \rangle$: As discussed above, some correlations could exist between the spectrum of the residence times τ_i and the time-interval Δt on which the smearing factor $\langle \Delta^2 \rangle$ is to be evaluated from MD trajectories of the polymer matrix (in the absence of solute molecules). It has been found that in some cases the value of $\langle \Delta^2 \rangle$ deduced from the maximum of the distribution of τ_i indeed gave [50] diffusion coefficients that favourably compared with experimental data. Nevertheless, in some situations this empirical rule failed to reproduce the solute's diffusion coefficients even to within an order-of-magnitude.

6 Finite Size Effects

In any simulation of large real systems by small models exists the possibility of effects of the small system size on the computed results. To check for such finite size effects, some computations on larger systems have been performed with TSA: micro-structures of edge-lengths 50 Å have been generated [75] for glassy

PC and the diffusion and permeability coefficients of Helium have been calculated. The transport coefficients in these 50 Å microstructures agreed with
those of 33 Å micro-structures to within 30%; it seems, therefore, that a length
scale of several dozen angstroms is sufficient for evaluating the solute's transport
coefficients. Similar conclusions can also be drawn for the MD simulations;
Changing from a 20 Å to a 30 Å simulation cell did not significantly alter
diffusion coefficients of O_2 in PIB [49].

7 Conclusions

Two atomistic approaches have been discussed in this article: MD and TSA.
They are complementary in several ways and can be used in order to understand
and, to a certain extent, predict the behavior of small gas molecules in polymers.
MD is the less coarse-grained of the two; its strength lies in the relatively few
arbitrary approximations: Except for the assumptions of an empirical force-field
and that of the applicability of classical dynamics on an atomistic level, no
further simplifications are implicit. Its main drawback is the computational cost
that, at present, prohibits simulations beyond 10 ns, these already being far from
routine. Here, the TSA is well suited to extend the time-scale of simulation,

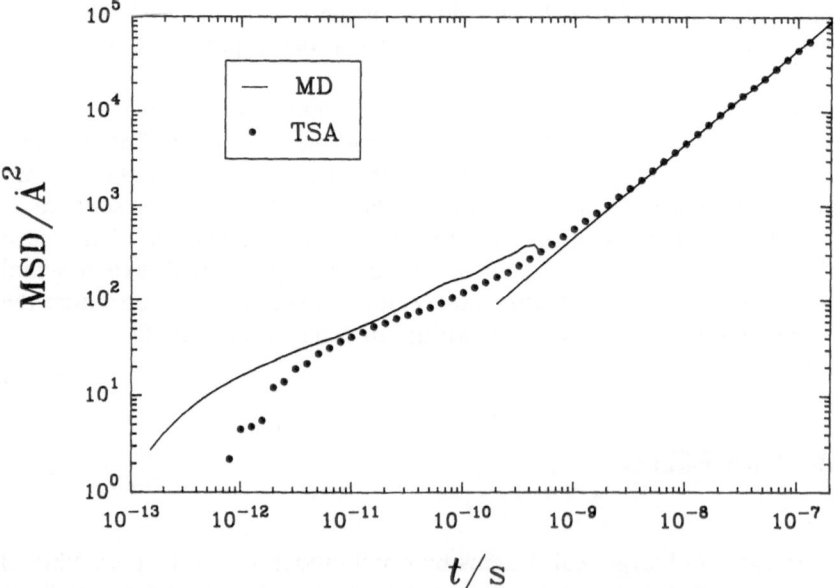

Fig. 12. Dynamics of Helium molecules in bisphenol-A-polycarbonate computed by MD and TSA
under otherwise identical conditions: T = 300 K, ρ = 1 bar, mean system density = 1.198 g·cm^{-3}.
The MD trace is the average of 10 He molecules, simultaneously present in the system; the *points*
represent the corresponding results of TSA, averages over 500 independent walkers

bringing new phenomena within reach. The TSA method does not require classical dynamics, but involves a number of other simplifying assumptions that must be justified.

The hopping mechanism of gas motion in dense polymers has been clearly established. It was observed many times in MD simulations. Abstraction from detailed short-time motions to jump events only, as is done in TSA, greatly enhances the computational efficiency, thus allowing considerably longer simulation times. MD has indicated that the solute's jumps are facilitated by an elastic response of the matrix ("the opening of channels connecting cavities") and that these elastic motions of the polymer atoms are not coupled to large-scale changes in the individual degrees of freedom of the polymer. This finding in turn allows for the assumption of an elastic matrix in TSA, a simplification that is at the basis of the high computational efficiency of the technique.

The finding of anomalous diffusion in gas transport through amorphous polymers by TSA was born out completely by MD. For a favourable case of fast diffusion, MD could detect the crossover from short-time motion inside cavities to the anomalous diffusion, and then to the Einstein regime. Anomalous diffusion must be expected for all transport of light gases through amorphous polymers.

It is important that MD and TSA are used in conjunction. As an example we mention the MD analysis of short-time rotational motion of small penetrants in dense polymers that provides the basis for the pre-averaging of these degrees of freedom in the TSA. Similarly, the duration of the hopping events, as estimated by MD (~ 2 ps), is an excellent guideline for selecting the smearing factor of TSA.

To date, the TSA is restricted to cases in which the matrix response to the guest molecule is elastic. This is an appropriate assumption for light gases, but breaks down in systems where the penetrants induce massive deformation in the polymer matrix, as with ions in complexing polymers or when the polymer is swollen by large amounts of penetrants. It remains to be seen if a similar coarse-grained approach can also be found in this situation.

Finally, we would like to point out that where the applicable time-scales overlap and when identical force-fields are used, both methods give practically identical results for diffusion coefficients and anomalies. When MD and the TSA are applied to identical systems with identical parameters, the dynamic behavior is almost identical, except for the shortest times ($<$ ca. 10^{-12} s) where the coarse-graining of TSA becomes evident. This is demonstrated in Fig. 12 [79].

8 References

1. Spillman RW (1989) Chem Eng Prog 85: 41
2. Matthes A (1944) Kolloid-Z 108: 79

3. Mears PJ (1954) J Amer Chem Soc 76: 3415
4. Barrer RN, Barrie JA, Slater (1958) J J Polym Sci 27: 177
5. Vieth WR, Sladek KJ (1965) J Colloid Sci 20: 1014
6. Paul DR (1969) J Polym Sci 7: 1811
7. Petropoulos JH (1970) J Polym Sci A2 8: 1797
8. Vieth WR, Howell JM, Hsieh JH (1976) J Membrane Sci 1: 177
9. Paul DR, Koros WJ (1976) J Polym Sci Polym Phys Ed 14: 675
10. Stern SA, Saxena VJ (1980) Membr Sci 7: 47
11. Stern SA, Frisch HL (1981) Ann Rev Mater Sci 11: 523
12. Frederickson GH, Helfand E (1985) Macromolecules 18: 2201
13. Cohen MH, Turnbull D (1959) J Chem Phys 31: 1164
14. Fujita H (1961) Fortschr Hochpolym Forsch 3: 1
15. Stern SA, Fang SM, Frisch HL (1972) J Polym Sci A2 10: 201
16. Vrentas JS (1977) J Polym Sci Polym Phys Ed 15: 403
17. Vrentas JS, Duda JL (1978) J Appl Polym Sci 22: 2325
18. Shah VM, Stern SA, Ludovice PL (1989) Macromolecules 22: 4660
19. Barrer RM (1937) Nature 140: 106
20. Barrer RM (1939) Trans Faraday Soc 35: 629, ibid (1939) 35: 644
21. Barrer RM (1947) Trans Faraday Soc 39: 3
22. Brandt WW (1958) J Phys Chem 63: 1080
23. DiBenedetto AT (1963) J Polym Sci A1: 3477, ibid (1963) A1: 3459
24. Pace RJ, Datyner A (1979) J Polym Sci Polym Phys Ed 17: 437, ibid (1979) 17: 453, ibid (1979) A1: 465
25. Jagodic F, Borštnik B, Ažman A (1983) Makromol Chem 173: 221
26. Theodorou DN, Suter UW (1985) Macromolecules 18: 1467
27. Boyd RH, Pant PVK (1991) Macromolecules 24: 6325
28. Sok RM, Berendsen HJC, van Gunsteren WF (1992) J Chem Phys 96: 4699
29. Felder RM, Huvart GS (1980) Methods of experimental physics p 16
30. Frisch HL (1980) Pol Eng Sci 20: 2
31. Crank J, Parks G S (1968) Diffusion in polymers. Academic, New York
32. Crank J (1975) The mathematics of difusion, Clarendon, Oxford
33. Allen MP, Tildesley DJ (1987) Computer Simulation of Liquids, Clarendon Press, Oxford
34. Roe RJ, Ed (1991) Computer Simulation of Polymers Prentice Hall, Englewood Cliffs
35. van Gunsteren WF, Berendsen HJC (1990) Angew Chem Int Ed Engl 29: 992
36. Car R, Parrinello M (1985) Phys Rev Lett 55: 2471
37. Ryckaert J-P, Ciccotti G, Berendsen HJC (1977) J Comput Phys 23: 327
38. Müller-Plathe F, Brown D (1991) Comput Phys Commun 64: 7
39. Evans DJ, Morriss GP, (1990) Statistical mechanics of nonequilibrium liquids, Academic Press London
40. Berendsen HJC, Postma JPM, van Gunsteren WF, DiNola A, Haak JR, (1984) J Chem Phys 81: 3684
41. Glasstone S, Laidler KJ, Eyring H (1941) The Theory of Rate Processes, McGraw-Hill
42. Takeuchi H, Okazaki K (1990) J Chem Phys 92: 5643
43. Sonnenburg J, Gao J, Weiner JH (1990) Macromolecules 23: 4653
44. Takeuchi H (1990) J Chem Phys 93: 2062, ibid 4490
45. Smit E (1991) Ph. D thesis, University of Twente, Enschede
46. Müller-Plathe F (1991) J Chem Phys 94: 3192
47. Müller-Plathe F (1991) Chem Phys Letters 177: 527
48. Müller-Plathe F, Rogers SC, van Gunsteren WF (1992) Chem Phys Lett 199: 237
49. Müller-Plathe F, Rogers SC, van Gunsteren WF (1992) Macromolecules 25: 6722
50. Gusev AA, Suter UW (1993) J Chem Phys 99: 2228
51. Debye P (1913) Verh Deut Phjysik Ges 15: 783
52. Dunitz JD (1979) X-ray analysis and the structure of organic molecules, Cornell University Press, London
53. Vainshtein BK (1966) Diffraction of X-rays by chain molecules, Elsevier, Amsterdam
54. Gusev AA, Suter UW (1991) Phys Rev A43: 6488
55. Landau LD, Lifshitz EM (1977) Course of Theoretical Physics, Pergamon, Oxford, Vol 5
56. Gusev AA, Suter UW (1992) ACS Polym Preprints 33: 631
57. Gusev AA, Arizzi SA, Moll DJ, Suter UW (1993) J Chem Phys 99: 2221

58. Gusev AA, Suter UW (1993) Computer-aided Mat Design 1: 63
59. Müller-Plathe F, van Gunsteren WF (1992) ACS Polym Preprints 33: 633
60. Müller-Plathe F, (1992) J Chem Phys 96: 3200
61. Smith W, Gillan MJ (1992) J Phys Condens Matter 4: 3215
62. Pant PVK, Boyd RH (1993) Macromolecules 26: 679
63. Müller-Plathe F, Laaksonen L, van Gunsteren WF (1993) J Mol Graph 11: 118
64. Müller-Plathe F, Rogers SC, van Gunsteren WF, (1993) J Chem Phys 98: 9895
65. Pant PVK, Boyd RH (1992) Macromolecules 25: 494
66. Trohalaki S, Rigby D, Kloczkowski A, Mark JE, Roe RJ (1989) ACS Polym Preprints 30(2): 23
67. El Amrani S, Kolb M (1993) J Chem Phys 98: 1509
68. Toxvaerd S (1990) J Chem Phys 93: 4290
69. Brandrup J, Immergut EH (1989) Polymer Handbook (eds.), 3rd Ed, Wiley, New York
70. Lundberg JL, Mooney EJ, Rogers CE (1969) J Polym Sci A 7: 947
71. Barrer RM, Chio HT (1965) J Polym Sci C10: 111
72. Crank J, Park GS, (1968) Diffusion in polymers, Academic Press, London
73. Tikhomirov BP, Hopfenberg HB, Stannett V, Williams JL (1968) Die Makromolekulare Chemie
    118: 177
74. Takeuchi H, Roe R-J, Mark JE (1990) J Chem Phys 93: 9042
75. Gusev AA, Moll DJ, Suter UW to be submitted
76. Insightll User Guide, Version 2.1.0, Copyright 1992, Biosym Technologies Inc., San Diego
77. Gusev AA, Suter UW to be submitted
78. Dauber-Osguthorpe P, Roberts VA, Osguthorpe DJ, Wolff J, Genest M, Hagler AT (1988)
    Proteins: Structure, functions and genetics 4: 31
79. Gusev AA, Tiller A, Suter UW to be submitted

Received: January 1994

Atomistic Monte Carlo Simulation and Continuum Mean Field Theory of the Structure and Equation of State Properties of Alkane and Polymer Melts

L.R. Dodd[1] and D.N. Theodorou
Department of Chemical Engineering, University of California, Berkeley, California 94720-9989, USA

A continuum mean field approach based on Generalized Flory Theory and the Polymer Reference Interaction Site Model is developed to describe the structural and equation of state properties of normal alkane liquids and linear polyethylene melts. Efficient Monte Carlo simulations based on a new algorithm that employs concerted rotations around up to seven consecutive skeletal bonds along a chain are also conducted on the same systems. A realistic united-atom model is chosen to describe the geometry and energetics of the molecules and used throughout the study. Comparisons between the simulations and experimental thermodynamic and structural results are good and those between the mean field theory and the exact simulation results are reasonable. A method is described for quickly sampling the conformation of unperturbed chains in continuous space. The statistics of these chains compare very well with conformationally equilibrated chain statistics from the bulk simulation; this provides a confirmation of Flory's Random Coil Hypothesis. The need for improving the mean field theory and for enhancing the equilibration rate of the Monte Carlo simulations are identified. A new neighbor-list scheme is introduced for use in polymer Monte Carlo simulations.

[1] Present address: Polytechnic University, Six Metrotech Center, Brooklyn, New York 11201, USA
E-Mail: dodd @ roebling.poly.edu

Advances in Polymer Science, Vol. 116
© Springer-Verlag Berlin Heidelberg 1994

1 Introduction

The prediction of the physical properties of amorphous polymers, above and below the glass temperature, from their fundamental chemical constitution is a very desirable goal. The importance of these properties in all technological applications involving polymers cannot be overstated. Phenomenological approaches have been developed to correlate the properties of polymer melts with the chemical constitution of the chains. However, the complexity of long-chain polymer systems limits the reliability of such correlations. There is a need for more fundamental understanding of the structure and properties of amorphous polymers. Statistical mechanics-based theory and molecular simulation can help fulfill this need by predicting macroscopic properties and at the same time illuminating underlying molecular processes.

A molecular theory is typically cast in closed form as a set of algebraic, integral, or differential equations that may be solved analytically or numerically. Theories are more general and less computer-intensive than simulations. Their major limitation is that, in order to afford a closed formulation, one has to introduce simplifying approximations in the mathematical treatment. Computer simulations, on the other hand, can be thought of as numerical solutions of the full statistical mechanics given a molecular model for the polymer. Advances in molecular theories of polymer systems that rely on realistic representations of molecular geometry and interaction energetics may only be possible if simulation methodologies are developed in parallel, to enable thorough testing of the theories. Designing reliable atomistic simulation approaches for long-chain liquids is in itself far from a trivial task, owing to the severe connectivity constraints and very long relaxation times that characterize these liquids.

In this paper we are concerned with the prediction of structure and equation-of-state properties of chain liquids. In recent years, several theoretical approaches have been proposed that have the potential of relating these properties to the detailed geometry and energetics of chains. Two of these approaches will be pursued in this paper: Generalized Flory Theory (GFT) and the Polymer Reference Interaction Site Model (PRISM). GFT [1, 2] relies upon the concept of the chain insertion probability, defined as the Boltzmann factor of the excess chemical potential of a chain relative to a hypothetical ideal gas state at the same density. The equation of state of the multichain system can be extracted from the temperature and density dependence of the insertion probability through an exact integral equation ("osmotic equation of state"). Conversely, the insertion probability can be derived given the equation of state. Following Flory's lattice-based mean field theory of polymer solution thermodynamics [3], GFT introduces an Ansatz for estimating the chain insertion probability as a product of insertion probabilities contributed by smaller sections (e.g., monomers, dimers) along the chain. The latter are computed from the equations of state of low molecular weight analogues of the chain (e.g., monomer, dimer fluid) which are more readily obtainable than that of the polymer. PRISM [4–6] is an extension of Chandler's RISM integral equation theory of molecular fluids [7]. In its non-self-consistent form, employed here, PRISM provides an estimate for the inter-molecular structure (site-site pair distribution functions) of the polymer liquid, using as input the intra-molecular pair density functions of chains. Thermodynamic predictions from GFT have been tested against computer simulations of simple model polymer fluids [8,9] and structural predictions from PRISM have been compared against simulations of simple chain models [10] and against experiment [6]. Very little has been done, however, towards implementing these theories for detailed atomistic models that are capable of reproducing the experimentally observed thermodynamic properties of melts and towards testing their predictions against molecular simulations based on the same models. One reason for this is the scarcity of atomistic simulations that can reliably predict polymer properties.

Two broad categories of simulation methods are available for extracting the structure and thermodynamic properties of liquids from molecular geometry and energetics: molecular dynamics (MD) and Monte Carlo (MC). MD tracks the temporal evolution of a microscopic model system through numerical integration of the equations of motion for all the degrees of freedom. Originally developed in the microcanonical ensemble, the method may be applied straight-forwardly in the canonical and isothermal-isobaric (NPT) ensembles as well. Constraints on the microscopic degrees of freedom, such as constancy of bond lengths and bond angles, can readily be handled [11]. The MD method has been applied rather extensively to united atom polyethylene-like systems [12–20]. A recent particularly thorough MD investigation of liquid tridecane (C_{13}) has demonstrated that thermodynamic, structural, and dynamical properties can be reproduced very accurately, provided care is taken in the potential representation used [21,22]. An asset of MD is that it provides a wealth of detailed information on short-time dynamical processes in the polymer. The major limitation of MD is that, since it faithfully mimics molecular motion in actual amorphous polymer systems, it is fully subject to the bottlenecks that limit this motion. Hundreds of CPU hours on a vector supercomputer are required to simulate a nanosecond of actual motion with atomistic MD. In view of the much longer relaxation times of actual long-chain polymer melts, the mere equilibration of a model melt by MD becomes a problem, and results are not trustworthy if one starts from an improbable initial configuration.

For the purpose of predicting structure and thermodynamics, MC is more promising than MD since it can bring about drastic reconfigurations of a model system and thereby speed up equilibration. Nevertheless, MC work on polymer melts has been much less extensive than MD, perhaps because of the challenge associated with designing moves that work well for polymers. Years of research on the simulation of small molecule systems does provide a foundation for polymer MC; it does not address all the difficulties that arise, however. In small molecule systems the degrees of freedom are not strongly coupled, and this allows simple elementary moves to be quite efficient. In liquids of rigid molecu-les, for example, translations and rotations are all that is required to sample configuration space reliably. In systems of long, flexible chains, the torsional (dihedral) angles along the chain backbones are the most important degrees of freedom. These degrees of freedom have to change cooperatively in a move if the constraints of chain geometry and excluded volume are not to be violated. A pioneering MC study of liquid n-triacontane (C_{30}) was presented by Vacatello et al. [23]. Boyd [24] undertook NPT MC simulations of n-tetracosane (C_{24}) using a united-atom model and a reptation-based algorithm. Very recently there has been much interest in atomistic MC simulation of polymers. In fact, one could say that this field is currently undergoing a renaissance, with some very promising methodological advances. Two of the most promising such advances are the Continuum Configurational Bias (CCB) method [25, 26] and the Concerted Rotation (CONROT) method [27]. An analysis of the two methods and of their combined use (COMBO) is presented in [28].

The purpose of the study reported in this paper is as follows:

to capture the structure, chain conformational characteristics, and equation of state properties of long-chain polymer liquids through MC simulations of models that take into account the detailed geometry and energetics of chains;
to assess the degree to which equilibration can be achieved with these simulations for different chain lengths; and
to test the ability of state-of-the-art theoretical approaches to predict structure and thermodynamics based on the same detailed molecular models.

Liquid hydrocarbons from n-decane to n-tetracosane and n-octaheptacontane (C_{78}) are used as test systems, the latter providing a reasonable model for molten monodisperse linear polyethylene. As shown below, the experimentally observed structure and thermodynamic properties of all these liquids are very well reproduced by MC simulations using the CONROT algorithm and a relatively simple united-atom representation of the chains. The degree to which chain conformations in our bulk MC runs follow Flory's random coil hypothesis [3] has been assessed in detail through Monte Carlo sampling of single unperturbed chains in continuous space, using the same model representation with the exception of nonlocal interactions. Our theoretical investigations are based on exactly the same model representation as our bulk MC work, thus enabling direct comparison with both simulation results and experiment. We develop a continuum mean field approach to the equation of state by combining the random coil hypothesis for intramolecular structure, GFT for repulsive, and PRISM for attractive contributions to pressure. Results are used to identify needs for improvement in the continuum mean field theory and in the MC simulations.

The paper is structured as follows. Our molecular model and MC methods for simulating the melt and sampling unperturbed chains are briefly outlined in Sect. 2. Results from the bulk MC simulations regarding the melt structure, chain conformation, and PVT behavior are presented in Sect. 3, where evidence is also provided of the degree of equilibration of the simulations. Structural predictions from the bulk MC are compared against PRISM estimates, also in Sect. 3. The formulation of the mean field theory for equation-of-state properties is explained in Sect. 4 with reference to Appendix B, and results from this theory are also discussed. Finally, general conclusions are drawn in Sect. 5.

2 Model and Simulation Methods

2.1 Molecular Model

The following molecular model was used for alkane and polyethylene chains throughout this study. Carbon–carbon bond lengths were held constant at 1.54 Å, the supplements of the carbon–carbon–carbon bond angles were fixed at 68°, and a united-atom description was used for the methylenes and methyls. In the following, the terms "site" or "mer" are used for a methylene or methyl segment. The configuration-space probability density simulated was that of the flexible model in the limit of infinite stiffness [27, 29, 30]. All inter-chain and intra-chain pairs of united-atom sites in the system interacted with the same nonbonded potential. This potential was of a Lennard-Jones (LJ) form, with parameters $\varepsilon/k_B = 49.3$ K and $\sigma = 3.94$ Å [31]. Finally, the torsional potential of Ryckaert and Bellemans was used [32]. Bulk MC simulations employed a finite-range modification of the LJ potential: the potential and its first and second derivatives were defined to be zero for $r > 2.3\ \sigma$, and a quintic spline was used between $r = 1.45\ \sigma$ and $r = 2.3\ \sigma$ to accomplish this with discontinuities occurring only in the third- and higher-order derivatives of the potential [33]. Attractive tail contributions to thermodynamic properties were dealt with through direct integration.

2.2 Bulk Monte Carlo

2.2.1 Attempted Moves

We have used the CONROT method in the melt simulations discussed here. Rather than going into the geometric and sampling subtleties of the CONROT move, we will confine ourselves to a brief description of the elementary moves, referring the reader to [27] for more details on the method.

Since one of our interests is to obtain PVT behavior, we perform isobaric-isothermal (constant NPT) simulations by setting the pressure and measuring the density and its fluctuations. As in the simulation of simple fluids, the dimensions of the simulation cell are changed periodically, requiring a displacement of all the molecules in the system. For chain systems, volume fluctuations can be implemented in several ways. In the first polymer NPT MC simulation, conducted by Boyd [24], all chains were displaced rigidly and affinely with respect to each other using the chain starts as points of reference. We adopt this strategy, although an alternative would be to use the center of mass of each chain as the reference point for the affine displacement.

As in a simulation of small-molecule systems, randomly selected single chains were: translated rigidly by a vector whose components follow a uniform distribution within a specified interval (e.g., -0.2 to 0.2 Å); pivoted rigidly about an axis passing through their center of mass by a rotation angle that follows a uniform distribution within a specified interval (e.g., -3 to $3°$). The direction of the rotation axis was chosen randomly among x, y, and z.

The torsional degrees of freedom of randomly selected chains were allowed to sample their configuration space using two types of moves. First, "reptations" of single chains were performed by randomly selecting one of either the chain end or the chain start, removing it, placing it at the other end of the chain with a torsion angle uniformly distributed between ($-\pi$, $+\pi$), and relabelling the mers of the chain appropriately. Clearly, reptations result in a net displacement of the chains. Second, concerted rotations of up to seven consecutive skeletal bonds along a randomly selected chain are performed using the algorithm presented in [27].

2.2.2 Acceptance

Rigid translations, pivots, and reptations of single chains were accepted with probability

$$p(m \rightarrow n) = \min\left[1, \frac{\exp(-v(n)/k_B T)}{\exp(-v(m)/k_B T)} \right] \tag{1}$$

where $v(i)$ is the total intermolecular potential of state i, T is absolute temperature, and k_B is Boltzmann's constant. Affine volume fluctuations were accepted with probability

$$p(m \rightarrow n) = \min\left[1, \frac{\exp[-v(n)/k_B T - PV(n)/k_B T + N\ln V(n)]}{\exp[-v(m)/k_B T - PV(m)/k_B T + N\ln V(m)]} \right] \tag{2}$$

where $V(i)$ is the volume of the simulation cell in state i. The $\ln V(i)$ terms in Eq. (2) result from changing the absolute chain start position degrees of freedom for all N chains when performing the volume fluctuation, while the scaled chain start coordinates (relative to the simulation cell edge length), orientational, and torsional degrees of freedom remain the same. Although not incorporated in [24], this term is necessary in order to sample the isothermal-isobaric probability density correctly. The acceptance criterion for the concerted rotations is discussed in detail elsewhere [27].

The initial configurations used to start all the Monte Carlo simulations were minimum energy structures obtained by the technique developed by Theodorou and co-workers [33, 34] using an initial guess for the density.

2.3 Continuous Unperturbed Chains via Fast Monte Carlo

In addition to the above melt Monte Carlo method for obtaining the configurations of chains in the bulk, we have developed a fast sampling technique for single unperturbed chains [35]. Flory's random coil hypothesis states that the conformations of polymer chains in the bulk are determined on average by local (short-range) intra-chain interactions and are unperturbed by non-local (long-range) interactions [3]. Flory's rotational isomeric state (RIS) model [36] is a discrete implementation of this hypothesis, wherein torsional angles along the chain can reside only in a finite number of states (e.g., trans, t; gauche plus, g^+; and gauche minus, g^- for polyethylene) and the conformational energy is expressed as a sum of terms depending on the states of two adjacent bonds. Only mers separated by four or fewer skeletal bonds are taken as contributing to the intra-chain energy in the RIS model; this accounts for the pentane-effect in polyethylene. Since there are only a finite number of torsion states, statistical weight matrices for the combinations of states assumed by successive angles may readily be constructed, and the model can be solved analytically or numerically for any conformation-dependent average property.

For an accurate comparison of our bulk simulation results against unperturbed conformational characteristics, it is imperative that we develop a model for single unperturbed chains based on exactly the same potential representation as used in our bulk simulation. To this end, we consider a continuous analogue of the RIS model, wherein each torsion angle is allowed to vary between $-\pi$ and π. Non-bonded and torsional interactions are only included between sites separated by at most four bonds. The torsional potential of Ryckaert and Bellemans [32] already takes into account all non-bonded and torsional interactions involving sites separated by three or less bonds. Therefore, for the continuous unperturbed chain model, only the non-bonded (LJ) interaction between sites four bonds apart (pentane-effect) is needed in addition to the torsional potential. Since overall translational and orientational degrees of freedom do not affect the energy of a single chain, only changes in the torsional degrees of freedom need be considered in calculating average properties of continuous unperturbed chains.

With the above potential representation, sections of the chain many bonds apart may cross and overlap and the conformation is literally unperturbed by long-range (non-local) interactions. In order to compute equilibrium average properties of our continuous unperturbed chains, we sample their configuration space through a fast Monte Carlo scheme where the only elementary move employed is simple reptation, performed exactly as described above for the bulk simulation. Note that reptations must be attempted in both directions along the chain in order for the simulation to be microscopically reversible.

The initial configuration for all single unperturbed chain simulations was all-trans, from which the conformation quickly departed. After permitting the chain to explore a sufficient number (~ 200 million) of its configurations, statistics were accumulated for the properties of interest. We will refer to the chains sampled by this method as "continuous unperturbed chains" (CUC). Due to the simplicity of the sampling method and the local nature of the unperturbed chain energy, reliable unperturbed chain statistics may be generated in a very small fraction of the CPU time that it takes to equilibrate the bulk NPT Monte Carlo simulation. More than ten thousand CUC configurations of a 24-mer alkane (n-tetracosane) could be sampled in one CPU second on an IBM 320H workstation. Furthermore, since the molecular model is identical in the bulk and unperturbed chain Monte Carlo simulations, one is given the opportunity to compare the chain statistics of the continuous unperturbed chains with the results for the chains from the bulk NPT simulation. This provides a direct test of Flory's random coil hypothesis.

3 Results

3.1 Tests for Equilibration

We used several criteria for monitoring equilibration of each of the NPT MC simulations we conducted. A necessary condition for equilibration is that the system energy fluctuates about a constant mean value, rather than drifting with simulation time. Since the initial configuration in all cases was a minimum energy configuration, one would expect the system potential energy to increase during the equilibration stage of the simulation as the configuration departs from the initial minimum energy well. Figure 1 shows a typical departure from the minimum energy well for a system consisting of 32 chains of 24 mers each (n-tetracosane) at 450 K and 0.1 MPa. The increase in the total energy is quite rapid and due mainly to increases in the rotational (torsional) potential. Clearly, the potential energy has stabilized beyond $\sim 3,000$ "cycles" (considerably less than would be required for a simulation without concerted rotations [27]). We define a "cycle" as the number of configurations that equals the number of degrees of freedom in the system. For example, in the n-tetracosane system each of the 32 chains has 21 torsional angle, 3 Eulerian angle, and 3 center of mass degrees of freedom, so one cycle equals $(21 + 3 + 3) \times 32 = 864$ configurations (attempted moves).

Another necessary condition for equilibration is that the average pressure calculated during the simulation be close to the set pressure of the NPT simulation. Since in this MC simulation of polymer chains the bond lengths and angles are infinitely stiff, the determination of the stress tensor from interatomic forces is non-trivial. There are in fact many techniques for computing the stress tensor; these are reviewed elsewhere [30]. Here we used both the so-called "Molecular Virial" method (see Appendix A of [37] and also [30]) and an "inter-chain force-based" method [30] to calculate the stress tensor and the pressure. We found the calculated pressure to be in excellent agreement with the set pressure for both methods: within 1% for the Molecular Virial method and within 10% for the inter-chain force-based method.

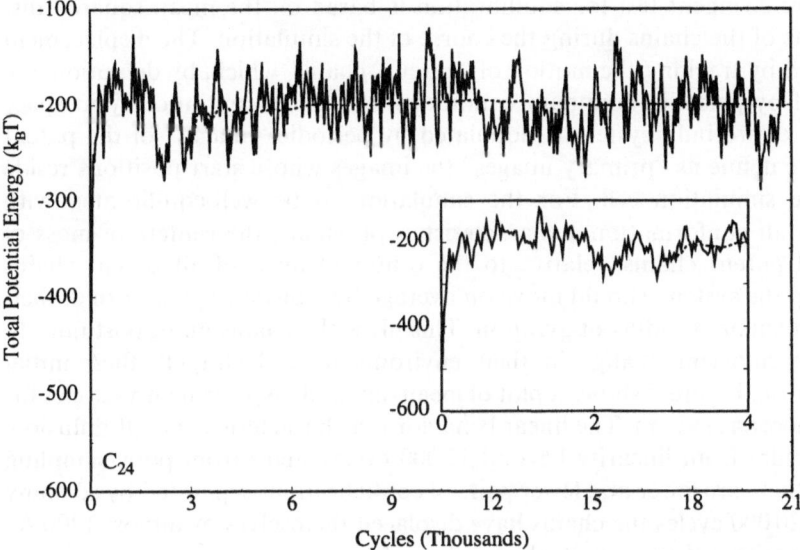

Fig. 1. Total potential energy vs Monte Carlo cycles for a 32-chain 24-mer (*n*-tetracosane) system at 450 K and 0.1 MPa starting from a minimum energy configuration. The running average is shown as a *dashed line*. The *inset* shows a closeup of the initial rise from the energy minimum

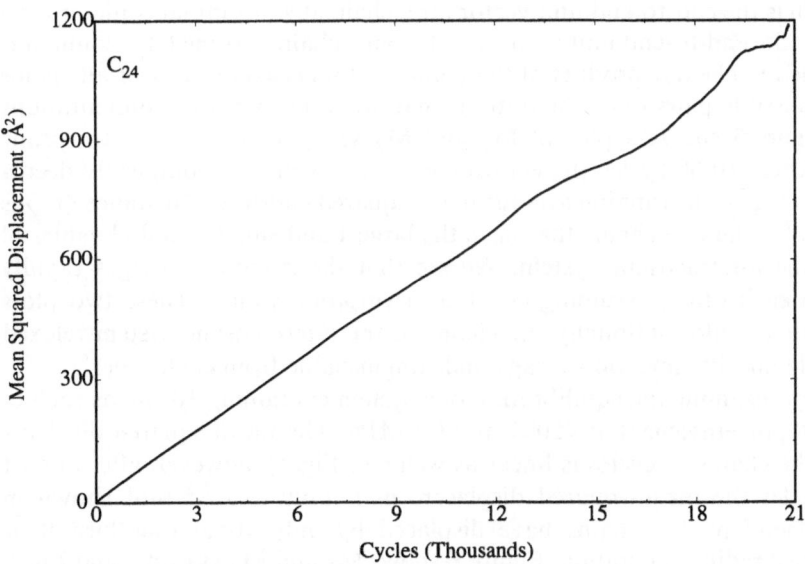

Fig. 2. Mean-squared displacement vs Monte Carlo cycles for a 32-chain 24-mer (*n*-tetracosane) system at 450 K and 0.1 MPa

A more stringent test for equilibration is based on the mean-squared displacements of the chains during the course of the simulation. The displacement is obtained by tracking the motion of "parent" chains, which, by definition, are unaffected by periodic boundary conditions and move continuously in space [30]. The entire bulk system is populated by periodic "images" of the parent chains. We define as "primary images" the images whose start positions reside within the simulation cell. For the simulation to be well-equilibrated with respect to all conformational characteristics of chains, the centers of mass of individual parent chains, relative to the center of mass of all parent chains making up the system, should move on average by a distance greater than their root mean-squared radius of gyration. This gives the chains an opportunity to experience sufficient change in their environment and "forget" their initial conformation. Figure 2 shows a plot of mean-squared displacement vs cycles for the n-tetracosane system. The linear behavior is a characteristic of self-diffusion. The departure from linearity beyond 15 000 cycles stems from poor sampling statistics: there are fewer and fewer pairs of configurations separated by so many cycles. In 20 000 cycles the chains have displaced themselves by almost 1200 Å^2 or over five times their unperturbed radius of gyration.

To quantify the extent to which the chains "forget" their initial conformation, we define the following autocorrelation functions for the end-to-end unit vectors of the chains

$$M_1(t) = \langle \mathbf{u}(t) \cdot \mathbf{u}(0) \rangle \tag{3}$$

$$M_2(t) = \tfrac{1}{2}[3\langle (\mathbf{u}(t) \cdot \mathbf{u}(0))^2 \rangle - 1] \tag{4}$$

where $\mathbf{u}(0)$ is the end-to-end unit vector for a chain at some initial configuration and $\mathbf{u}(t)$ is the end-to-end unit vector for the same chain at some later configuration labeled t. The dot-product of these unit vectors is averaged over all chains over all possible pairs of configurations that are a distance of t configurations apart. Figure 3 shows a plot of M_1 and M_2 vs cycles for the n-tetracosane system. After 10 000 cycles the end-to-end vector is almost completely decorrelated. In Fig. 4, the running average of the squared end-to-end distance $\langle r^2 \rangle$ is presented for the two chains that have the largest and smallest initial values of $\langle r^2 \rangle$ in the n-tetracosane system. We see that the running averages rapidly approach each other, assuming constant asymptotic values. These two plots indicate that conformationally the chains in the n-tetracosane system relaxed completely and became, on average, indistinguishable from each other.

We now examine the equilibration of a system containing 10 chains each of 78 units ("polyethylene") at 450 K and 0.1 MPa. The mean-squared displacement of the chains vs cycles is linear as well (see Fig. 5); however, after almost 31 000 cycles the mean-squared displacement is only 41.1 Å^2 (not shown in Fig. 5). Therefore, the chains have displaced by only about one-third their unperturbed radius of gyration. Figure 6 shows M_1 and M_2 vs cycles and Fig. 7 shows the running average of $\langle r^2 \rangle$ for the two chains that have the largest and smallest initial values in the polyethylene system. We see that at the end of the

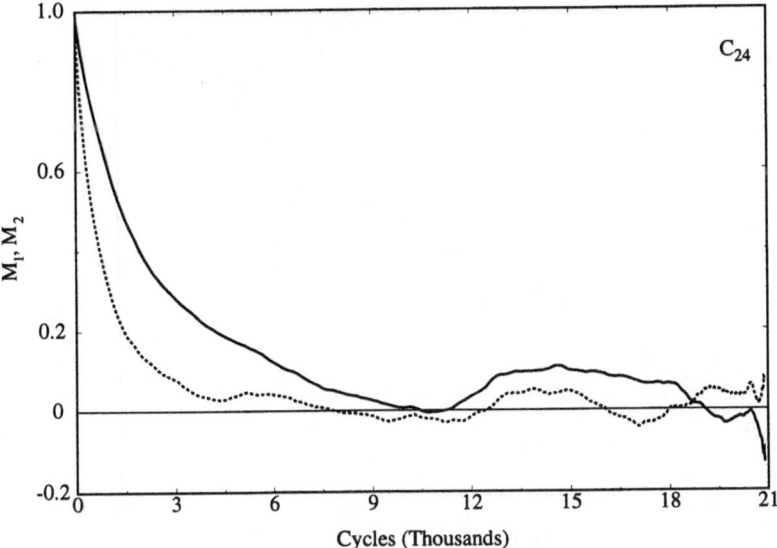

Fig. 3. Autocorrelation functions for the end-to-end unit vector vs Monte Carlo cycles for a 32-chain 24-mer (*n*-tetracosane) system at 450 K and 0.1 MPa. Solid curve: M_1; dashed curve: M_2

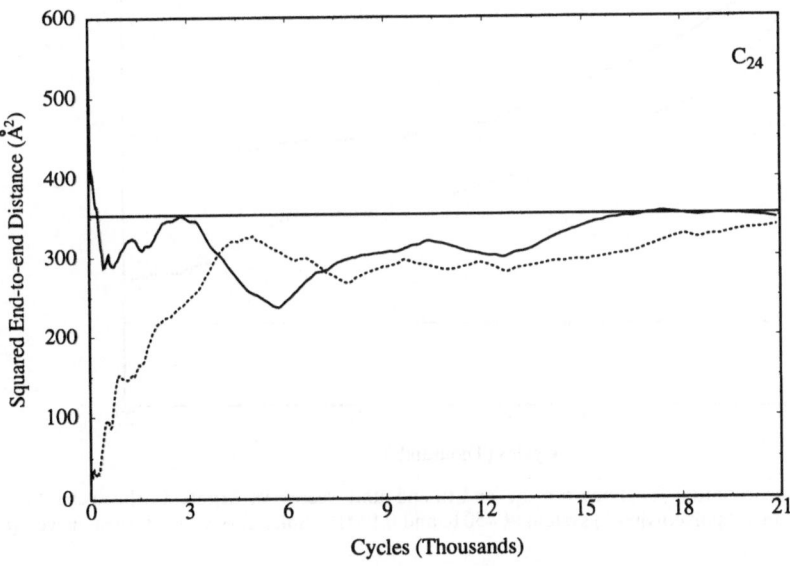

Fig. 4. Running average of the squared end-to-end distance vs Monte Carlo cycles for the chain that initially had the largest end-to-end distance and for the chain that initially had the smallest end-to-end distance in a 32-chain 24-mer (*n*-tetracosane) system at 450 K and 0.1 MPa. The *horizontal line* shows the value for continuous unperturbed chains

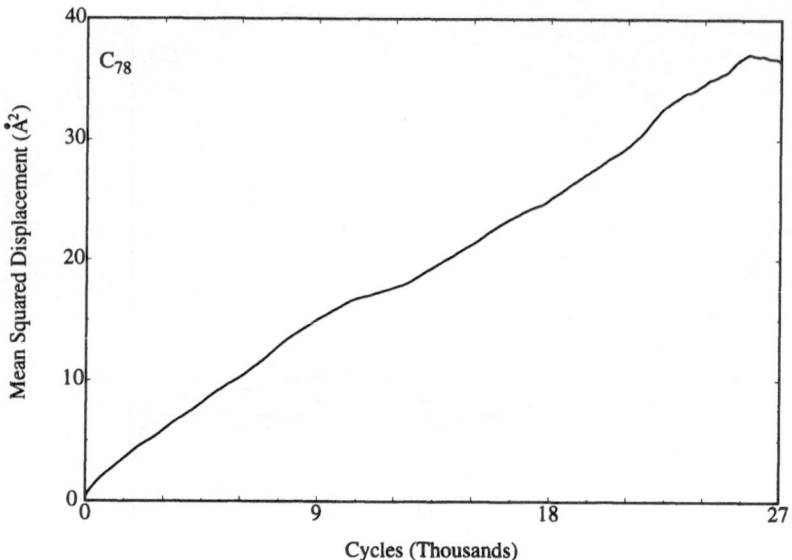

Fig. 5. Mean-squared displacement vs Monte Carlo cycles for a 10-chain 78-mer ("polyethylene") system at 450 K and 0.1 MPa

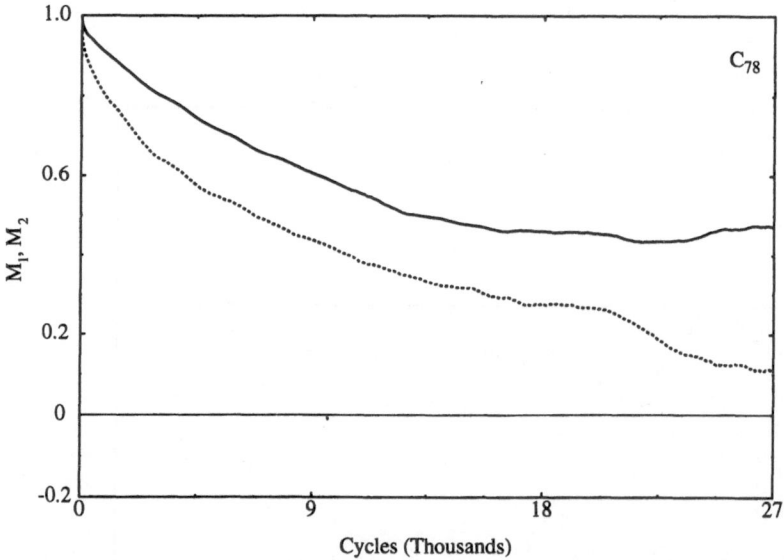

Fig. 6. Autocorrelation functions for the end-to-end unit vector vs Monte Carlo cycles for a 10-chain 78-mer ("polyethylene") system at 450 K and 0.1 MPa. Solid curve: M_1; dashed curve: M_2

simulation this longer chain system is still not completely relaxed conformationally.

Before moving on to the PVT and chain characteristics, some points should be made concerning the reptation elementary move. We found that if a reptation

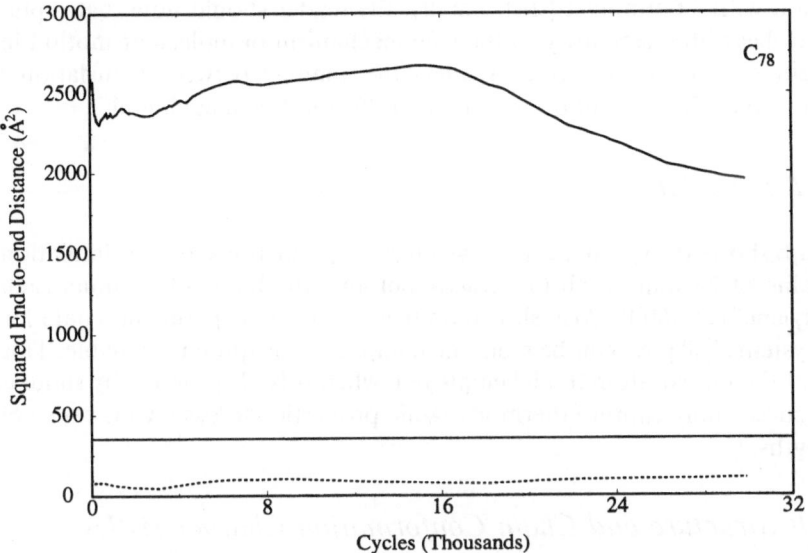

Fig. 7. Running average of the squared end-to-end distance vs Monte Carlo cycles for the chain that initially had the largest end-to-end distance and for the chain that initially had the smallest end-to-end distance in a 10-chain 78-mer ("polyethylene") system at 450 K and 0.1 MPa. The *horizontal line* shows the value for continuous unperturbed chains

move was successful in one direction for a given chain (e.g., the mer at the chain start was successfully placed at the chain end), then there was a very high probability that the next successful reptation for that chain would be in the opposite direction (e.g., the mer would be placed back at the chain start). This is because the pocket of free volume created near a chain end by a successful reptation is likely to be filled in the next reptation move involving the chain, if no significant redistribution has occurred between the two reptations. To quantify this, during the course of the simulation we accumulated a transition probability matrix **P** for the successful reptations performed on each chain

$$\mathbf{P} \equiv \begin{pmatrix} P_{ff} & P_{fb} \\ P_{bf} & P_{bb} \end{pmatrix} \qquad (5)$$

where, for example, P_{fb} is the conditional probability of a successful backwards reptation being followed by a successful forwards reptation. We found for the systems we studied that $P_{bf} \approx P_{fb}$ and were on average between 0.7 and 0.8 for all chains. This means that with the reptation elementary move the chains have a significant tendency to "shuttle" back and forth, which tends to reduce the rate at which the system explores its configuration space. One can reduce this tendency by introducing other elementary moves to rearrange the system between reptations and thereby "close" the pocket of free volume created by one reptation by the time a second one occurs. One of the advantages of the concerted rotation (CONROT) move is that it effects this rearrangement. A similar conclusion was reached with regards to using CONROT moves in

conjunction with CCB moves [28]. Finally, the reader should note that reptation moves bear little similarity to the true mechanism of molecular motion in
short chain polymer melts, which renders the analogy between translational
drift seen in the MC simulation and actual diffusion less meaningful.

3.2 PVT Behavior

Figure 8a,b shows the specific volume vs pressure predictions from a simulation
of 32 chains of 24 units each (n-tetracosane) and 10 chains of 78 units each
("polyethylene") at 450 K. Also shown on these plots are experimental data for
the two systems [38]. As can be seen, the comparison is quite reasonable. This
means that the united-atom model employed, when solved "exactly" by simulation, can successfully capture thermodynamic properties across a wide range of
chain lengths.

3.3 Melt Structure and Chain Conformation Characteristics

Having established the ability of the NPT Monte Carlo simulation to capture
PVT behavior, we now turn to the prediction of melt structure and single chain

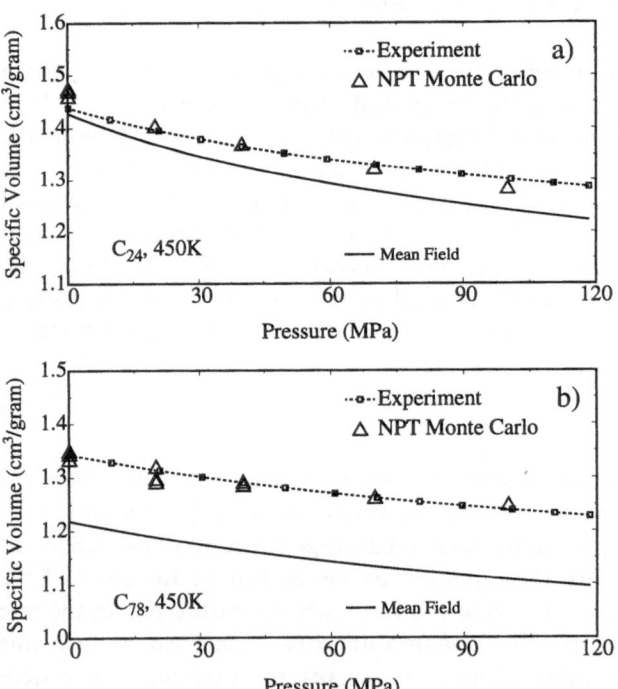

Fig. 8a, b. Specific volume (cm³/g) vs pressure (MPa) for: **a** n-tetracosane at 450 K; **b** 78-mer
"polyethylene" system; *squares* and *dotted line*: experimental data; *triangles*: predictions of NPT MC
simulation; *solid curve*; continuum mean field theory predictions

characteristics. We first look at the intra-chain pair density function $\omega(r)$, the inter-chain pair distribution function $g(r)$, and the total pair distribution function $g^{tot}(r)$. We calculate $\omega(r)$ as the average number of pairs of sites on the same chain that are between r and $r + dr$ apart, divided by the volume of a spherical shell of radii r and $r + dr$.

Figure 9a shows $\omega(r)$ obtained from the bulk NPT Monte Carlo simulation for the n-tetracosane system at 450 K and 0.1 MPa. The Dirac spikes at small distances are due to pairs that are separated by one fixed bond (1.54 Å) and by two fixed bonds (2.55 Å). Pairs separated by more than two bonds display sharp peaks that reflect the conformational preferences of the chains. For example, the peak at 3.93 Å is due to two mers separated by three bonds, the middle one of which is in a trans conformation and the peak at 3.05 Å is due to two mers separated by three bonds, the middle one of which is in a gauche conformation.

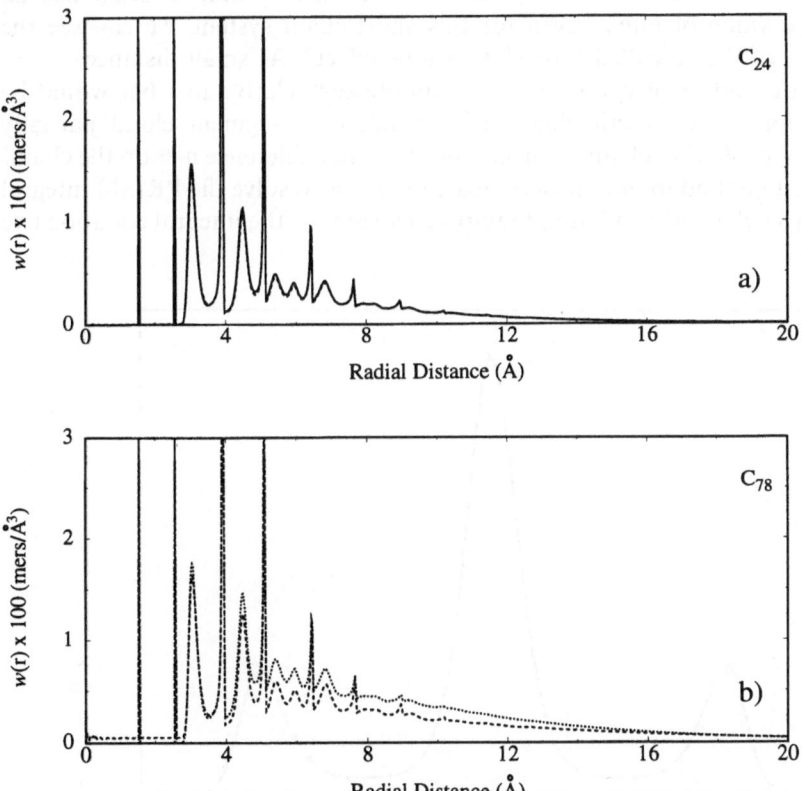

Fig. 9a, b. Intra-chain pair density function $\omega(r)$ vs radial distance: **a** *dotted curve*: result of bulk NPT Monte Carlo simulation of a 32-chain 24-mer (n-tetracosane) system at 450 K and 0.1 MPa; *dashed curve*: result from fast sampling of continuous unperturbed chains; curves are nearly identical; **b** *dotted curve*: result from bulk NPT Monte Carlo simulation of a 10-chain 78-mer ("polyethylene") system at 450 K and 0.1 MPa; *dashed curve*: result from fast sampling of continuous unperturbed chains

At longer radial distance, $\omega(r)$ drops to zero as a consequence of the finiteness of the chain's segment cloud; even if the chain were infinite, $\omega(r)$ would still do so, for a random walk cannot fill three-dimensional space [3].

Also shown in Fig. 9a is the $\omega(r)$ obtained from the sampling of continuous unperturbed chains at the same temperature using the same chain geometry and same potential representation for the short-range interactions. The agreement is quite striking. It should be pointed out that, while the bulk NPT simulation required some 80 CPU hours on a Cray X/MP-28 to generate the 18 million configurations used to determine the averages, the sampling of 200 million configurations of continuous unperturbed chains was accomplished on an IBM 320H workstation in 5 CPU hours (a speedup factor of ~ 300). A similar agreement is found for the torsion angle probability distribution Fig. 10.

The inter-chain pair distribution function g(r) from the bulk simulation for the *n*-tetracosane system at 450 K and 0.1 MPa is shown as the solid curve in Fig. 11a. Unlike the pair density function, the distribution function has an asymptotic value of unity. Even for this short-chain system, we can see the beginnings of the so-called "correlation hole effect." At small distances, compared to the radius of gyration, g(r) is suppressed relative to what would be observed for a monomeric fluid, as a chain's own segment cloud partially excludes mers of other chains from coming close to a reference mer on the chain.

One fast method to obtain an estimate for g(r) is to solve the PRISM integral equation [5]. Such a calculation requires a closure for the integral equation (we

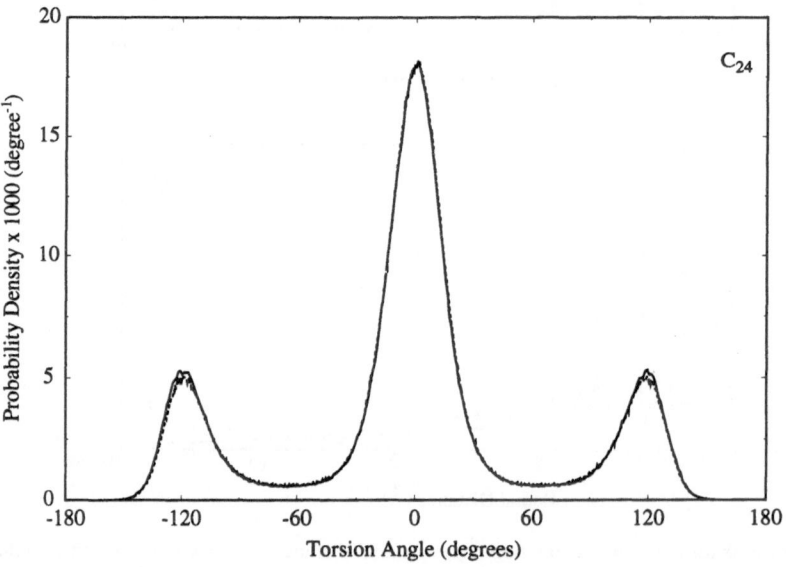

Fig. 10. Probability distribution of torsion angle relative to trans, $(\phi - \phi(t))$. *Solid curve*: result of bulk NPT Monte Carlo simulation of a 32-chain 24-mer (*n*-tetracosane) system at 450 K and 0.1 MPa. *Dashed curve*: result of fast sampling of continuous unperturbed chains. Curves are nearly identical

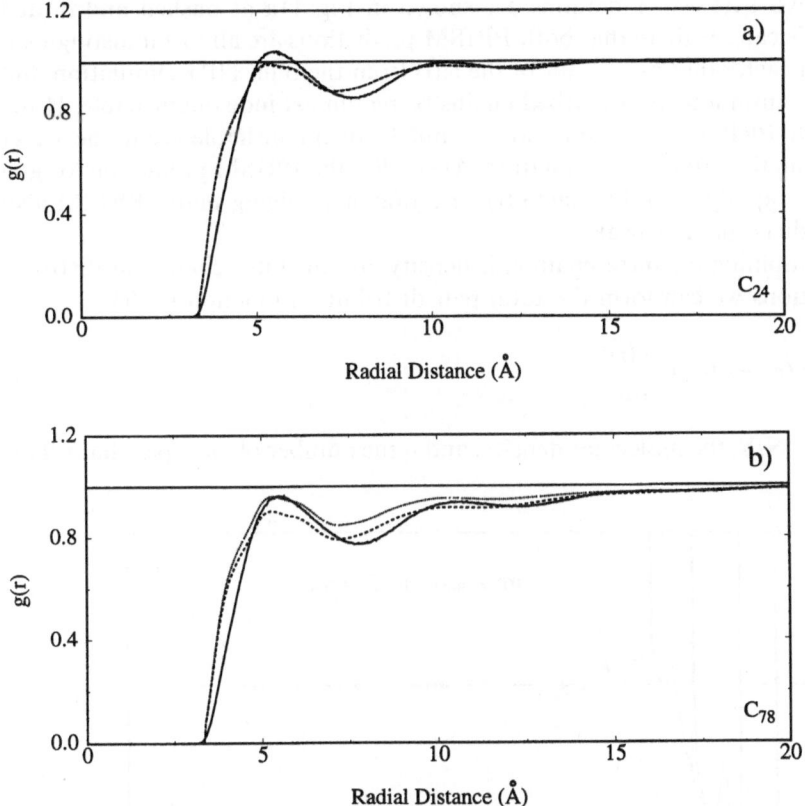

Fig. 11a, b. Inter-chain pair distribution function g(r). *Solid curve* is the result of bulk NPT Monte Carlo simulation. *Dashed curve* is the PRISM prediction using as input the exact intra-chain pair density $\omega(r)$ obtained from the NPT simulation. *Dotted curve* is the PRISM prediction using as input the approximate $\omega(r)$ obtained from fast sampling of continuous unperturbed chains: **a** 32-chain 24-mer (*n*-tetracosane) system at 450 K and 0.1 MPa; **b** 10-chain 78-mer (polyethylene) system at 450 K and 0.1 MPa

used soft Percus Yevick [39]), and, more importantly, an estimate for the intra-chain pair density function $\omega(r)$. We have incorporated the same potential model in these PRISM calculations as has been used throughout this work. Furthermore, we have studied the effects of two different estimates for $\omega(r)$. First, the exact $\omega(r)$ as obtained from our bulk NPT Monte Carlo simulations was used as input to the PRISM calculations. Comparing the g(r) predicted through PRISM by this method with the exact value from the bulk NPT simulation allows for a direct test of PRISM for model systems with realistic molecular geometry and energetics. Second, the $\omega(r)$ obtained from our sampling of continuous unperturbed chains was used as an input to the PRISM calculations. This second method (combined continuous unperturbed chains and PRISM) is of interest because it can generate estimates for the inter-chain pair distribution function extremely rapidly on a desktop workstation.

The two PRISM predictions are shown in Fig. 11a as dashed and dotted curves. We can see there that both PRISM predictions are almost indistinguishable from each other as a result of the ω(r) from the bulk NPT simulation and from the continuous unperturbed chains being almost indistinguishable. However, both PRISM predictions are definitely distinguishable from the exact results from the NPT MC simulation. At small r, the PRISM prediction for g(r) rises more rapidly than the exact g(r), resulting in a bulging shape; PRISM also underpredicts the first peak.

By combining the intra-chain pair density and the inter-chain pair distribution functions we can form the total pair distribution function $g^{tot}(r)$:

$$g^{tot}(r) = g(r) + \frac{\omega(r)}{\rho n} \qquad (6)$$

where $\rho = N/V$, the molecular density, and n the number of mers per chain. This

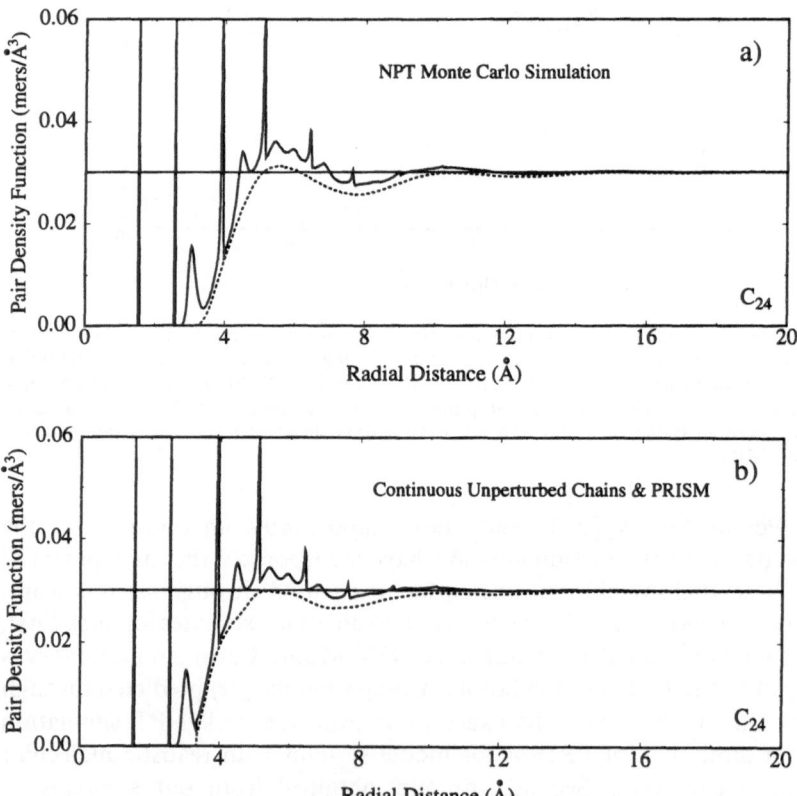

Fig. 12a, b. Total pair density function $\rho n \, g^{tot}(r)$ and the inter-chain pair density function $\rho n \, g(r)$ for 24-mer (n-tetracosane) system at 450 K and 0.1 MPa in units of mers per Å³: **a** results from 32 chain NPT Monte Carlo simulation; **b** results from combining fast sampling of continuous unperturbed chains and PRISM predictions

is displayed in Fig. 12 for the *n*-tetracosane system. Shown there are the exact results from the NPT simulation and the prediction from combining the $\omega(r)$ from the unperturbed chains and the $g(r)$ from the PRISM calculations. We can see that the comparison is reasonable, especially in terms of the intra-chain peaks captured by the continuous unperturbed chains. Again, it should be pointed out that the predictions of combined unperturbed chains and PRISM required less than six CPU hours on an IBM 320H workstation while the exact bulk NPT simulation required some 80 CPU hours on a Cray X/MP-28.

The ability of the bulk NPT simulation to predict the structure of real substances is demonstrated in Fig. 13. That figure displays the k-weighted structure factor $k \cdot S(k)$, where

$$S(k) - 1 = \rho n \int_0^\infty 4\pi r^2 \frac{\sin(kr)}{kr} [g^{tot}(r) - 1] dr \qquad (7)$$

for *n*-eicosane at 315 K and atmospheric pressure as obtained from X-ray studies of Habenschuss and Narten [40]. Also shown in the k-weighted structure factor obtained from a Fourier transform of $g^{tot}(r)$ from the bulk NPT Monte Carlo simulation of a system of 38 chains of 20 mers each. We can see that the first and most important inter-chain peak at low wave-vector is captured very well by the simulation. It is only at large wave-vectors that one sees some disagreement between prediction and experiment, with the simulation

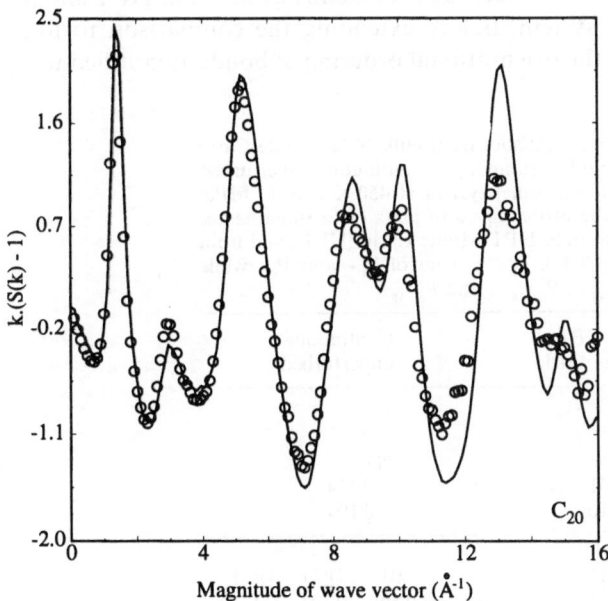

Fig. 13. *k*-weighted structure factor for *n*-eicosane (315 K, 0.1 MPa); *points*: X-ray diffraction data of Habenschuss and Narten; curve NPT MC simulation prediction

displaying more structure than the experiment; this is a consequence of using in our molecular model fixed bond lengths and bond angles and a united-atom description of the methylene groups.

Other chain characteristics of interest are the mean squared end-to-end distance $\langle r^2 \rangle$ and mean squared radius of gyration $\langle s^2 \rangle$. Table 1 shows the values of these (and other properties) for the n-tetracosane system obtained from the NPT simulation compared with the values predicted by the sampling of continuous unperturbed chains. Excellent agreement is obtained for $\langle s^2 \rangle$ and $\langle r^2 \rangle$ (see also line in Fig. 4). This agreement is also seen when comparing the torsion angle values of the NPT simulation and the unperturbed chains. The average gauche angle (i.e., $\langle |\phi(g) - \phi(t)| \rangle$) as well as the singlet probability $P_{g\pm}$ of a torsion angle ϕ being gauche (e.g., $|\phi - \phi(t)| \in (60°, 180°)$) of the continuous unperturbed chains agree very well with the exact bulk results. The comparison continues being very favorable when one examines the conformation probabilities for doublets and triplets of consecutive torsion angles. The value for the fraction of trans bonds P_t found in both our bulk NPT and in our CUC simulations is ~ 0.659. This is higher than values reported by Yoon and co-workers [21, 22] and higher than found experimentally. However, it is very difficult to identify a set of parameters to give good PVT and conformational properties (as was also discovered by Smith and Yoon [22]). No consideration was given to the conformational properties of C_{24} when ε and σ were determined.

Thus, with the fast sampling of continuous unperturbed chains, we are able to capture in great detail the characteristics of chains in the bulk NPT simulation for the n-tetracosane system. Before extending the comparison to long-chain systems, we examine the orientational ordering of bonds, quantified using

Table 1. Comparison between chain statistics from bulk NPT Monte Carlo simulation and chain statistics from fast sampling of continuous unperturbed chains for a 32-chain 24-mer (n-tetracosane) system at 450 K and 0.1 MPa. Also shown is a comparison of the probabilities of pairs of torsional states assumed by successive bonds from bulk NPT Monte Carlo (NPT) and from continuous unperturbed chains (CUC). $P_{g\pm}$ is sum of P_{g+} and P_{g-} while $P_{g\pm tg\pm}$ is the sum of P_{g+tg+}, P_{g+tg-}, P_{g-tg+}, and P_{g-tg-}.

Chain property	Bulk NPT Monte Carlo	Continuous unperturbed
$\langle r^2 \rangle$ Å^2	350	354
$\langle s^2 \rangle$ Å^2	43	44
$\langle |\phi(g) - \phi(t)| \rangle$	112°	111°
$P_{g\pm}$	0.341	0.334
$P_{g\pm tg\pm}$	0.110	0.104

$$
\text{NPT:} \begin{array}{c} \\ (t) \\ (g^+) \\ (g^-) \end{array} \begin{array}{ccc} (t) & (g^+) & (g^-) \\ \begin{pmatrix} 0.394 & 0.133 & 0.133 \\ 0.133 & 0.033 & 0.003 \\ 0.133 & 0.003 & 0.033 \end{pmatrix} \end{array} \quad \text{CUC:} \begin{array}{c} \\ (t) \\ (g^+) \\ (g^-) \end{array} \begin{array}{ccc} (t) & (g^+) & (g^-) \\ \begin{pmatrix} 0.406 & 0.131 & 0.131 \\ 0.131 & 0.031 & 0.003 \\ 0.131 & 0.003 & 0.031 \end{pmatrix} \end{array}
$$

the order parameter

$$S(r) = \tfrac{1}{2} \left[3 \langle \cos^2 \theta(r) \rangle - 1 \right] \tag{8}$$

where $\theta(r)$ is the angle formed between two "chords" whose mid-points are distance r apart. A "chord" is defined as a line segment connecting the mid-points of two adjacent skeletal bonds on a chain; chords may be paired inter-molecularly or intra-molecularly. The average squared-cosine of the angle between a pair of chords is accumulated for all chords distance r apart. Figure 14 shows the inter-chain and total order parameter for the n-tetracosane system (the results for the longer 78-mer chain system are nearly identical). We see here that the chords on different chains have a tendency to align themselves perpendicularly at the shortest radial distance and then align parallel at slightly longer distances before rapidly decaying to a random orientation as expected for an amorphous system. The total order parameter displays sharp peaks at small distances because of the conformational preferences of the chains but also becomes randomly oriented at large distances. The magnitude of the first peak of $S(r) \times g(r)$ for the n-tetracosane system agrees very well with results obtained by Yoon and co-workers using an explicit all-atom model of n-tridecane [22, 21].

Even though Monte Carlo is not a dynamic simulation, it is interesting to see whether it can provide estimates for the activation energy for chain self-diffusion. Figure 15 shows the mean-squared displacement vs cycles for a system of 32 chains of 14 mers (n-tetradecane) at 0.1 MPa, and 293, 353, and 450 K. The slope of these mean squared displacement curves is found to follow an Arrhenius temperature dependence from which one can extract an activation energy of $E_{act} \approx 15.61$ kJ/mole. The experimental value for the activation energy of self-diffusion on n-tetradecane is 14.99 kJ/mole [41].

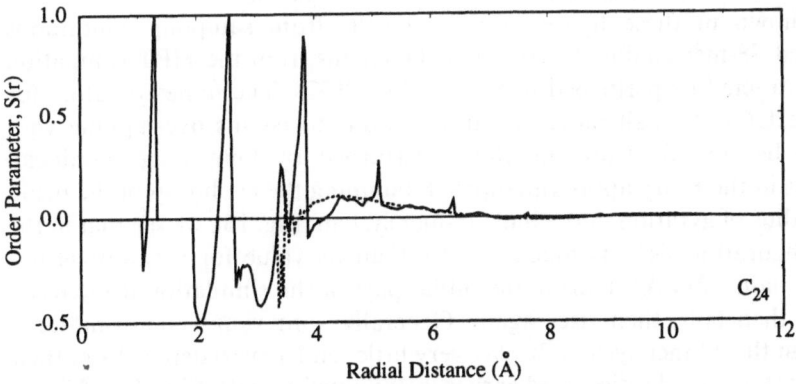

Fig. 14. Order parameter displaying bond orientational tendencies for 32-chain 24-mer (n-tetra-cosane) system at 450 K and 0.1 MPa. *Dashed curve:* inter-chain, *Solid curve:* total

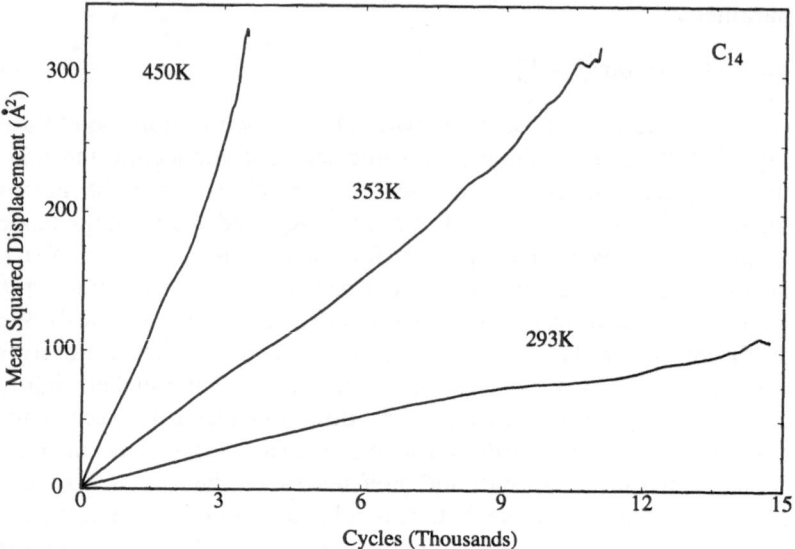

Fig. 15. Mean-squared displacement of chain centers of mass from simulations of 32 14-mer chains (*n*-tetradecane) at 0.1 MPa and 293 K, 353 K, and 450 K. The logarithm of the initial slopes, when plotted vs inverse temperature, gives an activation energy of 15.61 kJ/mol

3.4 Long-Chain System

Figures 9b and 11b show the intra-chain pair density function and the inter-chain pair distribution function, respectively, for a system of 10 chains of 78 units each ("polyethylene") at 450 K and 0.1 MPa. We see in Fig. 9b the increase in the size and density of the segment cloud with increase in chain length relative to the *n*-tetracosane system, Fig. 9a. In Fig. 11b the "correlation hole" now appears much stronger than for the *n*-tetracosane system.

Also shown in these figures are the results from sampling continuous unperturbed 78-mer chains. We see that the chains from the NPT simulation appear contracted or perturbed relative to the CUCs. The nonzero values for $\omega(r)$ from CUCs at small radial distance are due to chains overlapping with themselves because they are literally unperturbed by long-range (nonlocal) interactions in the Flory approximations. Examining the evolution of the mean squared radius of gyration in the bulk 78-mer system (Fig. 16), we see that in the initial configuration $\langle s^2 \rangle$ is much smaller than its value for continuous unperturbed chains, 219 Å^2. During the initial part of the simulation it increases somewhat, then falls, then rises again. Generally, long-range conformational properties in the 78-mer system display very little tendency to depart from their initial characteristics. As discussed earlier in conjunction with Figs. 6 and 7, the chains in this high-molecular weight system do not equilibrate conformationally even though almost 32 000 cycles have been performed (compared to only

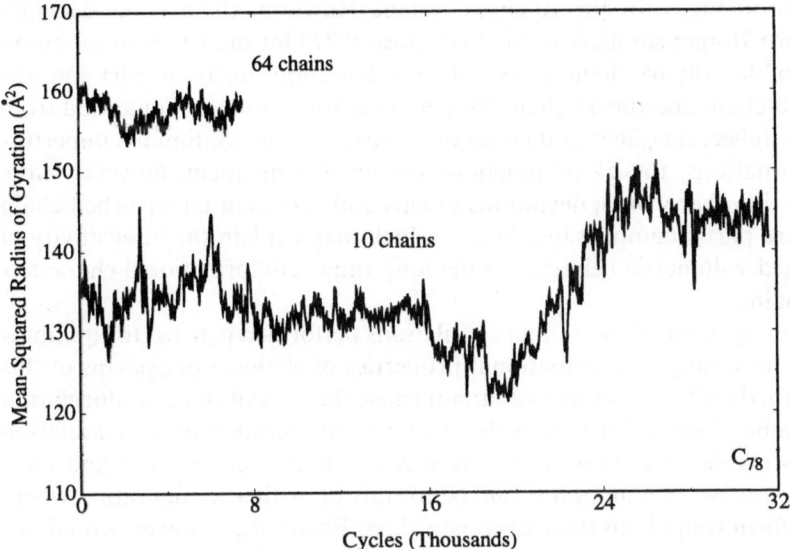

Fig. 16. Plot of the mean-squared radius of gyration vs Monte Carlo cycles for an NPT Monte Carlo of 78-mer ("polyethylene") system at 450 K and 0.1 MPa. The value from the sampling of continuous unperturbed chains is ≈ 219.2 Å^2

$\sim 20\,000$ cycles in the n-tetracosane system). This reflects the very sharp increase in the longest relaxation time of the melt with increasing chain length.

In order to explore conformational equilibration further and to study possible system size effects on the results of the long-chain simulation, we undertook a simulation of a system of 64 chains, each 78 mers long, at 0.1 MPa and 450 K. The edge length of the cell for this "superbox" system is approximately 53 Å as compared with 28 Å for the smaller-sized 10-chain system; a majority of the site pairs do not interact. The efficient simulation of such a large system requires the use of neighbor lists and cell lists. By implementing the algorithm outlined in Appendix A, we were able to reduce greatly the CPU time and memory requirements for the superbox Monte Carlo simulation. The initial configuration for this simulation was obtained by minimizing an 8-chain structure of 78-mer chains at the average density obtained from the NPT simulation of the 10-chain system at T = 450 K. An 8 million configuration NVT simulation was then performed on this system in order to remove it from its minimum energy well. Finally, this equilibrated 8-chain system was replicated 8 times to form the larger 64-chain superbox system. A 40 million configuration (7500 cycle) NPT simulation was undertaken on the superbox to allow it to approach equilibrium.

The evolution of $\langle s^2 \rangle$ in the superbox system is shown in Fig. 16. The values of $\langle s^2 \rangle$ are substantially larger than those of the 10-chain system but still short of the value for the continuous unperturbed chains; again, they show very little tendency to change away from the initial configuration. The autocorrelation function of the end-to-end unit vectors of the chains showed similar slow decay as was seen in Fig. 7 for the 10-chain system. However, the average densities from the two 78-mer simulations are very close: 0.773 for the 10-chain system vs 0.769 g/cm^3 for the 64-chain system. The distribution functions $\omega(r)$ and $g(r)$ from the 10-chain and the 64-chain 78-mer simulations differed somewhat from each other, reflecting different degrees of departure from continuous unperturbed conformations; the $g^{tot}(r)$ functions obtained from them, however, were quite close; in other words, deviations of $\omega(r)$ and $g(r)$ from unperturbed chain behavior are partly compensated in g^{tot}, which may explain the insensitivity of the predicted volumetric behavior to the long-range conformational characteristics of chains.

In summary, none of the 78-mer simulations performed here led to equilibration of the long-range conformational properties of chains. For systems of this chain length, the CONROT moves cannot cause the overall chain conformation to depart much from what it is in the starting configuration of the simulation within reasonable CPU time. The only way to obtain correct $\langle r^2 \rangle$ and $\langle s^2 \rangle$ values from the MC simulation would be to start off with a configuration where chains conform roughly to these average values. Even this, however, would not ensure adequate sampling of the entire distribution of chain conformations, which is extremely broad at these chain lengths. To sample this distribution adequately, one would need to conduct a very large number of simulations initiated at widely different starting configurations, or to devise a novel MC scheme that allows much more vigorous sampling of long-range chain conformational characteristics. We are currently working on such a new scheme. It should be noted, however, that properties depending primarily on total (inter- and intramolecular) short-range structure and interactions in the system, such as the melt density, are not significantly affected by the lack of long-range conformational equilibration; it is thus legitimate to use the MC simulations described above for the purpose of estimating such properties from a molecular-level description of the system.

4 Continuum Mean Field Theory

We now construct a continuum mean field theoretical approach for the PVT properties of bulk polymer melts that relies on the same detailed model for the chain geometry and energetics as was used in the simulations. Following the tradition of perturbation theory, we distinguish between repulsive and attractive contributions to the potential, recognizing that the former govern bulk structure. In our treatment of repulsive interactions we use an equivalent hard-sphere diameter $d^{HS}(T)$ for mers in place of the soft repulsive core of the Lennard-Jones potential; d^{HS} is determined from the parameters of the Lennard-Jones potential through the Barker-Henderson perturbation scheme [42]; other researchers have suggested that this may be more successful than the Weeks-Chandler-Andersen perturbation method for polymer systems [43, 44].

We start with the osmotic equation of state of Dickman and Hall [1], which expresses the pressure in terms of the insertion probability of a chain in the melt. While this equation is exact, it is not useful unless an expression for the insertion probability is available. In the definition of the insertion probability we separate attractive and repulsive contributions to the non-bonded interaction potential and treat the attractive contribution in a mean field sense (i.e., replace by its average over all configurations of the multichain melt). In this way, the pressure expression is straightforwardly separated into a repulsive and an attractive part. In calculating the repulsive part of the pressure we invoke Honnell and Hall's Generalized Flory-Dimer approximation [2], treating chains as collections of fused hard spheres of diameter $d^{HS}(T)$. Implementation of the Generalized Flory-Dimer approximation in our system requires the computation of the volume and excluded volume of molecules consisting of arbitrarily configured fused spheres. We have developed a fast, analytical algorithm that can address this problem in its full generality [45] and do not have to resort to approximate volume estimation methods based on Monte Carlo (cf. [43]). In calculating the attractive part of the pressure, knowledge of the inter-chain mer-mer pair correlation function is necessary. We determine this using PRISM [5] along with the intra-chain structure obtained from our sample of continuous unperturbed chains (CUC). Combining the attractive and repulsive parts of the pressure, we obtain a prediction for the pressure at given density and temperature, or, conversely, a prediction for the density at given pressure and temperature which we can compare against computer simulation results and experimental data. A detailed formulation of our continuum mean field theoretical approach is presented in Appendix B.

The mean field theory predictions for n-tetracosane and the 78-mer "polyethylene" systems at 450 K are shown in Fig. 8. For the short-chain system the agreement between simulation and mean field theory is reasonable. Note that, since the same model for chain geometry and energetics was used in the simulation and in the mean field theory, and since the simulation results are essentially exact, this comparison is a test of the assumptions incorporated in the mean field theory. For the longer-chain system the mean field theory shows the expected densification upon polymerization, but unfortunately overestimates this significantly. The isothermal compressibility (slope of the isotherms depicted in Fig. 8) is predicted rather well by the mean field theory for both chain lengths. We should emphasize that only the chain length changed in the mean field theory calculation in going from the 24-mer to the 78-mer system; the potential parameters were not changed.

5 Conclusions

We have presented continuum NPT Monte Carlo simulation results for alkane and polymer systems composed of chains up to 78 units long, using a realistic model for the chain geometry and energetics. We have found very good agreement between experimental data and simulation averages for PVT properties and structure factors.

In addition to the bulk simulation studies we have also developed a fast and efficient method of generating continuous unperturbed chains. This method is 300 times faster than the bulk MC simulation method. Comparisons between our bulk MC results and continuum unperturbed chains allowed for a direct test of the Flory random coil hypothesis. In those bulk MC simulations that equilibrate conformationally, chains in the melt were found to be indistinguishable from continuous unperturbed chains to an excellent approximation at all length scales.

The intra-chain pair density functions obtained from both the bulk simulation and the continuous unperturbed chains were used as input to the polymer-RISM integral equation for estimating the intermolecular pair distribution function g(r) (using a soft-Percus Yevick closure). We found that PRISM underpredicts the first peak in g(r), while also overpredicting the steepness of the rise to the first peak.

The long-range conformational properties of chains from the 10-chain 78-mer bulk simulation appeared to be perturbed relative to continuous unperturbed chains. To test whether this might be due to system size effects, we performed a simulation on a much larger system consisting of 64 78-mer chains. This required the development of a neighbor list algorithm suitable for Monte Carlo simulations. By comparing the results from the 10-chain and 64-chain 78-mer systems, we established that the apparent departure from unperturbed conformational statistics is due to a very slow relaxation of the long-range conformational characteristics of chains, which prevents average chain dimensions from departing significantly from their values at the beginning of the simulation. In other words, long-chain systems are not conformationally equilibrated in the course of a single bulk MC run, and new MC methodology is needed to achieve such equilibration. Short-chain systems (e.g., C_{24}), on the other hand, are thoroughly equilibrated at all length scales. The PVT properties from the large-box and the small-box 78-mer chain simulations agreed well with each other and with experiment and appeared unaffected by the lack of long-range conformational equilibration.

A mean field theoretical approach to the volumetric properties of polymer liquids that is usable with realistic chain models was developed by combining the random coil hypothesis for conformations in the bulk, Generalized Flory theory, and Polymer RISM. This approach estimates the contribution to pressure from repulsive forces by invoking the Generalized Flory Dimer Ansatz for the repulsive insertion probability of chains. The volumes and excluded volumes needed for this purpose were calculated exactly using a new, analytical algorithm. Attractive contributions to the pressure were obtained through the intermolecular pair distribution functions computed by Polymer RISM on the assumption of unperturbed chain conformations. Predictions of the mean field approach were reasonable for the short-chain system but overestimated the density significantly in the long-chain system. The major source of error in the mean field approach is thought to lie in the Generalized Flory Dimer Ansatz. Future work towards improvement of the theory should focus on this aspect, perhaps by using a fluid higher than the dimer as reference for estimating the chain insertion probability.

Acknowledgements. We are grateful for financial support from the B.F. Goodrich Company. DNT acknowledges the National Science Foundation for a Presidential Young Investigator Award, grant No. DMR-8857659. We thank the San Diego Supercomputer Center for a generous allocation of computer time.

Appendix A. Neighbor Lists

Grest and co-workers [46] have made a comprehensive study of the most efficient energy calculation method for use in MD of systems with a large number of particles. They discuss a merging of the two most popular list methods, neighbor lists and linked cell lists. Extending the algorithm of Grest and co-workers to MC simulations, however, requires care, since some of the conclusions they draw only apply to molecular dynamics simulations. We have developed the following algorithm for using neighbor lists in MC simulations of polymer molecules.

First, form the cell and neighbor lists.

1. The simulation cell is broken in sub-cells such that each has an edge length of at least the cut-off of the potential plus a "skin." For the 53 Å cell, $125 = 5^3$ sub-cells were used.

2. Form a list containing the sub-cell in which each site resides.

3. Form a list containing the sites that belong to each sub-cell.

4. Use these two cell lists to help accelerate the formation of a neighbor list for each site. This list includes all sites within the potential cut-off plus the skin.

Second, do the following for each attempted elementary move.

1. Find position of all displaced sites.
2. Find the maximum squared displacement of the sites, $(d_{max})^2$.
3. Compare $(d_{max})^2$ with squared skin thickness, $(r_{skin})^2$.
   - If $(d_{max})^2 < (r_{skin})^2$ then use existing neighbor list.
   - Else skin is broken, perform a full calculation of the energy (without using lists).
4. If, and only if, the elementary move is accepted and the skin was broken, then rebuild neighbor lists.

Appendix B. Formulation of the Mean Field Model

Consider a system of N n-mer chains in volume V at temperature T. To specify completely a chain i of length n in configuration space within our model representation, one needs to fix an (n + 3)-dimensional vector of degrees of freedom, q_i. As such, one can choose to use three Cartesian coordinates of a reference mer (e.g., the first mer) of the chain, three Eulerian angles specifying the overall orientation of the chain with respect to the laboratory frame of reference, and the sequence of all torsion angles of all skeletal bonds of the chain, except for the first and last. The set of configurational degrees of freedom of all chains will be collectively denoted as $\Omega_N \equiv \{q_1, q_2, ..., q_N\}$. The starting point for the formulation of the osmotic equation of state is the definition of the insertion probability (or factor) of a chain

$$p_n(N, V, T) \equiv \langle\langle \exp[-\beta v^{test}(q_{N+1}; \Omega_N)]\rangle_{\exp[-\beta v^{intra}(q_{N+1})]}\rangle_{\rho^{NVT}(\Omega_N)} . \qquad (9)$$

The conceptual process considered in Eq. (9) is one of inserting a "test" chain (chain number $N + 1$) into a system of N chains at V and T. v^{test} is the total potential energy felt by this test chain as a result of inter-chain interactions with the N chains already existing in the system; β stands for $1/k_B T$. The inner ensemble average in Eq. (9) is taken over all configurations $\{q_{N+1}\}$ of the test chain at fixed Ω_N; each configuration is weighted by the Boltzmann factor of its intra-molecular (bonded and nonbonded) energy. The outer ensemble average is taken over all configurations $\{\Omega_N\}$ of the N chains pre-existing in the system; each configuration is weighted by the Boltzmann factor of the total potential energy of the N chains. By its definition the insertion probability is related to the configurational integrals $Z_i(V, T)$ for i chains in volume V and temperature T through the equation

$$p_n(N, V, T) = \frac{Z_{N+1}(V, T)}{Z_N(V, T) Z_1(V, T)} \qquad (10)$$

with

$$Z_i(V, T) \equiv \int \ldots \int \exp[-\beta v(\Omega_i)] d\Omega_i \qquad (11)$$

where $v(\Omega_i)$ denotes the total (inter- and intra-chain) energy of a system of i chains in configuration Ω_i. By invoking the relation between configurational integral and Helmholtz energy, it is readily shown that, in a system of N chains in volume V at temperature T,

$$\beta P = \rho + \lim_{\substack{N, V \to \infty \\ \rho = const.}} \frac{\partial}{\partial V}\left[\sum_{i=1}^{N-1} \ln p_n(i, V, T)\right]. \qquad (12)$$

In the thermodynamic limit ($N, V \to \infty, \rho \equiv N/V =$ constant), $p_n(\rho, T)$ becomes an intensive property that is immediately related to the excess chemical potential, that is, to the chemical potential of a chain relative to an ideal gas of chains at the same temperature and density

$$\mu_n^{ex}(\rho, T) \equiv \mu_n(\rho, T) - \mu_n^{ig}(\rho, T) = -k_B T \ln p_n(\rho, T) . \qquad (13)$$

Through straightforward thermodynamic analysis, one can show that

$$\left(\frac{\partial \ln p_n}{\partial \rho}\right)_T = \frac{1}{\rho}\left[1 - \beta\left(\frac{\partial P}{\partial \rho}\right)_T\right]. \qquad (14)$$

The differential Eq. (14) relates the insertion probability with the pressure. When integrated for P, it gives a continuous analogue of Eq. (12) ("osmotic equation of state") that allows the calculation of PVT properties from a knowledge of the density-dependence of the insertion probability. When integrated for p_n, it provides an expression for the insertion probability if the equation of state is known.

Equation (12) is an exact expression for the pressure in terms of the insertion probabilities. It is a form of the "osmotic equation of state." The underlying physical picture is one of "building up" the system of N chains by adding chains one by one in the volume V. To make Eq. (12) practically useful, we need a way to calculate the insertion probabilities $p_n(i, V, T)$.

We separate the total potential energy function $v(\Omega_N)$ into a "repulsive" part $v^{rep}(\Omega_N)$ containing bonded interactions and the repulsive contribution to all nonbonded intra-chain and inter-chain interactions, and into an "attractive" part $v^{att}(\Omega_N)$, incorporating the attractive contributions to all nonbonded intra-chain and inter-chain interactions. Further-more, we introduce a mean field approximation in the treatment of $v^{att}(\Omega_N)$, in the following sense: in place of the configuration-dependent quantity $v^{att}(\Omega_N)$ we will use its value averaged over all configurations $\{\Omega_N\}$ at the prevailing density and temperature

$$v(\Omega_N) \approx v^{rep}(\Omega_N) + \langle v^{att}(N, V, T) \rangle .$$ (15)

Defining the repulsive insertion probability by exact analogy to Eq. (10),

$$p_n^{rep}(N, V, T) = \frac{Z_{N+1}^{rep}(V, T)}{Z_N^{rep}(V, T) Z_1^{rep}(V, T)}$$ (16)

with

$$Z_i^{rep}(V, T) = \int \cdots \int \exp[-\beta v^{rep}(\Omega_i)] d\Omega_i$$ (17)

we can write

$$p_n(i, V, T) = p_n^{rep}(i, V, T) \exp[-\beta(\langle v^{att}(i + 1, V, T) \rangle$$
$$- \langle v^{att}(i, V, T) \rangle - \langle v^{att}(1, V, T) \rangle)]$$ (18)

as a result of which the sum of logarithms of insertion probabilities appearing in the osmotic equation of state, Eq. (12), becomes

$$\sum_{i=1}^{N-1} \ln p_n(i, V, T) = \sum_{i=1}^{N-1} \ln p_n^{rep}(i, V, T)$$
$$- \beta[\langle v^{att}(N, V, T) \rangle - N \langle v^{att}(1, V, T) \rangle] .$$ (19)

Substituting Eq. (19) into Eq. (12) and taking the thermodynamic limit results in the pressure expression $P = P^{rep} + P^{att}$ where

$$\beta P^{rep} = \rho[1 - \ln p_n^{rep}(\rho, T)] + \int_0^\rho \ln p_n^{rep}(\rho', T) d\rho'$$ (20)

$$P^{att} = \frac{(\rho n)^2}{2} \int\limits_{\sigma}^{\infty} 4\pi r^2 u(r) g_{NB}^{tot}(r; \rho, T) dr \ . \tag{21}$$

In Eq. (21), g_{NB}^{tot} describes the distribution of separations of pairs of skeletal mers interacting through nonbonded forces. By definition, the number of non-boned skeletal mers we expect to find with their centers in a spherical shell of radii r and r + dr around a given mer is $4\pi r^2 (\rho n) g_{NB}^{tot}(r; \rho, T) dr$. (The mers in the shell interact with the central mer via the nonbonded interaction potential u(r); they may belong to the same chain as the central mer, or to other chains.) The density dependence of g_{NB}^{tot} is expected to be weak at the densities encountered in a bulk melt, and has been neglected in deriving Eq. (21). In view of Eqs. (20) and (21), the problem of calculating the pressure at given temperature and density reduces to one of finding the repulsive insertion probability $p_n^{rep}(\rho, T)$ and the total nonbonded pair distribution function g_{NB}^{tot}.

We express $p_n^{rep}(\rho, T)$ by invoking Honnell and Hall's Generalized Flory Dimer approximation for hard chains [2]. In this approximation, one considers a hard chain being inserted in a melt of hard chains one mer at a time. The probability of inserting a mer or pair of mers in the chain fluid is approximated by the corresponding probabilities of inserting a hard monomer in a fluid of hard monomers and a hard dimer in a fluid of hard dimers characterized by the same packing fraction as the chain fluid (GFD-A of [43])

$$p_n^{rep}(\rho, T) = p_1^{rep}(\eta) \left[\frac{p_2^{rep}(\eta)}{p_1^{rep}(\eta)} \right]^{Y_n + 1} \tag{22}$$

where $p_1^{rep}(\eta)$ is the insertion probability of a hard monomer in a sea of hard monomers at packing fraction η, $p_2^{rep}(\eta)$ is the insertion of a hard dimer in a sea of hard dimers at packing fraction η, and

$$Y_n + 1 \equiv \frac{\langle v_e(n) \rangle - \langle v_e(1) \rangle}{\langle v_e(2) \rangle - \langle v_e(1) \rangle} \tag{23}$$

$$\eta = \frac{\text{total volume of chains in chain fluid}}{V} = \rho \langle v(n) \rangle \ . \tag{24}$$

Thus, two important geometrical parameters of chains appear in the Generalized Flory Dimer approximation Eq. (22): the volume of a chain v(n) and the excluded volume of a chain $v_e(n)$. Both are weakly dependent on temperature. In defining v(n), the n-mer chain is considered as a succession of fused hard spherical segments of diameter $d^{HS}(T)$. The excluded volume $v_e(n)$ is the volume of the body formed by inflating all segments of a chain to a diameter $2 \times d^{HS}(T)$; because of the repulsive segment-segment interactions at short distances, this volume is inaccessible to the centers of segments belonging to other chains. These geometric parameters are obtained as averages over our sample of continuous unperturbed chains.

We now turn to the total nonbonded pair distribution functions $g_{NB}^{tot}(r, \rho, T)$ appearing in the attractive pressure term Eq. (21). Beyond a distance r_c that is sufficiently larger than σ, g_{NB}^{tot} in the bulk melt is indistinguishable from unity (Fig. 12). In practice, $r_c = 10$ Å is a satisfactory choice. For any distance r, g_{NB}^{tot} can be resolved into inter-chain and intra-chain terms (cf. compare Eq. (6))

$$g_{NB}^{tot}(r; \rho, T) = g(r; \rho, T) + \frac{\omega_{NB}(r)}{\rho n} . \tag{25}$$

The inter-chain pair distribution function $g(r; \rho, T)$ has been discussed in Sect. 3.3. The average nonbonded intermolecular pair density function $\omega_{NB}(r)$ has units of inverse volume. If one considers a spherical shell of radii r and $r + dr$ around a chain segment in the system, the quantity $4\pi r^2 \omega_{NB}(r) dr$ equals the number of centers of segments in the shell belonging to the same chain as the reference segment and interacting through nonbonded forces with it. By definition, $\lim_{r \to \infty} g^{inter}(r; \rho, T) = 1$ and $\lim_{r \to \infty} \omega_{NB}(r) = 0$. The intra-chain pair density function $\omega_{NB}(r)$ can, of course, be readily accumulated from our sample of continuous unperturbed chains.

Using Eq. (6) in Eq. (21), we obtain

$$P^{att} = \frac{(\rho n)^2}{2} \int_{\sigma}^{r_c} 4\pi r^2 u(r) g(r; \rho, T) dr + \rho \langle v_{NB}^{intra}(\sigma \to r_c) \rangle \tag{26}$$

$$+ \frac{(\rho n)^2}{2} \int_{r_c}^{\infty} 4\pi r^2 u(r) dr . \tag{27}$$

In the above equation, $\langle v_{NB}^{intra}(\sigma \to r_c) \rangle$ stands for the average nonbonded intra-chain potential energy of a chain due to attractive interactions at distances between ρ and r_c; it is readily calculable from our sample of unperturbed chains. The only unknown quantity is $g(r; \rho, T)$, needed between distances of σ and r_c, which we find using PRISM [5] as described in the text.

6 References

1. Dickman R, Hall C (1986) J Chem Phys 85: 4108
2. Honnell K, Hall CK (1989) J Chem Phys 90: 1841
3. Flory PJ (1953) Principles of polymer chemistry. Cornell University Press, Ithaca
4. Schweizer KS, Curro JG (1987) Phys Rev Lett 58: 246
5. Curro JG, Schweizer KS (1987) Macromolecules 20: 1928
6. Honnell KG, McCoy JD, Curro JG, Schweizer KS, Narten AH, Habenschuss A (1991) J Chem Phys 94: 4659
7. Chandler D (1982) In: Montroll E, Lebowitz J (eds) The liquid state of matter: Fluids, simple and complex, North Holland, p 275 (Studies in Statistical Mechanics, vol VIII)
8. Yethiraj A, Hall C (1991) J Chem Phys 95: 1999
9. Yethiraj A, Hall C (1991) J Chem Phys 95: 8494
10. Curro JG, Schweizer KS, Grest GS, Kremer K (1989) J Chem Phys 91: 1357
11. Allen M, Tildesley D (1987) Computer Simulation of Liquids, Oxford Science Publications, Oxford
12. Rigby D, Roe R-J (1987) J Chem Phys 87: 7285
13. Rigby D, Roe R-J (1988) J Chem Phys 89: 5280
14. Rigby D, Roe R-J (1989) Macromolecules 22: 2259
15. Brown D, Clarke JHR (1991) Comput Phys Commun 62: 360
16. Brown D, Clarke JHR (1991) Macromolecules 24: 2075
17. Takeuchi H, Roe R-J (1991) J Chem Phys 94: 7446
18. Takeuchi H, Roe R-J (1991) J Chem Phys 94: 7458
19. Boyd RH, Pant PVK (1991) Macromolecules 24: 6325
20. Pant PVK, Han J, Smith GD, Boyd RH (1993) J Chem Phys 99: 597
21. Yoon DY, Smith GD, Matsuda T (1993) J Chem Phys 98: 10037
22. Smith GD, Yoon DY (1993) J Chem Phys, submitted
23. Vacatello M, Avitabile G, Corradini P, Tuzi A (1980) J Chem Phys 73: 548
24. Boyd RH (1989) Macromolecules 22: 2477
25. de Pablo JJ, Laso M, Suter UW (1992) J Chem Phys 96: 2395
26. Siepmann JI, Frenkel D (1992) Molec Phys 75: 59
27. Dodd LR, Boone TD, Theodorou DN (1993) Molec Phys 78: 961
28. Leontidis EM, de Pablo JJ, Laso M, Suter UW (1993) Submitted to Advan Polym Sci
29. Gō N, Scheraga HA (1976) Macromolecules 9: 535
30. Theodorou DN, Boone TD, Dodd LR, Mansfield KF (1992) Makromol Chem, Theory Simul 2: 191
31. de Pablo JJ (1992) Personal Communication
32. Ryckaert JP, Bellemans A (1975) Chem Phys Lett 30: 123
33. Theodorou DN, Suter UW (1985) Macromolecules 18: 1467
34. Mansfield KF, Theodorou DN (1990) Macromolecules 23: 4430
35. Dodd LR, Theodorou DN (1992) Polym Prepr (Am Chem Soc, Div Polym Chem) 33: 645
36. Flory PJ (1969) Statistical mechanics of chain molecules. Wiley, New York
37. Ciccotti G, Ryckaert J (1986) Comput Phys Rep 4: 345, Appendix A by H.J.C. Berendsen
38. Dee G, Ougizawa T, Walsh D (1992) Polymer 33: 3462
39. Gillan M (1979) Molec Phys 38: 1781
40. Habenschuss A, Narten A (1990) J Chem Phys 92: 5692
41. Ertl H, Dullien F (1973) AIChE J 19: 1215
42. Barker J, Henderson D (1967) J Chem Phys 47: 4714
43. Yethiraj A, Curro JG, Schweizer KS, McCoy JD (1993) J Chem Phys 98: 1635
44. Curro JG, Yethiraj A, Schweizer KS, McCoy JD, Honnell KG (1993) Macromolecules 26: 2655
45. Dodd LR, Theodorou DN (1991) Molec Phys 72: 1313
46. Grest GS, Dünweg B, Kremer K (1989) Comput Phys Commun 55: 269

Received September 27, 1993

A Critical Evaluation of Novel Algorithms for the Off-Lattice Monte Carlo Simulation of Condensed Polymer Phases

E. Leontidis, J.J. de Pablo, M. Laso, and U.W. Suter
Institut für Polymere, Eidgenössische Technische Hochschule (ETH),
Zürich, Switzerland

Novel composite algorithms for efficient off-lattice simulations of dense polymer phases are implemented and evaluated in the case of polyethylene in the united and explicit atom approximations. The simulation algorithms are based on combinations of traditional methods with recently developed Monte Carlo moves. The classical Metropolis Monte Carlo (MMC) and Reptation techniques are supplemented with the Continuum-Configuration Bias (CCB) and the Concerted-Rotation (CONROT) methods. Several extensions of the CONROT method, involving the simultaneous coordinated displacement of four to seven chain backbone sites, are developed and tested in this paper. CONROT-based methods enhance the performance of the algorithm at the level of local segmental motions, whereas the CCB component is important for the convergence of global properties, such as the relaxation of end-to-end distance vectors. The present composite algorithms are able to reproduce quite efficiently equilibrium thermodynamic properties, such as the density and the radial distribution function in the liquid state. However, they fail to generate completely equilibrated melts of long chains (polyethylene with 70 carbon atoms) at the molecular level. In spite of this shortcoming, we believe that these methods constitute the most promising currently available tools for the off-lattice simulation of realistic models of polymer melts and glasses. A satisfactory treatment of relatively long polyethylene chains (with up to 40 carbon atoms in the backbone, according to our estimates) is possible at experimental melt densities, both in the NVT and NPT ensembles.

Advances in Polymer Science, Vol. 116
© Springer-Verlag Berlin Heidelberg 1994

1 Introduction

A fundamental goal of Materials Science and Technology is to relate the microscopic (molecular) structure of a material to its macroscopic (structural, mechanical, dynamic, thermodynamic) properties, and to be able to predict the change of these macroscopic properties upon the introduction of chemical modifications at the microscopic level. This is one of the key-problems of contemporary Materials Science, and in its solution computer simulation has been established as a powerful tool [1–4].

Polymers are an important class of modern materials. Computer simulation plays an ever-expanding role in the design of new polymers, and the understanding of the properties of existing polymers. Polymer simulations fall into the broad groups of minimum energy methods, the classical version of which is usually termed Molecular Mechanics [2, 3], and ensemble methods that are distinguished according to the techniques of ensemble creation as Monte Carlo (stochastic, MC) [2, 3, 6–9], and Molecular Dynamics methods (deterministic, MD) [4–6]. In the context of this report, only interactions specified in terms of classical mechanics are of interest. The mentioned classification is not exclusive, as many composite methods have appeared in recent years, which attempt to combine the advantages of more than one simulation method. Examples of such hybrid methods are Brownian Dynamics [5, 6, 10–12], Smart Monte-Carlo [13, 14, 37], Dynamic Monte-Carlo [15a, b], and Hybrid Monte-Carlo [16a, b, c].

Minimum energy methods generate static configurations or "snapshots", from which one can deduce equilibrium, thermodynamic and mechanical properties of realistic systems, when entropy does not play a significant role. This is believed by many to be the case for glassy materials, for which these techniques have provided a wealth of important information in the past decade [2, 3, 17–20]. In spite of some very successful attempts to study entropy-dependent material properties, such as the solubility and diffusion of small guest molecules into polymer matrices [21, 22], minimum energy methods alone cannot properly address the effects of thermal motion on important polymer properties, and do not provide accurate estimates for the partition function of the material.

Molecular Dynamics has recently become a popular method for polymer simulations [4–6, 23–26]. It generates successive configurations of a small volume of material by integrating the equations of motion starting from some initial configuration for all the degrees of freedom of the system. The method has now been extended to simulate the canonical [4, 6, 27, 28], NPT [4, 6, 27, 29], and grand-canonical ensembles [30]. Its main advantage is the realistic treatment of molecular motion, which allows the convenient study of dynamic properties. Because of current computer limitations however, the integration of the equations of motion can follow the temporal evolution of a complex polymeric system for a few nanoseconds at the most [4, 5]. While this is satisfactory for the study of certain fast relaxation processes, the majority of important dynamic processes in polymeric materials are in the microsecond to millisecond to second range [31, 32], thus rendering MD unsuitable for the study of many polymer properties.

Monte-Carlo methods provide an important alternative to minimum energy methods and MD, since they are in principle capable of providing the partition function, and thus entropy-dependent material properties [6–9, 27]. Being purely stochastic, these methods do not provide explicit real-time information. However, MC methods can in principle be "calibrated", using a variety of appropriate criteria and techniques [33]. An important advantage of MC methods is that the MC moves can be biased to place different emphasis on different degrees of freedom [13, 15]. This is impossible for MD, which treats all degrees of freedom on an equivalent basis. In dense polymer systems the goal of efficient phase-space sampling is a most difficult one, however, since polymer MC simulations suffer from frequent bottlenecks and very slow relaxations between successive configurations, especially at high densities [7, 9, 27, 33]. It is important to develop new MC methods that can sample the configuration space of a polymeric system efficiently, and are amenable to unequivocal calibration with respect to time, at least for selected dynamic properties of interest.

Significant advances in polymer MC simulations have been made in the past decade. In particular, the introduction of many new sampling techniques [7, 9, 34] has enhanced tremendously the efficiency of lattice polymer simulations, which were originally quite inefficient at high chain densities [7, 9]. Lattice-based MC polymer simulations of model systems, such as self-avoiding random walks (SAW's), offer many advantages [7, 9]. They are simple to implement, and often provide a startling variety of fundamental information on polymer static and dynamic behavior in dense phases. The main drawback of lattice-based simulations is that the lattice representation cannot emulate in detail realistic polymer structures and, consequently, these simulations cannot a priori become useful tools for the design of new materials. The recently introduced "Bond-Fluctuation Method" [35, 36] attempts to bridge the gap between lattice models and chemically realistic polymer structures, but it must still be demonstrated that this method can reproduce consistently a large variety of polymer properties with a minimum of parametrizations.

Little off-lattice polymer simulation work has been done to date, hence, efficient continuum MC methods are scarce. Simple Metropolis MC samples configuration space in a very inadequate way, because of the bottlenecks produced by high density and chain entanglements [7, 9, 41]. Reptation has been used successfully in the past for the simulation of short alkane chains in continuum space [37, 40], but is very ineffective for the treatment of long chains (with more than thirty backbone sites), since it relies on the existence of a relatively large concentration of chain ends. Force-Bias MC was shown to be inefficient for proteins [41].

Significant progress has been made in the last few years by a combination of methods originally developed for lattice simulations, and of geometric arguments. The Continuum-Configurational-Bias method (CCB) has attracted much attention, because of its great power and flexibility, and the very intuitive use of available "unoccupied" volume [42–45]. A number of promising results have already been obtained with this method, mostly for model systems [42, 45]. Configurational Bias methods can be applied to both lattice and off-lattice polymer systems. However, as is the case with simple Reptation, the CCB method also requires a relatively high concentration of chain ends to be efficient. It was found inefficient for the treatment of dense systems with long chains, and for chemically realistic polymers with bulky side groups (e.g., polypropylene, polystyrene) [46].

A novel Monte-Carlo algorithm for off-lattice systems was introduced very recently by Dodd, Boone, and Theodorou [47]. The Concerted-Rotation (CONROT) method is based on the classic Gō-Scheraga algorithm for the calculation of chain and ring closures in proteins [48]. Dodd et al. have reformulated the equations of the method and have created a powerful MC algorithm that satisfies microscopic reversibility [47]. Their original calculations have indicated that CONROT can simulate polyethylene and polypropylene melts, but, as originally formulated, it is not more efficient than Molecular Dynamics for the same systems [47]. However, these authors have suggested ways to improve the efficiency of CONROT. Probably the most attractive feature of this novel MC scheme is that it is quite suitable for simulation of glassy polymers, especially for long chains, in contrast to Reptation and CCB that operate on chain ends. In addition, the method involves very localized "pseudo-dynamic" moves of a few consecutive backbone sites at a time, hence it can be time-calibrated in a simple way [47].

The present work was motivated by these recent developments in off-lattice polymer simulations. It was undertaken in the hope that a suitable combination of improved variants of the CCB and CONROT methods may form the basis for a simulation algorithm that can overcome many of the present obstacles, and can sample configuration space efficiently, even for dense phases of long chains or for chemically realistic structures with bulky side-groups. Our goals in this paper are (1) to provide simple extensions of the newly developed CONROT method and integrate those in a composite algorithm consisting of a wide variety of possible MC moves, and (2) to compare a number of currently available methods for the simulation of dense polymer phases in continuum space (off-lattice), and prove that a combination of the many new methods can enhance our capabilities of simulating these complex systems.

In the next section we describe the various elementary MC moves that we have used in more detail. Section 3 contains information on the intermolecular potentials and simulation protocols used. Section 4 presents the results that we have obtained so far for a number of different systems. The results are discussed, the different methods compared, and conclusions drawn as to the most suitable method in each individual situation. Section 5 contains our final conclusions and suggested directions for future work.

2 Simulation Methodologies

2.1 Simulations in Torsion Angle (Generalized Coordinate) Space

In simulations of dense polymer phases one must face the problems posed by a high-dimensional configurational space. Luckily, the structural complexity of polymeric systems, which arises from geometry and connectivity, provides a way to simplify the simulation problem. The number of degrees of freedom may be reduced considerably by applying geometric constraints and working with generalized coordinates [49–51]. Typically, one assumes that the bond lengths and angles between consecutive backbone atoms are fixed (rigid constraint model). Molecular geometry is then determined by the position and orientation of the chain start, and by the successive torsion angles along the chain backbone. For a polymethylene chain of N methylene units (a polybead molecule) the degrees of freedom are thus reduced from $3 \times N$ to $N + 3$. For more complex structures the reduction in number of degrees of freedom is even more significant.

Ignoring the degrees of freedom associated with fluctuations of bond lengths and angles gives rise to an important conceptual and methodological problem, since it has been proved by theory [50, 52, 53] and simulation [54–60] that the rigid model is not equivalent to the infinitely rigid limit of the flexible constraint model, in which bond lengths and angles are constrained close to their equilibrium values with appropriate constraining potentials. However, analytical theory and simulation of short chain molecules have also demonstrated that the discrepancies between the two models are generally not large [53, 55, 59]. In fact, the main effect of fixing bond lengths and bond angles appears to be a reduction of the vibrational frequencies of the "soft" modes associated with torsion angles [57, 59].

Thus, it is argued that the rigid constraint model can provide exact results for important structural properties [49–51]. This assumption has been adopted by workers in the area of off-lattice polymer simulations [45, 47, 51], and will be retained in this work. However, we wish to point out that the validity of this assumption has been tested mostly on short chain molecules [55, 56, 58–60], and on a small protein [57]. As our computational facilities and methods improve, and it becomes possible to work with denser systems of longer chains, this assumption must be scrutinized closely and carefully, since the approximation might deteriorate with chain length.

2.2 Basic Requirements of Monte Carlo Methods

The simplest Monte Carlo simulations are carried out by sampling trial moves for the degrees of freedom from a uniform distribution. In the simplest case of a canonical ensemble simulation (see Sect. 2.3 below) a degree of freedom is chosen at random and then displaced randomly. The amplitude of the displacement is sampled from a uniform distribution. The trial move is accepted or rejected according to an importance sampling scheme [27, 33].

Nonuniform or biased distributions for sample moves may also be used in Monte Carlo simulations, provided proper acceptance/rejection criteria are formulated, satisfying microscopic reversibility (this is an unnecessarily strong condition) [13, 14, 27, 61]. State "new" of the system is generated from state "old" with a probability given by

$$P_{acc} = \min\left(1, \frac{T_{new \to old} P_{new}}{T_{old \to new} P_{old}}\right) \tag{1}$$

where $T_{new \to old}$ is the transition probability of going from state "new" to state "old", and P_{new} is the probability (Boltzmann weight) of being in state "new".

It is also possible to perform simultaneous random or biased moves on many degrees of freedom. This was shown not to be advantageous for the simulation of liquid phases of simple molecules [27, 33]. However, polymeric systems are expected to behave differently. For example, in simple Lennard-Jones systems it was found that the maximum efficiency of a Metropolis MC algorithm (fastest sampling of configuration space) corresponded to an overall acceptance rate of roughly 50%. This was found not to be the case in MC simulations of a model protein, where maximum efficiency corresponded to an acceptance rate of only 15% [41, 51].

2.3 Metropolis Monte-Carlo (MMC) [27, 33]

This is the simplest method that we have implemented. Random uniform sampling ensures that

$$T_{old \to new} = T_{new \to old} . \tag{2}$$

Combining Eqs. (1) and (2) and assuming that the generalized momenta provide an approximately constant term to the system Hamiltonian [49], we obtain the well-known Metropolis criterion [27, 33, 61] for the probability of going from state "old" to state "new":

$$P_{acc} = \min\left(1, \exp(-\beta[U_{new} - U_{old}])\right) \tag{3}$$

where U is the configurational potential energy of the system, and $\beta = (k_B T)^{-1}$.

The present simulations are carried out in torsional angle space (Fig. 1a). Random displacements of the position of the chain start correspond to random translations of the chain. Random displacements of the three Euler angles that define the orientation of the chain start correspond to rotation of the molecule as a whole. It was found advantageous to randomly displace all the configurational variables of a single chain within the same MC move. To achieve

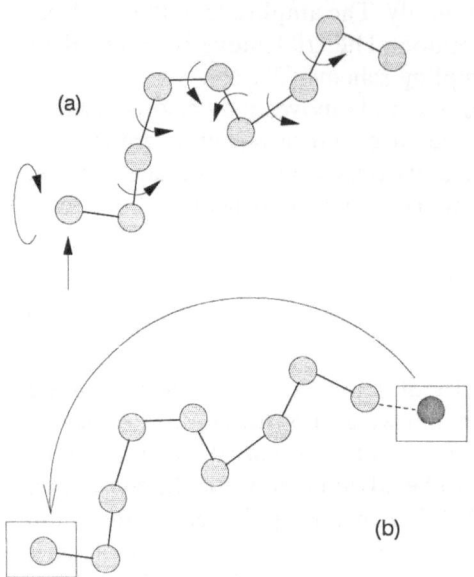

(a)

(b)

Fig. 1. a Schematic of random Metropolis Monte Carlo moves (MMC), corresponding to random translation, rotation of the chain as a whole, and changes in torsional angles. **b** Schematic of a reptation move

reasonable acceptance rates (20–30%), it was necessary to use very small displacements for the Euler and torsion angles of the chain backbone (the maximum displacement was set to 0.01–0.02 rad or 0.5–1 degree).

2.4 Simple Reptation [7, 9, 37–40]

Following De Gennes' theory of reptation in polymer melts [62], a reptation move was first introduced in lattice polymer simulations, and proved to be much more efficient than the simple Metropolis random move [63]. Reptation on a lattice is also more efficient than some existing kink-jump algorithms [7]. Reptation has been used successfully in off-lattice polymer simulations of model and realistic chains [37–40]. Implementation of reptation moves in continuum space simulations is straightforward (Fig. 1b). Each chain is considered in order, and one end is selected at random as the "head". The "head" site is erased and a new bond is generated at the other end of the chain with fixed bond length and angle. The torsion angle ϕ associated with the position of the new chain end can be drawn randomly from a uniform distribution or it can be chosen in a "smart way" [37, 40] by sampling the distribution

$$P_{sam}(\phi) = \exp(-\beta U_1(\phi)) \tag{4}$$

where U_1 is the part of the intramolecular potential energy that depends on torsion angle ϕ. If the new angle ϕ is chosen randomly from a uniform distribution, the acceptance/rejection criterion is identical to the standard

Metropolis criterion (Eq (3) above). If a biased choice is made, the bias must be removed by using a modified criterion [40]

$$P_{acc} = \min[1, \exp(-\beta \Delta U_2)] \tag{5}$$

where U_2 is the difference between the total potential energy and $U_1(\phi)$. Reptation can reproduce thermodynamic properties of a melt of tetracosane chains at realistic densities [39]. The efficiency of reptation methods decays fast as the chains become longer [40]. Strong cage effects appear in dense phases as well, to the effect that, while the acceptance rate is still considerable, the motions of a chain are restricted to a vibrational mode, in which only a limited number of chain end sites participate. In this work we have used simple reptation moves with random sampling from a uniform distribution.

2.5 Continuum Configurational-Bias (CCB) Monte Carlo [42–45]

The roots of this method can be traced back to the pioneering work of the Rosenbluths in the 1950s [64]. However, the CCB method in reality is a direct descendant of the Scanning method of Meirovich [65–68], in particular of the version for attractive random walks [68]. A related idea was introduced by Harris and Rice [69]. The method has recently attracted much interest, and has been fully developed as a simulation tool through the work of Siepmann [42], Frenkel et al. [43], and Siepmann and Frenkel [44]. de Pablo et al. [45] implemented the CCB method for the off-lattice treatment of realistic polymer systems. The initial off-lattice applications have demonstrated that the method can be used in a wide variety of important problems in polymer systems, most notably the determination of equilibrium thermodynamic properties, chemical potentials of polymers, solubilities of guest molecules in polymer melts, studies of phase transitions, and polymer-solvent interactions in supercritical fluids [70–72].

The CCB method is described in detail in the original references [42–45]. Here, we shall briefly state that a typical CCB move consists of the following steps (see Fig. 2):

(1) A chain is selected at random and "cut" at a random position.

(2) The chain is regrown step-by-step in a biased way. A number of possible positions (N_{sample}) for the insertion of the next site is examined. The new position of site "i" is chosen with a weight proportional to its normalized Boltzmann factor. A total of six potential insertion positions is sampled in the present work, since this number was found in previous work [45, 70] to be optimal for the systems considered here:

$$W_i = \frac{\exp(-\beta U_i)}{\sum\limits_{k=1}^{N_{sample}} \exp(-\beta U_k)} \tag{6}$$

$N_{CUT} = 4$

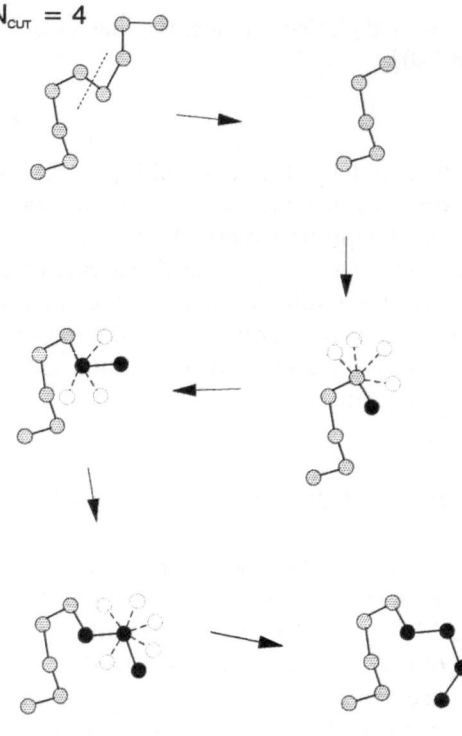

Fig. 2. Schematic of the Continuum Configurational Bias (CCB) move

(3) The bias introduced in step (2) must be removed in a proper way to ensure the satisfaction of the condition of microscopic reversibility. This is done by an appropriate choice for the transition probability. Equation (1) becomes:

$$P_{acc} = \min\left(1, \frac{W^{old}_{chain} P_{new}}{W^{new}_{chain} P_{old}}\right) \tag{7}$$

where the weight of the regrown part of the chain is given as a product of the weights of the individual segments

$$W_{chain} = \prod_{i=N_{cm}}^{N} W_i \tag{8}$$

In spite of its power, the CCB method has a number of shortcomings:

(1) In analogy to reptation techniques, it relies on a high concentration of chain ends. It becomes inefficient when long chains in dense systems are considered.

(2) Significant cage effects are observed in dense phases. A "tube" of empty space is artificially generated in the middle of a dense phase, when a large section of a chain is deleted. The regrown chain tends to develop in the direction of this tube.

(3) The method was demonstrated to give very good results for poly-methylene chains consisting of methylene beads in the united atom approximation [45, 70]. The presence of explicit hydrogen atoms, let alone bulky side groups, considerably reduces the efficiency of the method. By itself, it can only be applied in dense phases to polybeads and short polyethylene chains with explicit hydrogens less than 30 segments long. There is a rapid, almost exponential, decay of the acceptance rate as a function of the number of cut-and-regrown sites, which renders the method unsuitable for explicit continuum models of polymers such as polypropylene or polystyrene [46].

2.6 Concerted-Rotation Method (CONROT) [47]

Dodd, Boon, and Theodorou have very recently introduced the Concerted-Rotation method (CONROT), as a novel Monte Carlo algorithm for the atomistic simulation of polymer melts and glasses consisting of chains with fixed bond lengths and angles [47]. The algorithm is based on the seminal work of Gō and Scheraga on ring closures in polypeptides [48]. Gō and Scheraga demonstrated that joining two sites of an "interrupted" polymer chain through a bridge of four sites is a well-posed geometrical problem. Figure 3 demonstrates the basic concept on a decamer unit isolated from the interior of a longer polymer chain. Moving the four middle sites of the decamer involves a change of seven torsional angles (ϕ_3 to ϕ_9 in Fig. 3). However, six conditions (in the form of constraining equations) are available by the fixed geometry of the system. There are various ways of formulating these constraints [47, 48, 73], but we shall adhere to the intuitive choice of Dodd et al. [47]. Three constraint equations are provided by the constancy of the vector between sites 3 and 8 ($r_8 - r_3$). Two more conditions are obtained by the constancy of the unit vector of the bond

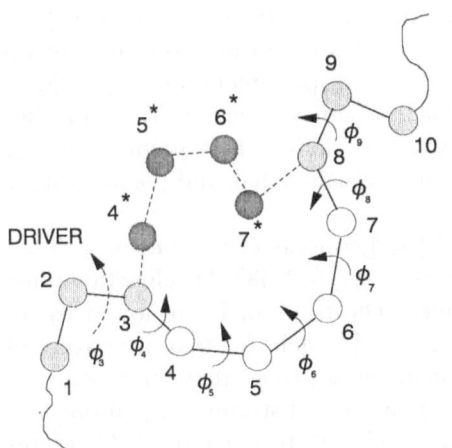

Fig. 3. Schematic of the basic Concerted-Rotation (CONROT) move. Beads indexed 1, 2, 3, and 8, 9, 10 remain fixed during the move

connecting sites 8 and 9 (u_9). Finally, since the orientation of sites 8, 9 and 10 must remain unchanged by the move, an appropriate Euler angle (γ) must be fixed.

Since seven consecutive torsional angles of the chain participate in the CONROT move, the first can be chosen as the independent variable, the rest being determined by solving the constraining equations. This is by no means an easy task, even after recasting the problem in the form of an implicit equation for the torsion angle ϕ_4

$$F[\phi_4, \phi_5(\phi_4), \phi_6(\phi_4), \phi_7(\phi_4)] = [u_9^{(4)}]^T T_4 T_5 T_6 T_7 e_1 - \cos\theta_8 = 0 \quad (9)$$

where $u_9^{(4)}$ is the unit vector of bond 9 in the local coordinate frame of bond 4 (Flory's notation [74] is used here for the definitions of the local coordinate frames), T_i is the rotation matrix of the i-th bond (as defined by Flory), and e_1 is the unit vector col(1, 0, 0). A complex numerical procedure finds all the roots of Eq. (9). One then obtains the six additional torsional angles by solving the constraint equations in a manner prescribed by Gō and Scheraga [48], and Dodd et al. [47]. Equation (9) may have no solutions, in which case one speaks of a "geometric failure", or multiple solutions, corresponding to different spatial arrangements of the four sites of the bridge and to different sets of torsional angles.

The complete description of the MC method based on Concerted-Rotations can be found in the paper of Dodd et al. [47]. Here we shall provide only a brief outline of the fundamental steps in the implementation of the CONROT algorithm.

(1) A chain is selected at random from a multi-chain system. The direction of the site numbering in the chain is chosen at random. A "decamer" is randomly selected from inside the selected chain. Several versions of the CONROT move for chain ends are discussed by Dodd et al. [47], but we have chosen to implement only the "internal CONROT move", since CONROT is less efficient for chain ends than reptation or CCB moves.

(2) Sites 4, 5, 6, and 7 of the decamer are cut, and the rotation angle ϕ_3 is turned by a random amount $\Delta\phi_3$. $\Delta\phi_3$ is sampled uniformly from the interval $[-\Delta\phi_3^{max}, +\Delta\phi_3^{max}]$. The maximum allowed displacement for angle ϕ_3 is restricted, since large changes would lead to unfavorable overlaps of sites 4 (mainly), 5, 6, and 7 with sites on adjacent chains. This point is further discussed in Sect. 4. The new position of site 4 is calculated, and the laboratory frame of bond 4 is set up.

(3) Equation (9) is solved. Function $F[\phi_4]$ consists of one or more closed loops, hence there is an even number of roots, N_{new} [47, 48]. This function must be evaluated on a dense grid of ϕ_4 values. The roots of F are found by an exhaustive search of the interval $[-\pi, +\pi]$. Each ϕ_4 that is a root of F generates a different sequence of torsion angles ϕ_5 to ϕ_7 that corresponds to a different possible site arrangement satisfying the constraining equations.

(4) One of the obtained sets of angles is chosen to enter the MC acceptance/rejection step. This choice was made randomly in the original Dodd et al.

formulation. In Sect. 4 we demonstrate that it is advantageous to choose one of the possible solutions to the CONROT problem according to its normalized Boltzmann weight, in direct analogy to the procedure adopted in the CCB method:

$$W_i^{new} = \frac{\exp(-\beta U_{[\phi_k]_i})}{\sum\limits_{j=1}^{N_{new}} \exp(-\beta U_{[\phi_k]_j})} \tag{10}$$

where $[\phi_k]_i$ denotes the torsion angle configuration associated with the i-th such solution.

(5) As pointed out by Dodd et al. [47], the solution of the CONROT problem entails a temporary change of the variables used to describe the configuration space. This is implicit in the use of the vector constraints discussed before. The Jacobian determinant, J_{new},

$$J_{new} = \frac{\partial(\phi_4, \phi_5, \phi_6, \phi_7, \phi_8, \phi_9)}{\partial(r_8 - r_3, u_9, \gamma)} \tag{11}$$

of the transformation from the variables $\{\phi_3, \phi_4, \phi_5, \phi_6, \phi_7, \phi_8, \phi_9\}$ to the variables $\{\phi_3, (r_8 - r_3), u_9, \gamma\}$ must be calculated and used in the acceptance criterion, otherwise detailed balance is not satisfied.

(6) The inverse problem is solved. One finds the solutions of a CONROT problem with the same constraints, but in which $\Delta\phi_3 = 0$ with respect to the original value. The weight, W_{old}, of the old (starting) configuration of the system is thus obtained. The Jacobian J_{old} of the starting configuration must also be computed.

(7) The proposed CONROT move is accepted with probability

$$P_{acc} = \min\left(1, \frac{J_{new} W_{old}}{J_{old} W_{new}} \exp(-\beta(U_{new} - U_{old}))\right) \tag{12}$$

Equation (12) ensures that microscopic reversibility is satisfied. Dodd et al. proved that erroneous results are obtained if the number of geometric solutions and the Jacobian determinants are not included in the acceptance criterion [47a]. Since we bias our selection of CONROT solutions, we have added the solution weights to the acceptance criterion. It must be pointed out however, that a successful removal of the bias requires long simulation runs, in order for some of the solutions that are not favored energetically to be selected.

Compared to other simulation methods the CONROT algorithm is quite complex, since it requires the complete solution of two geometric problems at each step. However, it has some very distinct advantages:

(1) Unlike the other methods discussed here, CONROT is not applied to chain ends, but works at a random position inside the chain. Its effectiveness does not depend critically on the length of the chains.

(2) Dodd et al. have demonstrated that CONROT moves have reasonable acceptance rates even at high densities [47].

(3) Concerted-Rotations are pseudo-dynamic in nature. They are highly localized, since they involve cooperative displacement of a few atoms only. There is in fact a close connection between the simple Concerted-Rotations described here, and the Schatzki process in glasses [31b, 32, 75]. There is also a direct correspondence between the CONROT move and various crankshaft moves used in polymer simulations on lattices [7, 9]. It is possible to base a dynamic Monte Carlo process on CONROT moves after appropriate calibration against a real-time method such as Molecular Dynamics. This procedure was carried out by Dodd et al. [47], who found that a MC method based on unbiased CONROT moves is not more efficient than Molecular Dynamics for a polypropylene melt. However, these authors argued that using reptation moves for the chain ends, and biasing the CONROT moves can alter this result significantly. It is of course not evident that a length-scale independent relation between time and MC steps can be obtained.

(4) Being highly localized motions, Concerted-Rotations appear to be particularly suitable for simulations of the glassy state.

2.7 Extended Concerted-Rotation (ECROT) Algorithms

Concerted Rotations can be components of more complex MC moves in off-lattice polymer simulations. Dodd et al. have discussed the possibility of using CONROT moves as components of a chain-identity exchange algorithm, similar to the pseudokinetic algorithm of Olaj and Lantschbauer [34a], Mansfield [34b], and Madden et al. [76] on lattices. Here we introduce and evaluate three simpler variants of the CONROT move, inspired by a discussion in a recent paper by Palmer and Scheraga [73]. The new methods are pictorially presented in Fig. 4.

2.7.1 First 9-Angle Method (ECROT1)

A dodecamer is selected at random from inside a chain in the first variant (Fig. 4a). There are nine degrees of freedom (torsional angles), three of which can be independently varied. Any three of the nine angles may be chosen according to Palmer and Scheraga, but we have chosen to use the first three for convenience. The first two angles are randomly turned by a small amount ($\pm 5°$ at the maximum), and the third is used as the driver angle for a simple CONROT move. This ECROT1 algorithm is a combination of two Metropolis random moves and one CONROT move in series. The acceptance probability for this move is given by Eq. (12), but the Jacobian determinant is modified to

$$J_1 = \frac{\partial(\phi_6, \phi_7, \phi_8, \phi_9, \phi_{10}, \phi_{11})}{\partial(\mathbf{r}_{10} - \mathbf{r}_5, \mathbf{u}_{11}, \gamma)} \tag{13}$$

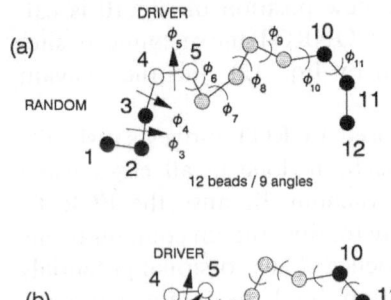

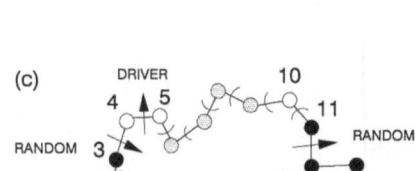

Fig. 4a–c. Schematic of the three Extended Concerted-Rotation moves discussed in the text: **a** ECROT1, **b** ECROT2, and **c** ECROT3. Beads represented by *filled circles* remain fixed in each move. Beads represented by *empty circles* are changed by random rotations of torsional angles. *Shaded beads* are moved by Concerted-Rotation

where u_{11} is the unit vector of the bond between sites 10, and 11, and together with γ defines the orientation of the plane specified by sites 10, 11, and 12.

2.7.2 Second 9-Angle Method (ECROT2)

An alternative method involving a dodecamer basic unit is shown in Fig. 4b. Instead of randomly turning two torsion angles before attempting the CONROT move, one now moves site 5 randomly on the surface of a sphere centered at site 3. The position of site 4 is subsequently calculated, and a CONROT move unites sites 5 and 10. The acceptance probability for this move is also given by Eq. (12), but the relevant transformation Jacobian is

$$J_2 = \frac{\partial(\phi_3, \phi_4, \phi_5, \phi_6, \phi_7, \phi_8, \phi_9, \phi_{10}, \phi_{11})}{\partial(r_5 - r_3, \phi_5, r_{10} - r_5, u_{11}, \gamma)} = \frac{\partial(\phi_3, \phi_4)}{\partial(r_5 - r_3)} J_1. \tag{14}$$

2.7.3 10-Angle Method (ECROT3)

The method of Fig. 4c uses a 13mer as its basic unit. Four torsional angles out of a total of ten are now varied independently. The move is identical to that of Fig. 4a, with the difference that the last torsion angle of the unit is also turned by a small fraction ($\pm 5°$ at the maximum). The new position of site 10 is calculated, as are the positions of sites 4 and 5. A CONROT move connects sites 5 and 10. The acceptance probability is given by Eq. (12), and the relevant Jacobian is J_1 of Eq. (13).

It is important to ascertain that the proposed ECROT moves satisfy the condition of microsopic reversibility. This has been done in all cases using simulations of a freely-rotating basic unit in vacuum. Because the ECROT methods do not move the chain ends, we generate totally random configurations of the basic unit every 1000 MC steps. In the absence of any torsional potentials and nonbonded interactions, a uniform distribution should be obtained for each individual torsion angle of the basic unit.

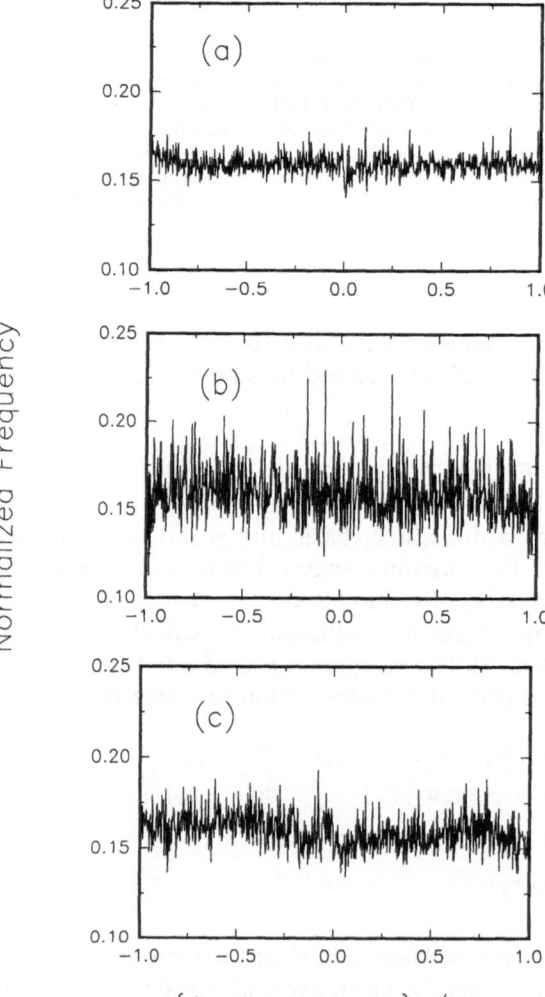

Fig. 5a–c. Distribution density of all angles that participate in CONROT moves obtained with freely-rotating chains in vacuum. The histograms were obtained after 10^6 moves. **a** ECROT1, **b** ECROT2, and **c** ECROT3

Figure 5 shows the distribution density, obtained by combining histograms, for the angles that are involved in the CONROT transformation, and demonstrates that uniform distributions are indeed sampled. The statistics for 10^6 MC steps are not so good for ECROT2 and ECROT3, since the higher fractions of geometric failures result in repeated countings of many configurations.

3 Computational Details

3.1 Interaction Energies

Most of the results of this work were obtained with polybead molecules in the united atom approximation. These are chains formed by methylene beads connected by rigid bonds of length 1.53 Å. The angle between successive bonds of a chain is also fixed at 112°. The torsion angles around the chain backbone are biased to favor trans or gauche states, using a torsional potential energy function originally introduced by Ryckaert and Bellemans in a Molecular Dynamics study of butane [77]

$$U_{tors}(\phi_i) = C \sum_{n=0}^{5} \alpha_n \cos^n(\phi_i) \qquad (15)$$

with constant coefficients $C = 9.27$ kJ/mol, $\alpha_0 = 1.0$, $\alpha_1 = 1.31$, $\alpha_2 = -1.414$, $\alpha_3 = -0.3297$, $\alpha_4 = 2.828$, and $\alpha_5 = -3.3943$ [24]. In Eq. (15), ϕ_i denotes the dihedral angle for rotation around the i-th bond. ϕ_i must be in the interval $[-\pi, +\pi]$. This torsional energy function has been used by many researchers in off-lattice polymer simulation work [24, 39, 40, 45, 47, 70].

Sites on different chains and sites on the same chain separated by more than three bonds interact through a "nonbonded" Lennard-Jones interaction function

$$U_{nb,ij} = 4\varepsilon_{ij} \left[\left(\frac{\sigma_{ij}}{r_{ij}} \right)^{12} - \left(\frac{\sigma_{ij}}{r_{ij}} \right)^{6} \right]. \qquad (16)$$

Here, all ε's and σ's are assumed equal, no distinction being made between end-methyl beads and middle-methylene beads. The energy depth parameter was set to $\varepsilon = 0.41$ kJ/mol, and the methylene bead size parameter to $\sigma = 3.94$ Å. These values had been found previously to reproduce PVT data of short polyethylenes successfully [45, 46, 70].

In the case of polyethylene with explicit hydrogen atoms, we have used a different model, keeping the assumption of fixed bond angles and bond lengths. A three-fold intrinsic torsional energy function with a barrier of 11.71 kJ/mol was assigned to each C–C bond [74]. Nonbonded interactions

Table 1. L-J interaction parameters and geometry for the polyethylene chains with explicit representation of H atoms

Atom	$r_i{}^a$	$\alpha_i{}^b$	$N_i{}^c$
C	1.8	0.93	5.0
H	1.3	0.42	0.9

Interacting pair	$a_{ij} \times 10^{-3}$ (kJ$\,$Å^{12}/mol)	c_{ij} (Å^6/mol)
C,C	1665	1531.3
C,H	238	535.6
H,H	30.5	196.6

Bond	Bond length (Å)
C–C	1.53
C–H	1.10

Bond angle	
∠CCC	112°
∠CCH	109°
∠HCH	109°

[a] Van der Waals atomic radii in Å
[b] Atomic polarizabilities in Å^3
[c] Effective number of electrons

between atoms separated by more than two bonds were calculated using the Lennard-Jones 6–12 pair function (Eq. (16)). Ketelaar's parameters [78, 79] a_{ij} and c_{ij} were used. These are related to the ε_{ij} and σ_{ij} of Eq. (16) by

$$\sigma_{ij} = \left(\frac{a_{ij}}{c_{ij}}\right)^{1/6}, \quad \varepsilon_{ij} = \frac{c_{ij}^2}{4a_{ij}}. \tag{17}$$

All atoms were treated individually. The London dispersion parameters c_{ij} were calculated according to the Slater-Kirkwood formula using Ketelaar's values [78, 79] for the atom polarizabilities α_i and the "effective number of electrons" N_i. The constants a_{ij} were so assigned as to minimize the potentials $U_{nb, ij}$ for a given pair of atoms when r_{ij} is set equal to the sum of the corresponding adjusted van der Waals radii r_i^0. The values used are shown in Table 1. The geometry of the polyethylene chain is fully determined by the geometry of the methylene group. This geometry is specified by the assumption that all ∠HCH are equal.

3.2 Computer Experiments-General Considerations

All the simulations have been performed using cubic simulation boxes in the NPT ensemble, following the procedure of Boyd [39], and de Pablo et al. [45]. Volume fluctuation moves were performed every 500, 1000 or 2000 moves, depending on the overall length of the simulation. The maximum change of the edge of the simulation box in a volume fluctuation move was 0.2 Å for the polybead simulations (acceptance ratio of 20–30% for volume fluctuation moves), and 0.1 Å for polyethylene with explicit hydrogens (acceptance ratio ca. 60%).

The minimum image convention was used in all these simulations, and no tail corrections to the thermodynamic properties were applied, since we were investigating the relative performance of various algorithms and did not concentrate on the precise values of thermodynamic properties.

Most of the results presented here were obtained using a Silicon Graphics CRIMSON workstation. We have also used the NEC SX-3 supercomputer of the Swiss Center for Scientific Computations (CSCS).

3.3 Systems Studied

3.3.1 C_{71} Polybead Melt

The first system investigated is a melt of ten polybead chains of seventy one carbon atoms, at a temperature of 240 °C (513 K) and pressure of 1 bar. The experimental density of the C_{71} system is 0.71 g/cm^3 [80]. From a previous study it was known that Reptation performs poorly in this dense system, because of the relative sparsity of chain ends [81]. CCB was an improvement, yielding reasonable convergence for PVT properties [81]. However, the end-to-end distance vector of the chains relaxes slowly, and the diffusion of the center-of-mass of the chains is quite small. Closer scrutiny of configurations obtained from these simulations revealed that large portions around the middle of the chains are not moving at all, because very few deep CCB cuts can be accepted. CCB is hence rather inefficient for this system, both at the segmental and at the chain-level.

The simulations of this system started from a pre-equilibrated configuration obtained from a previous CCB-Reptation study [81]. The evolution of this system in configuration space was followed to gain insights as to the relative strengths and weaknesses of the methods described in Sect. 2.

3.3.2 C_{24} Polybead Melt

The second system is a melt of 20 polybead chains of C_{24} (tetracosane) at a temperature of 200 °C (473 K), and a pressure of 1 bar. Both, Reptation and CCB are known to perform adequately in this system [39, 45, 47, 70, 81]. The experimental density of C_{24} is 0.68 g/cm^3 at the conditions chosen [80]. The simulation starts from a preequilibrated configuration, obtained after 10^6 moves of pure Reptation.

3.3.3 C_{71} Polyethylene Melt

Polyethylene with explicit hydrogen atoms is used for the third test system. This is also a melt of 10 chains of C_{71} at a temperature of 240 °C and a pressure of 1 bar. Reptation is known to fail for this system, and CCB was found to be effective in moving only the 10 to 15 skeletal atoms near the chain end, depending on temperature [46]. Deeper cut-and-regrow moves have such low acceptance probabilities that the method is not useful in practice. It was hoped that biased CONROT moves may provide a much needed breakthrough. The simulation started from an equilibrated configuration obtained from the corresponding polybead system. Although having a reasonably low energy, this starting configuration did not correspond to a situation of high probability of occurrence at equilibrium for the polyethylene chains. 5.5×10^5 random MMC steps were taken to equilibrate the system to a reasonably low constant energy. The evolution of the system was subsequently followed for 7.5×10^5 steps. Volume fluctuation moves were performed every 1000 steps. The calculation of the interaction energy was implemented in a straightforward way. The overall performance of this program can be improved substantially if neighbor-lists or linked-cell techniques are used [85], but we decided not to implement such techniques for this preliminary investigation.

3.4 Criteria of Simulation Performance

Because systems of long chain molecules are characterized by a wide range of relaxation times, there are many different criteria of performance of a simulation. The ones employed here can be grouped in two well-defined classes, namely criteria for short time and distance behavior, and criteria for long time and distance behaviour.

Many additional criteria have been defined. For example, in the context of a temperature-jump simulation one can study the rate of approach of various thermodynamic and molecular properties to their final equilibrium values. Such properties are the configurational internal energy and the density of the system (for NPT simulations), as well as the average end-to-end distance and radius of gyration of the chains. Such a study was not undertaken in this work.

3.4.1 Criteria of Short-Time, Short-Range Performance

These provide information as to how fast the local properties change as the simulation proceeds.

Important information on local rearrangements is provided by the bond orientation autocorrelation functions (BCF's). The 1st and 2nd degree BCF's are given by:

$$f_{1;bcf}(n) = \langle \mathbf{u}_i(k) \cdot \mathbf{u}_i(k+n) \rangle_{i,k} \tag{18}$$

$$f_{2;bcf}(n) = \tfrac{3}{2} \langle (\mathbf{u}_i(k) \cdot \mathbf{u}_i(k+n))^2 \rangle_{i,k} - \tfrac{1}{2} \tag{19}$$

where a double averaging is performed over all the bonds in the system (subscript i) and a large number of Monte Carlo steps (subscript k) for improved statistics. Here, $\mathbf{u}_i$ is the unit vector of the i-th bond, and the number of Monte Carlo steps effectively replaces time. The quantities $f_{p;bcf}(n)$ are measures of how fast the tumbling motion of the bonds erases the memory of the previous configuration. These functions are equal to 1 when $n = 0$, and decay to zero when $n \to \infty$. Bond autocorrelation functions are usually fitted by stretched exponentials [31, 47, 82], or by even more complicated functions [86], but the long-n tail may also be fitted by simple exponentials. This procedure was used by Dodd et al. to calibrate the original CONROT algorithm vs Molecular Dynamics [47].

The bond correlation functions of Eqs. (18) and (19) are not appropriate when Reptation is used, because this method continuously changes the identity of the end bonds. As a result, $f_{p;bcf}(n)$ decays rapidly for small n; however, this is a spurious decay, associated with renumbering-related decorrelation and the very rapid random changes at the chain ends, while the overall shape of the chain does not change much. The final long-n slopes of the bond correlation functions can still be useful [81].

The mean-square displacement of chain sites, often called site-autocorrelation function [25, 82, 83], is also a useful measure of simulation efficiency at the local scale:

$$g_{site}(n) = \frac{1}{N_{site}} \sum_{i=1}^{N_{site}} \langle (\mathbf{r}_i(n) - \mathbf{r}_i(0))^2 \rangle \tag{20}$$

where the average in Eq. (20) is taken over all chains in the system, and over a number of configurations for improved statistics.

3.4.2 Criteria of Long-Time, Long-Range Performance

These criteria examine the ability of the algorithm to probe the longest relaxation times, or the greatest spatial relaxations of the studied systems. A measure of chain overall translation is the mean-square displacement of the center of mass of the chains:

$$g_{CM}(n) = \langle (\mathbf{r}_{CM}(n) - \mathbf{r}_{CM}(0))^2 \rangle \tag{21}$$

The chain self-diffusion coefficient can be extracted only when $g_{CM}(n)$ becomes proportional to n. Usually, one requires that the chain center-of-mass moves first over a distance that is large compared to the radius of gyration [25].

A measure of chain rotation is provided by the relaxation function of the end-to-end distance vector:

$$f_{1;eed} = \frac{(\mathbf{r}_N(n) - \mathbf{r}_1(n)) \cdot (\mathbf{r}_N(0) - \mathbf{r}_1(0))}{|\mathbf{r}_N(n) - \mathbf{r}_1(n)| \; |\mathbf{r}_N(0) - \mathbf{r}_1(0)|} \tag{22}$$

This is similar to the autocorrelation function of the end-to-end distance [82–84], and decays to zero much more slowly than the bond correlation functions of Eqs. (18) and (19) [82–84].

Estimates of equilibrium statistical inefficiencies [27, 33] can also be useful, since they provide a measure of the sampling efficiency of a method. Different values of statistical inefficiencies are obtained for different thermodynamic properties [27]. In this way, more insights are gained as to the way that a method operates to provide uncorrelated configurations. The statistical inefficiency $I(\Omega)$ of a property Ω is calculated according to the method described by Allen and Tildesley [27]. Roughly $I(\Omega)$ simulation steps are needed to obtain a configuration completely uncorrelated from the starting configuration with respect to property Ω.

4 Simulation Results

4.1 An Equilibrium Melt of Ten C_{71} Polybead Chains

Our first goal was to optimize the Concerted-Rotation method in a dense phase. The only adjustable parameter of the method is the maximum displacement $\Delta\phi_3^{max}$ of the driver angle ϕ_3 (see Fig. 3). While a smaller value of $\Delta\phi_3^{max}$ will certainly lead to higher acceptance rates of the CONROT moves, a larger value produces larger steps through configuration space. There is a further subtlety involved in CONROT moves, because of their geometric nature. Trial runs with a decane chain in vacuum have demonstrated that the number of geometric failure increases substantially with $\Delta\phi_3^{max}$, since it becomes more difficult on the average to generate a rigid four-site bridge, when the new value of ϕ_3 differs substantially from that of the initial configuration. This is an effect that has also been observed in the use of the Gō-Scheraga algorithm for loop-closure problems in proteins, and has prompted the development of modified procedures that involve CONROT [73]. The optimum value of $\Delta\phi_3^{max}$ in dense systems is, of course, highly system-dependent. Dodd et al. have used values ranging from $10°$ to $30°$ in their simulations of polybead and polypropylene melts.

We have made test runs of 10^5 MC steps to find the optimum value of $\Delta\phi_3^{max}$ in this system. Each run consisted of a random sequence of 80% Concerted-Rotation moves and 20% CCB moves. The CCB moves were restricted to cutting and regrowing a maximum of 15 sites from both ends of each chain. CONROT moves were performed in both chain directions by reversing the indexing direction of each chain for every other move on the average; they were biased in the manner described in Sect 2. The results of these simulations are summarized in Table 2. The CONROT algorithm has not been fully-optimized, therefore the execution times in Table 2 are of qualitative value only.

Acceptance rate drops considerably with increasing $\Delta\phi_3^{max}$. Counteracting this reduction in efficiency is, at least partially, the fact that execution time is also reduced, although not as dramatically. Because of the higher fraction of geometric failures, the program spends less time performing costly geometric calculations. Similar reductions of necessary CPU time were observed for trial runs of 1000 steps on the NEC SX-3 supercomputer, with a semi-vectorized version of the CONROT code (see Table 2). Execution of the algorithm on a vector machine is only 15–20 times faster than that on a fast scalar workstation, since the CONROT subroutines are inherently quite scalar. The end-to-end distance relaxation, the radial distribution function, and the torsion angle distribution are similar in all cases. Performance at the polymer segment level is evaluated with the bond correlation relaxation functions $f_{1;bcf}(n)$ and $f_{2;bcf}(n)$, plotted in Fig. 6. The performance of the CONROT moves for this system does not change substantially for $\Delta\phi_3^{max}$ up to 30°! This is a rather surprising result, since the acceptance rate is only 19% when $\Delta\phi_3^{max}$ is equal to 30°. We have adopted values of $\Delta\phi_3^{max}$ ranging from 10° to 30° for the rest of this work. However, our results suggest that a higher $\Delta\phi_3^{max}$ might produce better results in less dense phases.

Figure 7 presents results of the comparison between runs with different fractions of CCB and CONROT moves, and also demonstrates the effect of biasing CONROT moves. For this series of runs CCB cuts were made until the middle of the chain from both ends (cutting and regrowing a maximum of 36 chain sites). The cases examined are:

(1) 20% CCB-80% CONROT moves without bias (as prescribed by Dodd et al. [47]).
(2) 80% CCB-20% biased CONROT moves,

Table 2. Summary of results of runs performed for the determination of the optimum maximum step size of the driver angle for the Concerted-Rotation

$\Delta\phi_3^{max}$ (deg)	Execution time on a silicon graphics CRIMSON workstation (cpu s/10^3 moves)	Fraction of geometric failures	Fraction of accepted moves (based on total moves)	Fraction of accepted moves (after subtracting geometric failures)	Execution time on a NEC, SX-3 supercomputer (cpu s/10^3 moves)
10°	544	0.179	0.377	0.450	34.5
20°	507	0.240	0.250	0.330	33.4
30°	518	0.274	0.186	0.257	32.8
60°	463	0.404	0.110	0.184	28.3
90°	427	0.509	0.076	0.156	24.1

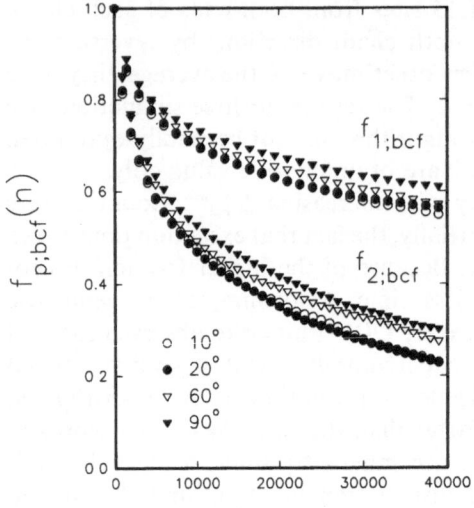

n (# of steps)

Fig. 6. Effect of the maximum change in CONROT driver angle change ($\Delta\phi_3^{max}$). First and second degree bond correlation functions plotted vs the number of Monte Carlo steps. A melt of ten C_{71} polybead chains was simulated with algorithms composed of 20% CCB moves ($N_{CUT} = 15$) and 80% CONROT moves. (○) $\Delta\phi_3^{max} = 10°$, (●) $\Delta\phi_3^{max} = 20°$, (▽) $\Delta\phi_3^{max} = 60°$, and (▼) $\Delta\phi_3^{max} = 90°$

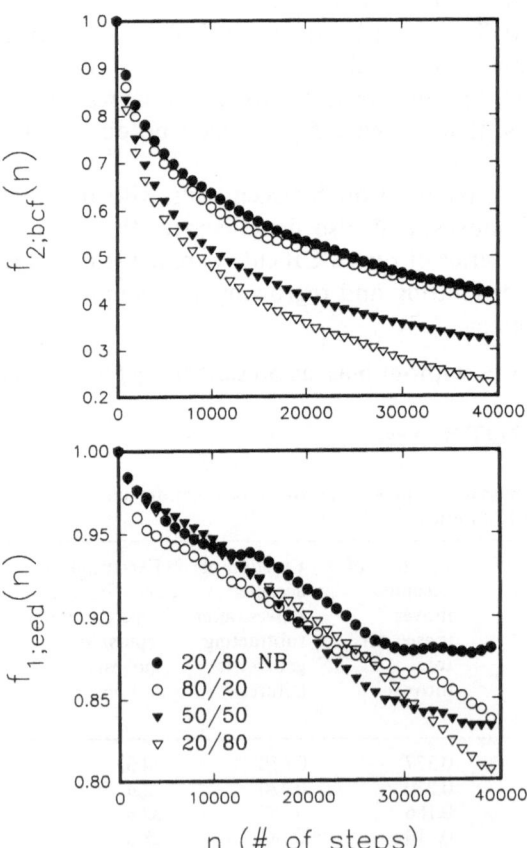

n (# of steps)

Fig. 7. Effect of different combinations of CCB and CONROT moves for the C_{71} chain system. Second degree bond correlation function (*top*) and end-to-end distance vector relaxation function (*bottom*) as functions of the number of simulation steps. The results shown are for (●) 20% CCB ($N_{CUT} = 36$) − 80% CONROT with $\Delta\phi_3^{max} = 10°$ with random choice of the CONROT solutions, (○) 80% CCB − 20% CONROT with biased choice of solutions, and similarly (▼) 50% CCB − 50% biased CONROT, and (▽) 20% CCB − 80% biased CONROT

(3) 50% CCB-50% biased CONROT moves, and
(4) 20% CCB-80% biased CONROT moves.

The diffusion of the center of mass is insignificant in all these cases. The bond orientation correlation function demonstrates that (a) unbiased CONROT moves do not constitute a significant improvement over the CCB method in this dense melt (as can be seen by comparing cases (1) and (2)), and (b) biasing the CONROT moves produces a dramatic improvement of simulation efficiency at the segmental level. It should be pointed out that the CONROT subroutines, which perform the solution of the CONROT geometric problem, are computationally very expensive. Biasing the CONROT moves does not tax the algorithm significantly. Since the end-to-end distance relaxation does not change essentially, it can be concluded that the introduction of CONROT moves is necessary for an algorithm with improved overall sampling ability. A ratio of CCB/CONROT equal to 20/80 was subsequently used in the investigation of this system.

Figure 8 demonstrates that an optimum value exists for the length of the chain that is cut and regrown in a CCB move (N_{cut}). These runs were performed

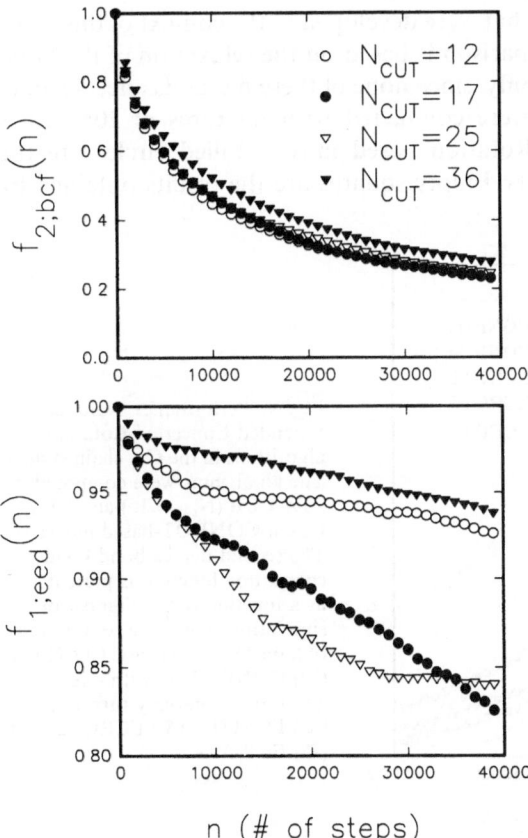

Fig. 8. Determination of optimum N_{CUT} for CCB moves in the C_{71} chain system. The second degree bond correlation function (*top*), and the end-to-end vector relaxation function (*bottom*) are plotted as functions of the number of simulation steps. The algorithms were composed of 20% CCB-80% biased CONROT with $\Delta\phi_3^{max} = 10°$. The results shown are for ($\bigcirc$) $N_{CUT} = 12$, ($\bullet$) $N_{CUT} = 17$, (∇) $N_{CUT} = 25$, ($\blacktriangledown$) $N_{CUT} = 36$

with a CCB/ECROT1 ratio of 20/80, and the independent angles in the ECROT1 move were turned by a maximum of 5° for the first two angles and by 10° for the driver angle of the Concerted-Rotation (see Fig. 4a for the meaning of these angles). The relaxation of the bond orientation correlation function is a weak function of N_{CUT}. Figure 8 (top) demonstrates that only cuts to the middle of the chain give distinctly slower relaxation, because of the much lower acceptance rates of the deep cuts. Figure 8 (bottom) however shows that N_{CUT} has a strong impact on the relaxation of the end-to-end distance vector. Shallow cuts do not change the end-to-end vector fast, but deep cuts are also disadvantageous, because of low acceptance rates. Figure 8(bottom) implies that the optimum value for N_{CUT} in this system is approximately 20. Probably the most interesting conclusion that may be drawn from Fig. 8 is that different components of the composite simulation algorithm influence local and global relaxation properties of the chains in different, well-defined ways. Thus, CONROT-based moves lead to rapid relaxation at the segmental level, but do not affect overall chain rotation directly. On the other hand, CCB does not have a strong impact on local relaxation properties, but its effect on global chain dynamics is very important. Both these moves are necessary for an algorithm with optimum efficiency for the present system.

The extensions of CONROT that were developed in the context of this work are compared in Fig. 9. The comparison is based on the relaxation of the bond orientation correlation function only, since none of these methods affects global properties directly. These runs were conducted with mixtures of 20% CCB moves and 80% of Concerted-Rotation based moves. Filled circles are the standard biased CONROT results. Empty squares are the results obtained by

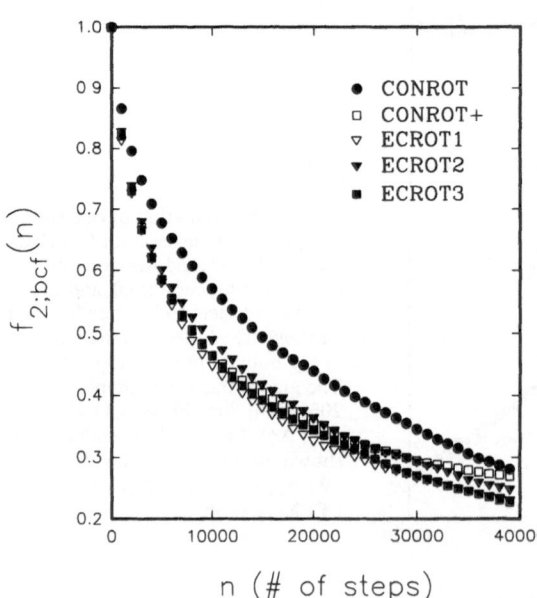

Fig. 9. Performance evaluation of Extended Concerted-Rotation algorithms in the C_{71} chain system. The algorithms were composed of 20% CCB ($N_{cut} = 15$)-and 80% biased CONROT-based moves. The second degree bond bond correlation function is plotted as a function of the number of simulation steps. The results were obtained with (●) basic CONROT, (□) CONROT + with one additional randomly turned angle, (▽) ECROT1, (▼) ECROT2, and (■) ECROT3

turning randomly one torsional angle before initiating a CONROT move (the basic unit of the move is thus an 11-mer). Filled and empty triangles are the results of ECROT2 and ECROT1 respectively, and filled squares are the results of ECROT3. In all these cases, the maximum displacement of randomly turned angles was set to $\pm 5°$, and of CONROT driver angles to $\pm 10°$. For ECROT2, the new position of vector $\mathbf{r}_5 - \mathbf{r}_3$ was also assumed to form a maximum angle of $5°$ with the old position of the vector. To reduce the fraction of geometric failures in the case of ECROT2, a maximum of four attempts is made for each driver angle with different $\Delta\phi_5$. The results in Fig. 9 demonstrate that additional degrees of freedom can improve the efficiency of the CONROT move, but that limits exist to the improvements that can be obtained. Turning one additional torsion angle randomly leads to an initial improvement, although the long-time behaviour may actually be worse. Turning two additional degrees of freedom leads to further improvements, but turning three does not produce better results. ECROT2 performs worse that ECROT1 because of smaller acceptance rates. Acceptance rates and other useful information about these runs are listed in Table 3.

Figure 10 presents bond correlation functions, and end-to-end relaxation functions obtained from longer simulations. Both runs were made with 20% CCB end moves ($N_{CUT} = 15$), 20% reptation moves, and 60% biased ECROT1 moves with $\Delta\phi_3^{max}$ equal to $10°$. The first run was started from a system configuration where the constituent chains have conformations close to the unperturbed ones; the second run departed from a set of relatively collapsed chain conformations. It is demonstrated that, even with the present algorithm, it is still not possible to obtain fully-equilibrated melts for these long-chain systems. From the different relaxations of bond correlation functions and end-to-end distance vectors (Fig. 10 (top)), it is clear that the results obtained depend strongly on the initial conditions chosen. An examination of the fluctu-

Table 3. Summary of results of runs performed for the determination of the optimum Extended Concerted-Rotation method for a melt of C_{71} polybead molecules. The maximum step size for randomly turned angles was $5°$; that for CONROT "driver" angles was $10°$

Method	Execution time on a silicon graphics CRIMSON workstation (cpu s/10^3 moves)	Fraction of geometric failures	Fraction of accepted moves (based on total moves)
CONROT	544	0.179	0.377
CONROT + [a]	—	0.197	0.294
ECROT1	485	0.212	0.250
ECROT2	580[b]	0.151[b]	0.172
ECROT3	470	0.218	0.237

[a] This is a restricted ECROT1 method, in which only one torsional angle is turned randomly before initiating the CONROT move
[b] The longer time, lower fraction of geometric failures, and lower fraction of accepted moves stems from the fact that FOUR (instead of the standard ONE) attempts to complete a CONROT move were made for each selected driver angle. Solutions are thus found more often, but most of the accepted ones are close to the starting configuration

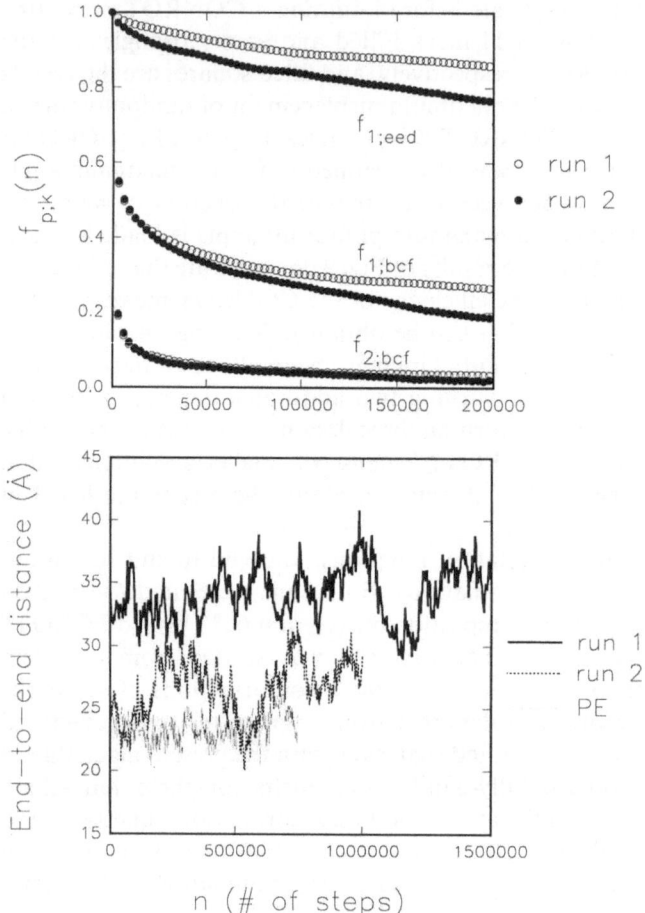

Fig. 10. Demonstration of bottlenecks in long runs with the C_{71} chains system. The algorithm was composed of 20% reptation moves − 20% CCB ($N_{CUT} = 15$) 60% biased ECROT1 with $\Delta\phi_3^{max} = 5°$ and $\Delta\phi_5^{max} = 10°$. (*Top*) The first-degree bond correlation function and the end-to-end vector relaxation function as functions of the number of simulation steps. (○) Run 1 (normal chains), and (●) run 2 (collapsed chains). (*Bottom*) Fluctuation of the average of the square of the end-to-end distance of the chains for run 1 (*continuous curve*), run 2 (*medium-dashed curve*), and the explicit H-polyethylene run (*dotted curve*)

ations of the average end-to-end distance (Fig. 10 (bottom)) reveals that the algorithm cannot escape rapidly from "bottlenecks". The very slow relaxation of autocorrelation functions of long polymer chains is a well-known problem, even when less realistic polymer models or less dense phases are simulated; Smith and Rapaport have recently illustrated this point using molecular dynamics on simplified polymer models in a solvent [84].

The motion of the center of mass of the C_{71} chains is very slow. The results from the first run are plotted in Fig. 12, together with the results of the C_{24} simulations. Statistical inefficiencies for these runs are reported in Table 4.

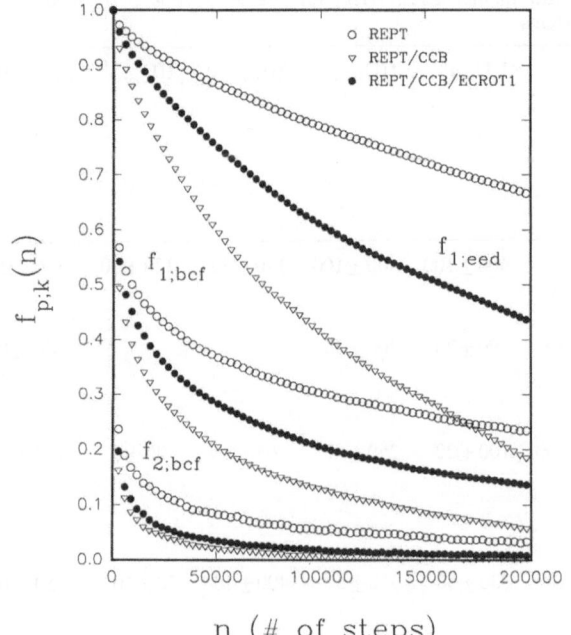

Fig. 11. Comparison of methods in the C_{24} chain system. End-to-end distance relaxation function (*top*), and first (*middle*) and second (*bottom*) degree bond correlation functions plotted vs the number of simulation steps. The results were obtained with the following algorithms: (○) 90% reptation − 10% MMC moves, (▽) 70% CCB ($N_{CUT} = 14$) − 20% reptation − 10% MMC moves, and (●) 50% biased ECROT1 ($\Delta\phi_3^{max} = 5°, \Delta\phi_5^{max} = 10°$) − 20% CCB ($N_{CUT} = 14$)-20% reptation − 10% MMC moves

The results for the C_{71} chains also show that the performance of off-lattice simulations must be assessed by looking at all possible aspects of the simulation. Often used criteria, such as internal energy variation, torsional angle distribution function, and radial distribution function [38, 39, 45] do not by themselves provide reliable performance indicators.

4.2 An Equilibrium Melt of Twenty C_{24} Polybead Chains

Three runs (2×10^6 steps each) were made and analyzed. The first consisted of 90% reptation and 10% random MMC moves. The second run consisted of 70% CCB moves with cuts of the last 14 bonds from both of the chains, 20% reptation, and 10% random moves. The last run consisted of 50% ECROT1 moves as described in Sect. 4.1, 20% CCB moves, 20% Reptation and 10% random moves. Average acceptance rates of the various methods did not change significantly over all these runs, being 42% for MMC moves, 12% for Reptations, 17% for CCB moves, and 17% for ECROT1 moves. The results of these runs are plotted in Figs. 11 and 12. Figure 11 contains the bond orientation correlation functions of first and second order, and the end-to-end relaxation function for all three runs. It can be seen that the combination of CCB with Reptations is optimal for this system. Concerted-Rotations appear to reduce the efficiency of the algorithm, but still produce good results at the level of the chain

Table 4. Time requirements and statistical inefficiencies, $I(\Omega)$[a], for the long runs performed on the C_{24} polybead and the C_{71} polybread systems

System/ Method	Time required on a CRIMSON workstation (cpu s/10^3 moves)	$I(\rho)$[b]	$I(U_{tot})$[b]	$I(U_{inter})$[b]	$I(U_{intra})$[a]	$I(U_{tors})$[b]	$I(e\text{-}e\text{-}d)$[b]
C_{24} 10% MMC 90% REPT	78	350 + 50	400 ± 100	400 ± 100	170 ± 50	320 ± 50	150 ± 20
C_{24} 10% MMC 20% REPT 70% CCB	136	300 ± 50	90 ± 20	200 ± 50			70 ± 10
C_{24} 10% MMC 20% REPT 20% CCB 50% ECROT1	187	400 ± 100	100 ± 20	250 ± 50	65 ± 15	20 ± 5	120 ± 30
C_{71} 20% REPT 20% CCB 60% ECROT1	486	500 ± 100	330 ± 50	330 ± 50	400 ± 100	20 ± 10	450 ± 100

[a] Statistical inefficiency (in units of 10^3 simulation steps)
[b] ρ is the density, U_{tot} the total configurational energy, U_{inter} the contribution from LJ intermolecular interactions, U_{intra} the contribution from LJ intramolecular interactions, U_{tors} the torsional potential energy, e-e-d the end-to-end distance

segments. Pure reptation moves sample configuration space very slowly, even for this system of relatively short chains. The reason that CCB is the best overall method for the C_{24} polybead system is the following: Cut-and-regrow moves of 14 bonds from a chain end have an acceptance rate of approximately 8% [46]. These moves change drastically the configuration of a significant part of a C_{24} chain.

Figure 12 shows the diffusion of the center of mass of C_{24} during these simulations. Reptation alone provides little towards displacing the center of mass of the chains; the data indicates that the diffusion regime has not yet been attained, even though the mean-square displacement of the center of mass is larger than s^2 (square of the radius of gyration; s of the C_{24} chains is 7.0 Å, that of the C_{71} is 13.5 Å). Introducing CCB dramatically improves performance: the C_{24} chains diffuse over a distance of 5s during 10^6 steps. The mean-square displacement of the center of mass becomes approximately proportional to the number of simulation steps after 10^6 steps if CCB is primarily used, indicating that the diffusion regime for self-diffusion of the C_{24} chains is attained. By contrast the center of mass of C_{71} chains moves only by a distance s in 10^6 moves. Substituting roughly 2/3 of CCB moves for ECROT1 moves proves to be detrimental to the progress of the center of mass through space.

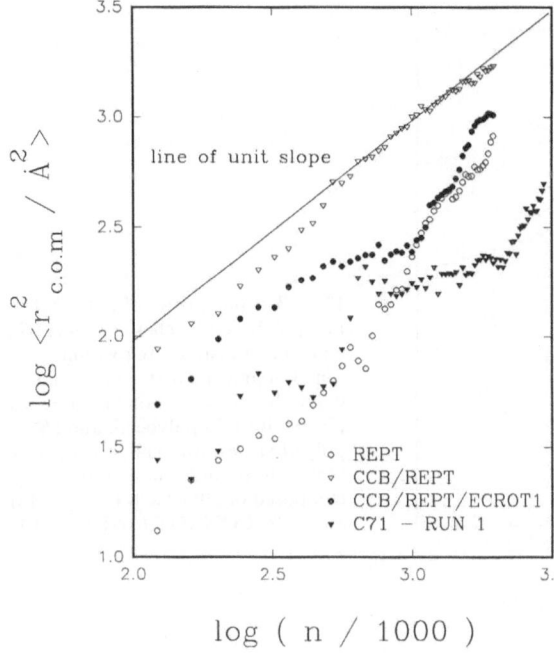

Fig. 12. Logarithm of the mean-square displacement of the center of mass of the C_{24} chains as a function of the logarithm of the number of simulation steps. The symbols are as in Fig. 11. *Filled triangles* are results of the first C_{71} run of Fig. 10

Time requirements and statistical inefficiencies are reported in Table 4. The relative speed of the move types Reptation:CCB:CONROT is roughly $3:2:1$. Inefficiency results indicate that different properties decorrelate at different rates from their starting values. CONROT moves produce drastic decorrelation of the torsional energy of the chains, and CCB leads to faster decorrelation of the end-to-end distance, as expected. For the C_{71} system it is rather disconcerting that 3×10^5 moves are needed to produce a configuration that is totally statistically uncorrelated from the starting configuration. Table 4 also indicates that Reptation alone is a poor method for the treatment of C_{24} chains, since the autocorrelations of energy and end-to-end distance decay very slowly. This is a restatement of the results of Fig. 11. Density relaxation and the radial distribution function are rather insensitive measures of simulation efficiency, since they are practically identical for the three algorithms applied to the C_{24} chains. It can be concluded that simulations comprising CCB and CONROT moves provide a good basis for the treatment of C_{24} chains.

4.3 A Melt of Ten C_{71} Polyethylene Chains with Explicit H-Atoms

A run of 7.5×10^5 steps was performed consisting of 20% CCB moves ($N_{CUT} = 14$), and 80% biased CONROT moves with a maximum driver angle

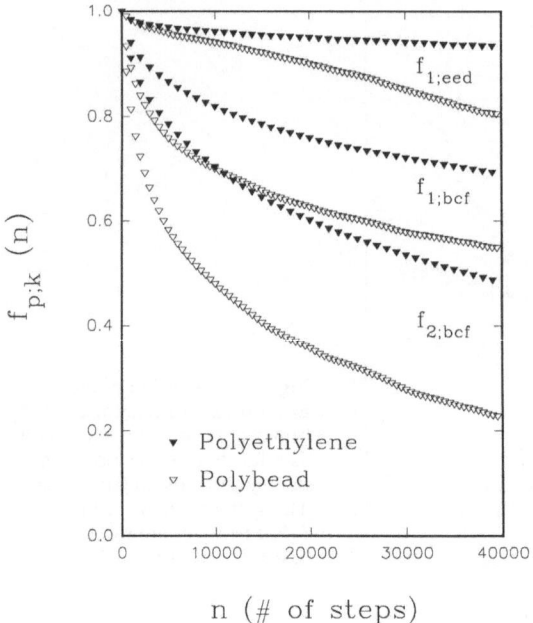

$f_{p;k}(n)$

n (# of steps)

Fig. 13. Comparison of polyethylene and polybead C_{71} chain systems. The end-to-end distance relaxation function and the first and second degree bond correlation functions are plotted for (∇) polybead, and ($\blacktriangledown$) polyethylene with explicit hydrogen atoms. Both algorithms were composed of 20% CCB ($N_{CUT} = 15$), and 80% CONROT ($\Delta\phi_3^{max} = 10°$)

"wedge" of 10°. No reptation moves were applied. The first and second degree bond correlation functions and the end-to-end relaxation function are plotted in Fig. 13. The corresponding results for the polybead system (inverted empty triangles from Fig. 7) are also included for direct comparison. Besides rendering the calculations more expensive (since for the same number of degrees of freedom there are now three times more interacting sites in the system), the presence of explicit hydrogen atoms reduces relaxation rates considerably, as evidenced by the much slower decay of the bond correlation and end-to-end distance relaxation functions.

It takes clearly more than 10^6 steps to simulate an equilibrated melt of C_{71} polyethylene chains with CCB or CONROT. Evidence for this is also provided in Fig. 10 (bottom), where the end-to-end distance fluctuations are found to be much smaller than the corresponding fluctuations of the polybead model, and the system remains close to the original (collapsed-chain) configuration. Statistical inefficiencies for energy and end-to-end distance could not be calculated from this exploratory run, so that the decorrelation length must be at least equal to the length of the run. One therefore needs more than 10^6 moves to generate a new uncorrelated configuration! It is also evident that the efficiency of CONROT, as measured from the relaxation of autocorrelation functions, is inferior for polyethylene than for the corresponding polybead system. Nevertheless, the bond correlation function $f_{2;bcf}(n)$ drops to 0.2 after 2×10^5 steps (results not shown here). Such a relaxation rate, although lower than that observed in the polybead system, is in itself quite satisfactory and cannot be achieved with other methods.

5 Discussion and Concluding Remarks

Our results from off-lattice simulations of dense polymer phases, obtained with a combination of newly available simulation tools, demonstrate that the Concerted-Rotation MC algorithm, introduced recently by Dodd et al. [47], is a valuable tool for the simulation of realistic polymer models. While CONROT does not constitute an improvement over existing methods for systems with many chain ends, CONROT-based algorithms considerably extend the range of systems and problems that can be studied with MC methods. Such algorithms are particularly powerful when the solutions to the geometric CONROT problem are selected with bias.

A dense system of C_{71} polyethylene chains can now be employed to simulate certain aspects of realistic polymer models, although very long simulations are needed to sample the phase space of these systems in a satisfactory way, especially for models with explicit hydrogen atoms or (probably) side groups. Dense phases of long chains (of 50 carbon atoms or more) can still not be fully explored, but CONROT algorithms provide rapid relaxations at the segmental level. It is thus possible to apply such methods to the study of problems where the relevant interactions occur at the segmental level, e.g. of polymers interacting with small guest molecules.

It is rather unexpected that the presence of the modestly-sized explicit hydrogen atoms dramatically reduces the efficiency of seemingly robust methods such as CCB and CONROT, that perform reasonably well on polybead models. The increased probability of overlaps due to the small hydrogen apparently has a strong effect on acceptance rates for both CCB and CONROT. We conclude that the polybead chain, although a very useful polymer model for development work, might, due to its "smoothness", lead to optimistic conclusions when assessing the suitability of a given MC method for the simulation of dense phases of chemically detailed polymers. Bulkier side groups exacerbate the severity of the problem.

The basic CONROT move developed by Dodd et al. can form an important component of more complex MC moves, involving, for example, serial combinations of different techniques. The extended CONROT (ECROT) moves presented in this paper are simple examples of the possibilities that become available with CONROT algorithm. These moves could be extended even further to encompass more degrees of freedom, by performing Metropolis moves on additional torsional angles, and augmenting the operational unit. However, there is a limit to the advantages obtainable in this way. Acceptance rates drop as more chain sites become involved. This effect counterbalances the greater configurational space change per move achieved by the extended algorithms.

Dodd et al. have also discussed possible extensions of the basic CONROT methodology that they developed. They suggest that CONROT may allow chain-identity-exchange moves, in the spirit of the "pseudokinetic" algorithms of Olaj et al. [34a], Mansfield [34b], and Madden and coworkers [76]. These are bold, "unphysical", two-chain moves that may be expected to alter dramatically the convergence of thermodynamic properties, if reasonable acceptance rates are achieved. However, any "dynamic information" inherent in the CONROT move will be lost, and time calibration might not be possible any more when two chains exchange parts. In spite of this drawback, we believe that this will be a very welcome addition to the arsenal of new MC moves for off-lattice polymer simulation. Extended CONROT moves, such as ECROT1, may actually facilitate the design of pseudokinetic moves, since they allow spanning of larger distances in the simulation unit cell.

The more elaborate MC moves have been contemplated with two long-range aims in mind:

The first goal is to improve the efficiency of MC algorithms to such a point, that it will be possible to conveniently simulate dense phases (primarily melts and glasses) of chemically realistic polymers of arbitrary length. Current lattice-based polymer models, such as the Bond-Fluctuation model [35], do offer such a promise. However, we believe that there exists a large potential for improvement of MC algorithms for off-lattice polymer models in generalized coordinate space. Novel methods, such as CCB and CONROT, constitute such an improvement. In addition, CCB and CONROT offer incentives to abandon half-century old model systems, such as Gaussian chains [7, 9], in a number of applications, since it is now possible to attack some problems with more realistic chain models, such as the polybead model.

The second goal is to attempt to employ multi-chain step methods for many-chain polymer systems. We believe that the often-heard argument against simultaneous displacement of many degrees of freedom, based on results obtained with Lennard-Jones particles in a dilute phase [27], is not valid in dense many-chain polymer systems. Molecular and Stochastic Dynamics simultaneously update all degrees of freedom [4–6, 10–12, 15, 16], and important relaxation processes in dense polymer systems involve coordinated motions of many segments of many chains [31, 32]. However, multiple displacements in a MC context must be carried out with bias, to ensure efficient sampling of configuration space. CCB and CONROT go a long way in this direction, since they involve biased combined moves of many degrees of freedom, although the elementary moves are single-chain moves. Multiple-chain CCB and CONROT moves, or many-chain moves that contain CCB and CONROT as important components, appear to hold considerable promise.

Acknowledgements. Financial support from the german Bundesministerium für Forschung und Technologie (grant No. 03M4028) is gratefully acknowledged. We would like to thank Professors Larry Dodd and Doros Theodorou for many enlightening discussions on the CONROT method and on polymer melt simulations in general. We would also like to acknowledge the Swiss Center for Scientific Calculations (CSCS) for generous allocation of Supercomputer time on the NEC SX-3 Supercomputer of Manno, Switzerland.

References

1. Abraham FF (1986) Adv Phys 35: 1
2. Sommer K, Batoulis J, Jilge W, Morbitzer L, Pittel B, Plaetschke R, Reuter K, Timmermann R, Binder K, Paul W, Gentile FT, Heermann DW, Kremer K, Laso M, Suter UW, Ludovice PJ (1991) Adv Mater 3: 590
3. Ludovice PJ, Suter UW (1989) In: Encyclopedia of polymer science and engineering, Supplement Vol. 2nd edn. John Wiley, New York, p 11
4. Van Gunsteren WF, Berendsen HJC (1990) Angew Chem Int Eng 29: 992
5. McCammon JA, Harvey SC (1987) Dynamics of proteins and nucleic acids. Cambridge Univ Press, Cambridge
6. Heermann DW (1990) Computer simulation methods, 2nd edn. Springer, Berlin Heidelberg New York
7. Binder K (1992) In: Bicerano J (ed) Computational modeling of polymers. Marcel Dekker, New York
8. Roe RJ (ed) (1991) Computer simulation of polymers. Prentice Hall, Englewood Cliffs
9. Baumgartner A (1987) In: Binder K (ed) Applications of the Monte Carlo method, 2nd edn. Springer, Berlin Heidelberg New York
10. Van Gunsteren WF, Berendsen HJC, Rullmann JAC (1981) Mol Phys 44: 69
11. Helfand E, Wasserman ZR, Weber TA (1980) Macromolecules 13: 526
12. Toxvaerd S (1987) J Chem Phys 86: 3667
13. Rossky PJ, Doll JD, Friedman HL (1978) J Chem Phys 69: 4628
14. Pangali C, Rao M, Berne BJ Chem Phys Let 55: 413
15. (a) Kotelyanskii MJ, Suter UW (1992) J Chem Phys 96: 5383; (b) (1992) ACS Polymer Preprints 33: 663
16. (a) Mehlig B, Heermann DW, Forrest BM (1992) Phys Rev B 45: 679; (b) (1992) Mol Phys 76: 1347; (c) Forrest BM, Suter UW (1994) Mol Phys (in press)
17. (a) Theodorou DN, Suter UW (1986) Macromolecules 18: 1467; (b) (1985) ibid 19: 139; (c) (1986) ibid 19: 379
18. Hutnik M, Argon AA, Suter UW (1991) Macromolecules 24: 5970
19. (a) Rutledge GC, Suter UW (1992) Polymer 32: 2179; (b) (1991) Macromolecules 25: 1546
20. Boyd RH, Krishna Pant PV (1991) Macromolecules 24: 4073
21. Gusev AA, Suter UW (1991) Phys Rev A 43: 6488
22. (a) Gusev AA, Suter UW (1992) ACS Polymer Preprints 33: 631; (b) (1993) J Chem Phys 99: 2221
23. Mansfield KF, Theodorou DN (1991) Macromolecules 24: 6283
24. (a) Rigby D, Roe R-J (1989) J Chem Phys 87: 7285; (b) (1987) ibid 89: 5280; (c) (1988) Macromolecules 22: 2259
25. Kremer K, Grest G (1990) J Chem Phys 92: 5057
26. Smith GD, Boyd RH (1992) Macromolecules 25: 1326
27. Allen MP, Tildesley DJ (1989) Computer simulation of liquids. Clarendon, Oxford
28. (a) Berendsen HJC, Postma JPM, Van Gunsteren WF, DiNola A, Haak JR (1984) J Chem Phys 81: 3684; (b) (1984) Nosé, S Mol Phys 52: 255
29. Andersen HC (1980) J Chem Phys 72: 2384
30. (a) Çagin T, Pettitt M (1991) Mol Phys 72: 169; (b) (1991) Mo Simul 6: 5
31. (a) Bailey RT, North AM, Pethrick RA (1981) Molecular motion in high polymers. Clarendon, Oxford; (b) Sperling LH (1992) Introduction to Physical Polymer Science. Wiley-Interscience, New York
32. Boyer RF (1992) In: Bicerano J (ed) Computational modeling of polymers. Marcel Dekker, New York
33. Binder K (1986) Monte Carlo methods in statistical physics, Springer, Berlin Heidelberg New York
34. (a) Olaj OF, Lantschbauer W, Macromol (1987) Chem Rapid Commun 3: 847; (b) Mansfield ML (1982) J Chem Phys 77: 1554; (c) Pakula T (1982) Macromolecules 20: 679; (d) Pakula T, Geyler S (1987) ibid 20: 2909
35. (a) Carmesin I, Kremer K (1992) Macromolecules 21: 2819; (b) Wittmann HP, Kremer K, Binder K (1988) J Chem Phys 96: 6291
36. Baschnagel J, Binder K, Paul W, Laso M, Suter UW, Batoulis I, Jilge W, Burger T (1991) J Chem Phys 95: 6014
37. Bishop M, Ceperley D, Frisch HL, Kalos MH (1980) J Chem Phys 72: 3228
38. Vacatello M, Avitabile G, Corradini P, Tuzi A (1980) J Chem Phys 73: 548

39. Boyd RH (1989) Macromolecules 22: 2477
40. Almarza NG, Enciso E, Bermejo FJ (1992) J Chem Phys 96: 4625
41. Northrup SH, McCammon JA (1980) Biopolymers 19: 1001
42. Siepmann JI (1990) Mol Phys 70: 1145
43. Frenkel D, Mooij GCAM, Smit B (1991) J Phys Condens Matter 3: 3053
44. Siepmann JI, Frenkel D (1992) Mol Phys 75: 59
45. de Pablo JJ, Laso M, Suter UW (1992) J Chem Phys 96: 2395
46. Widmann A unpublished results
47. (a) Dodd LR, Boone TD, Theodorou DN (1992) Mol Phys in print; (b) ACS Polymer Preprints 33: 645; (c) Dodd LR, Theodorou DN (1994) article in this volume
48. Gō N, Scheraga HA (1970) Macromolecules 3: 178
49. Flory PJ (1974) Macromolecules 7: 381
50. Gō N, Scheraga HA (1976) Macromolecules 9: 535
51. Kitao A, Gō NJ (1991) J Comput Chem 12: 359
52. Fixman M (1974) Proc Natl Acad Sci USA 71: 3050
53. Helfand E (1979) J Chem Phys 71: 5000
54. (a) Gottlieb M, Bird RB (1977) J Chem Phys 65: 2467; (b) Gottlieb M (1976) Comp & Chem 1: 155
55. (a) Pear MR, Weiner JH (1980) J Chem Phys 71: 212; (b) (1979) ibid 72: 3939
56. Van Gunsteren WF (1980) Mol Phys 40: 1015
57. Van Gunsteren WF, Karplus M (1982) Macromolecules 15: 1528
58. Perchak D, Weiner JH, (1982) Macromolecules 15: 545
59. Perchak D, Skolnick J, Yaris R (1985) Macromolecules 18: 519
60. Almarza NG, Enciso E, Alonso J, Bermejo FJ, Alvarez M (1990) Mol Phys 70: 485
61. Valleau JP, Whittington SG (1977) In: Berne BJ (ed) Statistical mechanics a equilibrium techniques. Plenum Press, NY
62. (a) De Gennes PG (1971) J Chem Phys 55: 572; (b) De Gennes PG (1979) Scaling concepts in Polymer Physics. Cornell Univ Press, Ithaca
63. Wall FT, Mandel F (1975) J Chem Phys 63: 4592
64. Rosenbluth MN, Rosenbluth AW (1955) J Chem Phys 23: 356
65. Meirovich H (1984) Macromolecules 17: 2038
66. Öttinger HC (1985) Macromolecules 81: 93
67. Meirovich H (1985) Phys Rev A 32: 3699
68. Livne S, Meirovich H (1988) J Chem Phys 88: 4498
69. Harris J, Rice SA (1988) J Chem Phys 88: 1298
70. (a) de Pablo JJ, Laso M, Suter UW (1992) J Chem Phys 96: 6157; (b) Laso M, de Pablo JJ, Suter UW (1992) ibid 97: 2817; (c) de Pablo JJ, Laso M, Suter UW Cochran HD (1993) Fluid Phase Equilibria 83: 323
71. Frenkel D, Smit B (1992) Mol Phys 75: 983
72. Siepmann JI, McDonald IR (1992) Mol Phys 72: 255
73. Palmer KA, Scheraga HA (1991) J Comput Chem 12: 505
74. Flory PJ (1989) Statistical mechanics of chain molecules. Hanser, Munich
75. Boyd RH, Breitling SM (1974) Macromolecules 7: 855
76. (a) Madden WG (1987) J Chem Phys 87: 1405; (b) (1988) ibid, 88: 3934; (c) Lastoskie CM, Madden WG, in Roe RJ, (ed) (1991) Computer simulation of polymers. Prentice Hall, Englewood Cliffs
77. Ryckaert JP, Bellemans A (1975) Chem Phys Let 30: 123
78. Abe A, Jernigan RL, Flory PJ (1966) J Am Chem Soc 88: 631
79. Ketelaar J (1958) Chemical constitution. Elsevier, NY
80. Dee GT, Ougizawa T, Walsh DJ (1992) Polymer 33: 3462
81. de Pablo JJ, Laso M, Suter UW, Siepmann (1993) Mol Phys 80: 55
82. Wittmann H-P, Kremer K, Binder K (1992) Macrom Chem Theory Simul 1: 275
83. Kolinski A, Skolnick J, Yaris R (1986) J Chem Phys 84: 1922
84. Smith W Rapaport DC (1992) Mol Simul 9: 25
85. Dodd LR, personal communication
86. Bahar I, Erman B, Monnerie L (1989) Macromolecules 22: 431

Received: November 1993

PRISM Theory of the Structure, Thermodynamics, and Phase Transitions of Polymer Liquids and Alloys

K.S. Schweizer[1] and J.G. Curro[2]
[1] Departments of Materials Science & Engineering and Chemistry, University of Illinois, Urbana, Illinois 61801, USA
[2] Sandia National Laboratories, Albuquerque, New Mexico 87185, USA

The recent development of a microscopic theory of the equilibrium properties of polymer solutions, melts and alloys based on off-lattice Polymer Reference Interaction Site Model (PRISM) integral equation methods is reviewed. Analytical and numerical predictions for the intermolecular structure and collective density scattering patterns of both coarse-grained and atomistic models of polymer melts are presented and found to be in good agreement with large scale computer simulations and diffraction measurements. The general issues and difficulties involved in the use of the structural information to compute thermodynamic properties are reviewed. Detailed application of a hybrid PRISM approach to calculate the equation-of-state of hydrocarbon fluids is presented and found to reproduce accurately experimental PVT data on polyethylene. The development of a first principles off-lattice theory of polymer crystallization based on a novel generalization of modern thermodynamic density functional methods is discussed. Numerical calculations for polyethylene and polytetrafluoroethylene are in good agreement with the experimental melting temperatures and liquid freezing densities. Generalization of the PRISM approach to treat phase separating polymer blends is also discussed in depth. The general role of compressibility effects in determining small angle scattering patterns, the effective chi-parameter, and spinodal instability curves are presented. New theoretical concepts and closure approximations have been developed in order to describe correctly long wavelength concentration fluctuations in macromolecular alloys. Detailed numerical and analytical applications of the PRISM theory to model athermal and symmetric blends are presented, and the role of nonmean field fluctuation processes are established. Good agreement between the theory and computer simulations of simple symmetric polymer blends has been demonstrated. Strong, nonadditive compressibility effects are found for structurally and/or interaction asymmetric blends which have significant implications for controlling miscibility in polymer alloys. Recent generalizations of PRISM theory to treat block copolymer melts, and nonideal conformational perturbations, are briefly described. The paper concludes with a brief summary of ongoing work and fertile directions for future research.

Advances in Polymer Science, Vol. 116
© Springer-Verlag Berlin Heidelberg 1994

1 Introduction

Understanding the intermolecular packing, conformation, thermodynamics, and phase transitions of high polymer fluids and alloys is both a challenging problem in statistical mechanics and a subject of immense technological importance. Historically, theoretical progress has been possible only by introducing extreme approximations of both a chemical structural and statistical mechanical nature. Simple lattice models solved at the lowest mean field level represent the classical approach [1]. Recently, Freed, Dudowicz and collaborators have pioneered the development of statistical thermodynamic theories of sophisticated lattice models which can deal approximately with monomer shape effects [2]. Fluctuation corrections to the simple random mixing approximation are perturbatively computed, and many interesting results have emerged. However, the lattice restriction is still, potentially, a major limitation, especially if local, chemically specific packing effects are important. Structural properties are also not explicitly addressed. Phenomenological field theoretical and/or Landau expansion approaches have been extensively developed [3] and provide considerable qualitative insight into the general aspects of phase behavior and polymer conformation. However, such approaches heavily coarse-grain over chemical structure and often introduce a locally unrealistic incompressibility approximation. Thus, packing correlations on spatial scales of the order of and less than the statistical segment length are ignored. From a practical point of view, the presence of empirical parameters of uncertain relation to the microscopic intermolecular forces and chemical structure of specific materials limit the detailed predictive power of such theories.

A distinctly different approach, which has witnessed much progress recently, is large scale Monte Carlo and molecular dynamics computer simulations [4]. These studies provide many insights regarding the physics of model polymer fluids, and also valuable benchmarks against which approximate theory can be tested. However, an atomistic, off-lattice treatment of high polymer fluids and alloys remains immensely expensive, if not impossible, from a computational point of view.

Over the past several years we and our collaborators have pursued a continuous space liquid state approach to developing a computationally convenient microscopic theory of the equilibrium properties of polymeric systems. Integral equations methods [5–7], now widely employed to understand structure, thermodynamics and phase transitions in atomic, colloidal, and small molecule fluids, have been generalized to treat macromolecular materials. The purpose of this paper is to provide the first comprehensive review of this work referred to collectively as "Polymer Reference Interaction Site Model" (PRISM) theory. A few new results on polymer alloys are also presented. Besides providing a unified description of the equilibrium properties of the polymer liquid phase, the integral equation approach can be combined with density functional and/or other methods to treat a variety of inhomogeneous fluid and solid problems.

The outline of the paper is as follows. In Sect. 2 we describe the basic RISM and PRISM formalisms, and the fundamental approximations invoked that render the polymer problem tractable. The predictions of PRISM theory for the structure of polymer melts are described in Sect. 3 for a variety of single chain models, including a comparison of atomistic calculations for polyethylene melt with diffraction experiments. The general problem of calculating thermodynamic properties, and particularly the equation-of-state, within the PRISM formalism is described in Sect. 4. A detailed application to polyethylene fluids is summarized and compared with experiment. The development of a density functional theory to treat polymer crystallization is briefly discussed in Sect. 5, and numerical predictions for polyethylene and polytetrafluoroethylene are summarized.

The second general part of this paper describes the PRISM theory of phase-separating polymer mixtures. This aspect is less well developed than the one-component melt problem, especially with regards to its atomistic implementation for real materials. However, construction of a microscopic theory of polymer alloys is presently of great interest both scientifically and due to the technological desire to "molecularly engineer" miscibility and phase structure. The general theoretical issues for polymer blends within the PRISM framework are discussed in Sect. 6. Specific applications to model athermal polymer mixtures are summarized in Sect. 7. Section 8 treats thermally-induced phase separating blends, and the subtle question of the appropriate closure approximation is discussed. New "molecular-based" closure approximations are described, and representative results for model symmetric polymer blends are presented and compared with recent Monte Carlo simulations. Analytical results are also presented for stiffness and interaction potential asymmetric blends where non-additive compressibility effects are found to be very important. PRISM theory of periodic block copolymers is briefly described in Sect. 9. Section 10 discusses the generalization of PRISM methodology to allow an ab initio assessment of "nonideal" conformational effects. The paper concludes with a brief overview of ongoing and future directions.

2 PRISM Theory

The theoretical approach we take to describe amorphous polymer liquids is based on integral equation theory which has its roots in the theory of mon-atomic liquids [5]. Consider for a moment a uniform system of n spherical particles of density $\rho = n/V$. A convenient measure of the degree of order in such a system is the radial distribution function $g(r)$ defined as

$$\rho^2 g(r) = \left\langle \sum_{i \neq j = 1}^{n} \delta(\vec{r}_i) \delta(\vec{r} - \vec{r}_j) \right\rangle. \tag{2.1}$$

Physically $\rho g(r)$ is the density of particles at distance r from a given particle. Most thermodynamic properties of interest can be computed from a knowledge of $g(r)$ and the interparticle pair potential $v(r)$. The starting point in calculating the radial distribution function is the well known Ornstein-Zernike equation [5]:

$$h(r) = C(r) + \rho \int C(|\vec{r} - \vec{r}'|)h(r')d\vec{r}' \tag{2.2}$$

where $h(r) = g(r) - 1$ is frequently called the total correlation function and approaches zero at large r. Equation (2.2) serves as a definition for the direct correlation function $C(r)$ which plays a central role in liquid state physics.

Unfortunately $C(r)$ does not have a readily apparent physical interpretation. However, some insight [5, 6] can be gained by iterating Eq. (2.2) in explicit order of density

$$h(r) = C(r) + \rho C^*C(r) + \rho^2 C^*C^*C(r) + \rho^3 C^*C^*C^*C(r) + \cdots \tag{2.3}$$

where the * operator denotes the convolution integral following standard notation. We may interpret this equation as follows: two given particles at distance r apart are correlated "directly" from the first term on the RHS of Eq. (2.3); subsequent higher order terms in density represent "indirect" correlations between the two given particles mediated by the remaining particles in the system. At low density, high temperature, or large separation the direct correlation function is exactly related to the pair potential as $C(r) = \exp[-\beta v(r)] - 1 \cong -\beta v(r)$ where $\beta = 1/k_B T$ and k_B is Boltzmann's constant. At higher densities $C(r)$ is strongly modified due to many-particle effects. However, it can be argued [5, 6] that $C(r)$ still has roughly the same range as $v(r)$ itself and is a simpler object to approximate than $h(r)$. Since Eq. (2.3) shows that $h(r)$ can be expressed as an expansion in $C(r)$, one can view the direct correlation function, for qualitative purposes, as an effective or renormalized pair potential which accounts for the many body contribution to $h(r)$.

The expectation that $C(r)$ is a short range, relatively "simple" function can be exploited to develop a second approximate relationship, or closure relation, between $h(r)$ and $C(r)$. Based on graph theoretical techniques [5] or functional expansions [5, 7], one can deduce the Percus-Yevick approximation:

$$C(r) \cong \{1 - \exp[\beta v(r)]\}g(r) \equiv \{\exp[-\beta v(r)] - 1\}y(r) \tag{2.4a}$$

where the "indirect" correlation function, $y(r)$, is a continuous and finite function for all values of r. For hard spheres of diameter d, Eq. (2.4a) takes the simple form

$$\begin{aligned} g(r) &= 0 \quad r < d \\ C(r) &\cong 0 \quad r > d \; . \end{aligned} \tag{2.4b}$$

The first condition on $g(r)$ inside the hard core is an exact statement of the impenetrability of hard spheres. The second condition on $C(r)$ outside the hard core is approximate and emphasizes that $C(r)$ and $v(r)$ have roughly the same spatial range. Equations (2.4) plus the Ornstein-Zernike equation, Eq. (2.2), lead to the well-known Percus-Yevick integral equation theory that has been successful [5] in describing the structure of monatomic liquids with strongly repulsive and weakly attractive interactions.

These integral equation ideas of monatomic liquids were generalized and applied to molecular liquids by Chandler and Andersen [6, 8] to formulate the Reference Interaction Site Model or RISM theory of molecular fluids. In the RISM approach, each molecule is subdivided into spherically symmetric, interaction sites. The intermolecular pair structure of a uniform molecular liquid of M molecules is now specified through a site-site radial distribution function matrix $g_{\alpha\gamma}(r)$:

$$\rho^2 g_{\alpha\gamma}(r) = \left\langle \sum_{i \neq j = 1}^{M} \delta(\vec{r}_i^\alpha)\delta(\vec{r} - \vec{r}_j^\gamma) \right\rangle . \tag{2.5}$$

In Eq. (2.5), ρ is the number density of molecules and $\vec{r}_i^\alpha$ is the position vector of site α on molecule i. It can be seen from Eq. (2.5) that $g_{\alpha\gamma}(r)$ is the intermolecular radial distribution function for sites α and γ on different molecules. Chandler and Andersen generalized the Ornstein-Zernike equation to reflect the fact that in molecular liquids, unlike monatomic liquids, there exist intramolecular correlations. This generalized Ornstein-Zernike-like, or RISM, equation has the form [6, 8]

$$\underline{h}(r) = \iint d\vec{r}_1 d\vec{r}_2 \underline{\omega}(|\vec{r} - \vec{r}_1|)\underline{C}(|\vec{r}_1 - \vec{r}_2|)[\underline{\omega}(r_2) + \rho\underline{h}(r_2)] \tag{2.6}$$

where $\underline{h}(r)$, $\underline{C}(r)$, and $\underline{\omega}(r)$ are $N \times N$ matrices (for molecules consisting of N sites) with matrix elements $h_{\alpha\gamma}(r) = g_{\alpha\gamma}(r) - 1$, $C_{\alpha\gamma}(r)$, and $\omega_{\alpha\gamma}(r)$, respectively. The functions $h_{\alpha\gamma}(r)$ and $C_{\alpha\gamma}(r)$ are intermolecular correlations functions between sites on different molecules. In contrast, $\omega_{\alpha\gamma}(r)$ is the intramolecular probability density between a pair of sites on the same molecule. For molecules composed of hard sphere sites the closure for RISM theory is carried over by analogy with the Percus-Yevick closure of monatomic liquids.

$$\begin{aligned} g_{\alpha\gamma}(r) &= 0 \quad r < d_{\alpha\gamma} \\ C_{\alpha\gamma}(r) &\cong 0 \quad r > d_{\alpha\gamma} . \end{aligned} \tag{2.7}$$

Equations (2.6) and (2.7) form a set of nonlinear integral equations which can be solved by standard numerical techniques [5, 9]. Thus, by solution of these integral equations, one can obtain a quantitative description of the short range, intermolecular packing in the liquid state which includes the effect of the intramolecular bonding constraints. Chandler and coworkers used this RISM formalism to study various small, rigid polyatomic liquids [9, 10] and found good agreement with computer simulation and scattering experiments for the intermolecular structure.

We have generalized the RISM formalism to describe the structure of flexible polymer chain liquids [11–15]. The application of RISM theory to polymers is referred to as polymer RISM, or PRISM, theory. (Note that there is no connection with the well known *rotational isomeric state model* commonly employed to calculate the intramolecular structure of isolated polymer chains [16].) In the case of rigid molecules, the intramolecular probability functions $\omega_{\alpha\gamma}(r)$ are simply delta functions which specify the positions of the atoms or sites on the molecule. When the PRISM theory is applied to flexible polymers, a difficulty arises because in general the intramolecular functions $\omega_{\alpha\gamma}(r)$ and the intermolecular radial distribution functions $g_{\alpha\gamma}(r)$ depend on each other [6]. A rigorous calculation would thus require that the intramolecular and intermolecular structure be determined self consistently. This general problem is discussed in Sect. 10. For one-component polymer melts, we can avoid this difficult self-consistent calculation, at least nominally, by invoking the Flory *ideality hypothesis* [17]. The *intra*molecular excluded volume forces that tend to expand a polymer chain in dilute solution are nominally cancelled in the melt by *inter*molecular excluded volume interactions. The net result of these two opposing effects is that the polymer melt acts as a theta solvent for itself and the average intramolecular structure of a chain is ideal. There is ample evidence that these ideas are substantially correct based on computer simulations [18] and neutron scattering experiments [19] on polymer melts. Within this approximation the intramolecular structure can be calculated from an "ideal" single chain model without long range excluded volume interactions [16]. The result of this single chain calculation is then used as input to the PRISM theory to compute the intermolecular correlation functions $g_{\alpha\gamma}(r)$ and $C_{\alpha\gamma}(r)$ in the polymer melt.

Inspection of the generalized Ornstein-Zernike-like equations reveals that, for an ideal chain molecule fluid consisting of N(even) identical (but still symmetry nonequivalent) sites, the matrix equation, Eq. (2.6), consists of $N(N + 2)/8$ independent integral equations. The large number of coupled integral equations results from the fact that the correlation functions $g_{\alpha\gamma}(r)$ depend on the specific locations of site α on one chain and site γ on another chain. For a polymer the number of equations would obviously become unmanageably large. We can argue [13], however, that, for long polymer chains, end effects can be neglected, and thus all the sites on a homopolymer chain can be treated as equivalent. This simplification is exact for a ring homopolymer [12]. Such an equivalent-site approximation leads to a considerable mathematical simplification and renders the calculation of the structure of polymer melts a tractable problem. A computationally tractable scheme for computing "chain end" corrections has also been formulated [13]. Taking all the sites as equivalent on average allows us to write

$$h(r) = g(r) - 1 = h_{\alpha\gamma}(r) ,$$

$$C(r) = C_{\alpha\gamma}(r) . \tag{2.8a}$$

With this simplification, and subsequent preaveraging over the intramolecular structure, the generalized Ornstein-Zernike-like matrix equations reduce to a single scalar equation of the form

$$h(r) = \iint d\vec{r}_1 d\vec{r}_2 \omega(|\vec{r} - \vec{r}_1|) C(|\vec{r}_1 - \vec{r}_2|)[\omega(r_2) + \rho_m h(r_2)] \qquad (2.8b)$$

where ρ_m is the site density $N\rho$, and $\omega(r)$ is defined as the intramolecular distribution averaged over all pairs of sites on a single chain, and

$$\omega(r) = \frac{1}{N} \sum_{\alpha, \gamma = 1}^{N} \omega_{\alpha\gamma}(r) . \qquad (2.9)$$

Since it is the second relation in Eq. (2.8a) which is the fundamental approximation, $h(r)$ in Eq. (2.8b) can be rigorously interpreted as the "average" correlation function, i.e. $h(r) = N^{-2} \sum_{\alpha\gamma}^{N} h_{\alpha\gamma}(r)$. Recent work has demonstrated that the equivalent site approximation is remarkably accurate for $h(r)$, even for very short alkanes such as propane [20] ($N = 3$) and butane [21] ($N = 4$).

For hard core homopolymers, $d = d_{\alpha\gamma}$ and the Percus-Yevick/RISM closure of Eq. (2.7) applies. Using the convolution theorem for Fourier transforms, Eq. (2.8b) can be conveniently written as

$$\hat{h}(k) = \hat{\omega}^2(k)\hat{C}(k) + \rho_m \hat{\omega}(k)\hat{C}(k)\hat{h}(k) \qquad (2.10)$$

where the caret denotes Fourier transformation with wave vector k. $\hat{\omega}(k)$ is the Fourier transform of Eq. (2.9) and can be identified with the single chain structure factor:

$$\hat{\omega}(k) = \frac{1}{N} \sum_{\alpha, \gamma = 1}^{N} \hat{\omega}_{\alpha\gamma}(k) . \qquad (2.11)$$

Equations (2.8) together with the closure in Eq. (2.7) make up a single nonlinear integral equation for $g(r) = 1 + h(r)$. It can be seen that all pair correlation information about the chemical structure of the polymer enters the theory through $\hat{\omega}(k)$. An important and unique feature of the PRISM theory that should be emphasized is that one has the ability to account for the effect of both local and global structural details, through $\hat{\omega}(k)$, on the intermolecular packing and thermodynamic properties of polymer liquids.

3 Intermolecular Packing in Homopolymer Melts

In PRISM theory the intramolecular architecture of the polymer is specified through $\hat{\omega}(k)$. The level of detail needed in the $\hat{\omega}(k)$ calculation depends on the particular question that one is addressing. For example, if one is interested in long range, universal aspects, a Gaussian model, in which the monomers are heavily coarse grained, would suffice. The effect of structural features like chain stiffness, operating on intermediate length scales, could be accounted for by a semiflexible chain model. If one is interested in making detailed comparisons with experiment, the local monomeric structure details presumably must be included. In this section we give examples of PRISM calculations covering this range of detail.

3.1 Gaussian Chains

For chains consisting of sites connected by harmonic springs, the probability density between two intramolecular sites is a Gaussian distribution which can be written in Fourier transform space as

$$\hat{\omega}_{\alpha\gamma}(k) = \exp[-|\alpha - \gamma|k^2\sigma^2/6] \qquad (3.1)$$

where σ is the statistical segment length. For simplicity we will take the statistical segment length equal to the hard core diameter d. The summation in Eq. (2.11) can easily be performed to yield

$$\hat{\omega}(k) = \frac{(1 - f^2 - 2f/N + 2f^{N+1}/N)}{(1 - f)^2} \qquad (3.2)$$

where $f = \exp(-k^2\sigma^2/6)$. Equation (3.2) reduces to the well known Debye function in the limit of large N and $k\sigma \leqslant 1$. Then $\hat{\omega}(k)$ from Eq. (3.2) can be inserted into the generalized Ornstein-Zernike-like equation, Eq. (2.10) for the homopolymer, and the resulting equation can be solved numerically [11–15] subject to the Percus-Yevick hard sphere closure condition, Eq. (2.7). The numerical solutions were obtained using a variational method of Lowden and Chandler [9].

Results from this solution are shown for a melt of 2000 unit Gaussian chains in Fig. 1 as the points. Note that g(r) is a small number at contact (r = σ = d) and increases monotonically to unity on a scale of the radius of gyration R_g of the Gaussian chains. Such behavior is an example of the correlation hole predicted by deGennes [22]. The correlation hole is a consequence of the shielding of intermolecular interactions due to chain connectivity when two polymers are brought within R_g of each other. The "negative" correlation hole behavior (i.e., g(r) < 1) is a universal aspect of polymer melts and is a necessary consequence of chain connectivity and repulsive interactions between inter-molecular segments. Note in Fig. 1 that the contact value of g(r) increases as the density increases at fixed chain length, reflecting the fact that the chain domains are pushed together as the density increases.

An analytical approximation [23] can be found for the Gaussian chain melt by taking either the so-called thread or string limits. The "thread-like chain" model has been discussed in depth elsewhere [23]. Mathematically, it corres-ponds to the limit that all microscopic length scales approach zero but their ratios remain finite. In particular, the site hard core diameter d → 0, but the site density ρ_m → ∞ such that the reduced density, $\rho_m d^3$, is non-zero and finite. In

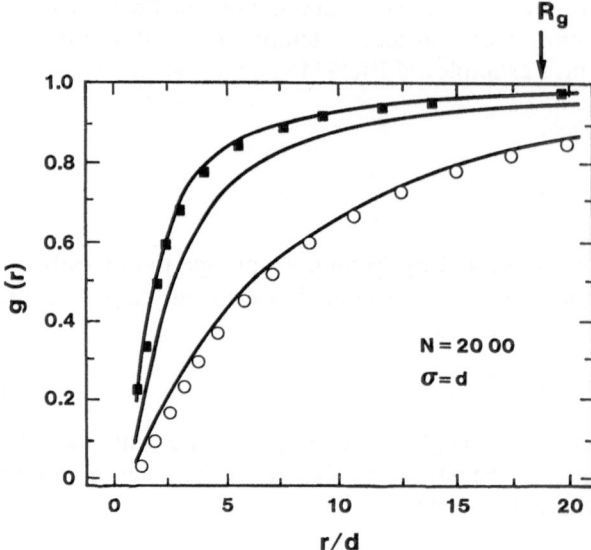

Fig. 1. Site-site intermolecular pair correlation function g(r) for 2000 unit Gaussian chains plotted as a function of reduced separation r/d. The solid curves are analytical results from the string model [23]. The *points* refer to the predictions from numerical solution of the PRISM theory for two reduced densities $\rho_m d^3$ of 0.2 (*open circles*) and 1.0 (*solid squares*). The *middle curve* is the analytical result for a reduced density of 0.6

the spirit of the Edwards' pseudopotential model[3a], the hard core direct correlation function is effectively replaced by a density-dependent delta-function. Hence the thread model does not represent the continuous $d \to 0$ limiting case of a finite range hard core repulsion. The latter model would have trivial ideal gas properties. Gaussian statistics are assumed to describe the single chain structure factor $\hat{\omega}(k)$, which is thus characterized solely by the statistical segment length σ, and the number of segments N. Moreover, in the thread limit the intramolecular structure factor can be accurately represented by the simple form[3a]: $\hat{\omega}(k) \cong [N^{-1} + (k\sigma)^2/12]^{-1}$.

With the above simplifications the PRISM equation can be analytically solved and yields [23] a site-site radial distribution of the Yukawa, or screened Coulomb, form:

$$g(r) = 1 + \frac{3}{\pi\rho_m\sigma^3} \frac{[\exp(-r/\xi_\rho) - \exp(-r/\xi_c)]}{r/\sigma} . \tag{3.3}$$

Note the presence of two screening lengths ξ_ρ and ξ_c. ξ_c is of macromolecular dimension and is simply related to the radius of gyration $\xi_c = R_g/\sqrt{2}$. ξ_ρ, on the other hand, is short range at high density and is determined from the core condition. Note also that at high densities there is a wide "intermediate" region of intersite separation, $\xi_\rho \ll r \ll \xi_c$, where power law correlations occur, i.e. $-h(r) \propto (\sigma/r)$. In the thread limit the hard core diameter is shrunk to a point and the core condition in Eq. (2.7) is approximated as $g(r = 0) = 0$, for which ξ_ρ becomes [23]

$$\xi_\rho^{-1} = \xi_c^{-1} + \frac{\pi\rho_m\sigma^3}{3\sigma} \quad \text{(thread limit)} . \tag{3.4a}$$

A related model, the string model, takes the hard core d as finite, but the core condition is only satisfied on average, that is

$$\int_0^d r^2 g(r) dr = 0 .$$

This condition can be justified as an optimized perturbative treatment of the $d \neq 0$ polymer using the thread as a reference system [23]. In this case ξ_ρ is given by the transcendental equation

$$\frac{\xi_\rho}{\sigma}\left(\frac{\xi_\rho}{\sigma} + \Gamma^{-1}\right)\exp(-\sigma/\xi_\rho\Gamma) - \left(\frac{\xi_\rho}{\sigma}\right)^2 = \frac{\pi\rho_m d^3}{9} - \left(\frac{\xi_c}{\sigma}\right)^2$$

$$+ \frac{\xi_c}{\sigma}\left(\frac{\xi_c}{\sigma} + \Gamma^{-1}\right)\exp(-\sigma/\xi_c\Gamma) \quad \text{(string limit)} \tag{3.4b}$$

where $\Gamma = \sigma/d$ is the aspect ratio of the chain.

The analytical predictions from Eq. (3.3) in the string limit are shown as the solid curves in Fig. 1. It can be seen that the analytical approximation shows a remarkable agreement with the corresponding exact numerical PRISM calculations for a finite hard core diameter. Implicit to the thread or string approach is a coarse-graining of molecular structure over the segmental length scale. Hence, local chemical information is lost (except in an average manner) and detailed structural features such as oscillatory g(r) functions indicative of local solvation shells due to the non-zero space-filling volume of real monomers are not captured. However, the thread idealization does yield analytical results which for many systems and properties are in excellent qualitative, or semi-quantitative, agreement with numerical PRISM predictions for more realistic non-thread polymer models. This achievement is the primary reason for constructing and studying the thread and string models.

3.2 Semiflexible Chains

More realism can be introduced into the intramolecular calculation of $\hat{\omega}(k)$ by the freely jointed chain model in which the chains are made up of rigid bonds connected by freely rotating joints. In this model $\hat{\omega}(k)$ is given by Eq. (3.2) with $f(k) = \sin(k\sigma)/k\sigma$ where σ represents both the bond length and hard core diameter for the model of present interest. Because of the rigid bond nature of the model, an additional short range structure is superimposed on the long range correlation hole aspect of g(r) that is not generally evident in Gaussian chains at the same density [14, 15]. Although the freely jointed chain model includes the constant bond length constraint similar to the very stiff chemical bonds present in real polymers, it assumes the chain is completely flexible and thus neglects local stiffness and short range excluded volume which are important features of real polymers.

The exact intramolecular distribution function for a semiflexible chain characterized by a local bond bending energy $\varepsilon_b(1 + \cos\theta)$ cannot be computed analytically. A convenient approximation, however, is given by the Koyama distribution [24, 25]

$$\hat{\omega}_{\alpha\gamma}(k) = \frac{\sin(B_{\alpha\gamma}k)}{B_{\alpha\gamma}k} \exp(-A_{\alpha\gamma}^2 k) \tag{3.5}$$

where

$$A_{\alpha\gamma}^2 = \langle r_{\alpha\gamma}^2 \rangle (1 - C_{\alpha\gamma})/6$$

$$B_{\alpha\gamma}^2 = C \langle r_{\alpha\gamma}^2 \rangle$$

$$C_{\alpha\gamma}^2 = \frac{1}{2}\left[5 - 3\frac{\langle r_{\alpha\gamma}^4 \rangle}{\langle r_{\alpha\gamma}^2 \rangle^2}\right].$$

Equation (3.5) can be shown to preserve the correct second $\langle r_{\alpha\gamma}^2 \rangle$ and fourth $\langle r_{\alpha\gamma}^4 \rangle$ moments of the intramolecular separation $r_{\alpha\gamma}$. Thus for a semiflexible chain of specified stiffness (through the second and fourth moments) the intramolecular structure function $\hat{\omega}(k)$ can be computed numerically from Eqs. (3.5) and (2.11). The second and fourth moments have been calculated by Honnell et al. [24] for the discrete version of the *wormlike chain* model of Porod from a knowledge of the chain stiffness parameters $\langle \cos\theta \rangle$ and $\langle \cos^2\theta \rangle$. The persistence length ξ of the chain is given by

$$\xi = \frac{\ell}{1 + \langle \cos\theta \rangle} \tag{3.6}$$

where ℓ is the bond length and θ is the bond angle. The semiflexible, wormlike chain is thus completely characterized by two parameters ξ (or equivalently the reduced bond bending energy $\beta\varepsilon_b$) and N. Another, more realistic alternative, is to employ the standard rotational isomeric state model [16] for the required moments. From a knowledge of the bond length and angle, and the rotational states and potentials, the rotational isomeric state model can be used to compute the moments $\langle r_{\alpha\gamma}^2 \rangle$ and $\langle r_{\alpha\gamma}^4 \rangle$, although calculation of the fourth moment is tedious.

An example [24] of the effect of chain stiffness on the radial distribution function can be seen in Fig. 2. In Fig. 2a, g(r) is plotted for a completely flexible,

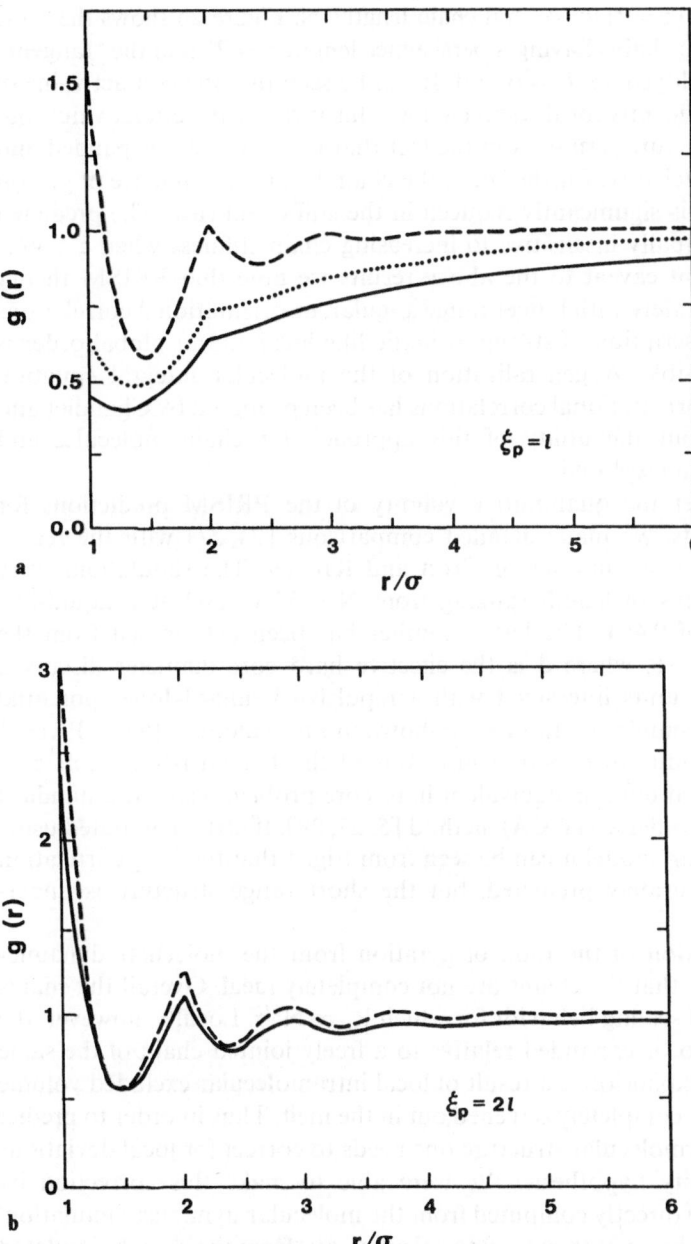

Fig. 2. a Intermolecular site-site radial distribution function g(r) for tangent hard core, freely jointed chains for a packing fraction $\eta = 0.5$ at three different chain lengths: $N = 20$ (*dashed*), $N = 200$ (*dotted*), $N = 2000$ (*solid*). The persistence length, $\xi = 1 = \sigma$, where 1 is the bond length and σ the hard core diameter. **b** Intermolecular site-site radial distribution function g(r) for tangent hard core, semiflexible chains [24] for a packing fraction $\eta = 0.5$ at two different chain lengths: $N = 20$ (*dashed*), and $N = 2000$ (*solid*). The persistence length is $\xi = 21$, where 1 is the bond length which equals the hard core diameter σ

freely jointed chain ($\xi = 1$) for several chain lengths N. Figure 2b shows the same plot for semiflexible chains having a persistence length $\xi = 2\ell$ and the "tangent" bead model is employed, i.e. $\ell = \sigma = d$. It can be seen that the contact value of g(r) increases significantly for the stiff chains. This increase in contact value and solvation shell structure results from the fact that the chains are expanded and thus pack more efficiently. Furthermore the chain length dependence of g(r) on local length scales is significantly reduced in the stiff chain case. The predicted results become virtually insensitive to increasing chain stiffness when $\xi \geqslant 4\ell$.

As an important caveat to the above results we note that PRISM theory does not deal accurately with longer range angular, or orientational correlations [26]. Thus, the description of strong, nematic-like local and/or global order is not generally possible. A generalization of the molecular integral equation approach to treat orientational correlations has been proposed by Chandler and co-workers [26] but the utility of this approach for chain molecules and polymers remains unexplored.

In order to test the quantitative validity of the PRISM predictions for homopolymer melts, we made detailed comparisons [24, 27] with the recent molecular dynamics simulations of Grest and Kremer. The simulations were performed on chains of length ranging from N = 50 to 200 at a liquid-like packing fraction of 0.464. The latter number has been determined from the relation $\eta = \pi \rho_m d^3 / 6$, where d is the effective hard core diameter discussed below. The repeat units interacted with a repulsive Lennard-Jones potential with $\beta \varepsilon = 1.0$. The simulation results are shown in Fig. 3 along with our PRISM predictions. The radial distribution functions of the Lennard-Jones soft core system were mapped onto an equivalent hard core problem using the standard Weeks, Chandler, Andersen (WCA) method [5, 27, 28]. If g(r) is computed using a freely jointed chain model it can be seen from Fig. 3 that the long correlation hole regime is accurately predicted, but the short range structure is underestimated.

Close examination of the radii of gyration from the molecular dynamics simulations reveals that the chains are not completely ideal. Overall the chains exhibit nearly ideal scaling behavior for which $R_g \propto N^{1/2}$. Locally, however, the chains are found to be expanded relative to a freely jointed chain of the same length. This local expansion is a result of local intramolecular excluded volume which has not been completely screened out in the melt. Thus in order to predict accurately the intermolecular structure one needs to correct for local deviations from Flory's ideality hypothesis. We were able to make this correction by employing the $\hat{\omega}(k)$ directly computed from the molecular dynamics simulation. When the PRISM theory is then used to calculate g(r) from the actual, simulated $\hat{\omega}(k)$, excellent agreement is seen in Fig. 3 between the theory and the simulation [27]. This agreement suggests that the PRISM theory can predict intermolecular structure with about the same accuracy as corresponding RISM calculations on small, rigid molecules. If only the radius of gyration, or mean square end-to-end distance, is available from simulation, an approximation can be made by using the semiflexible chain model in Eq. (3.5) where the persistence

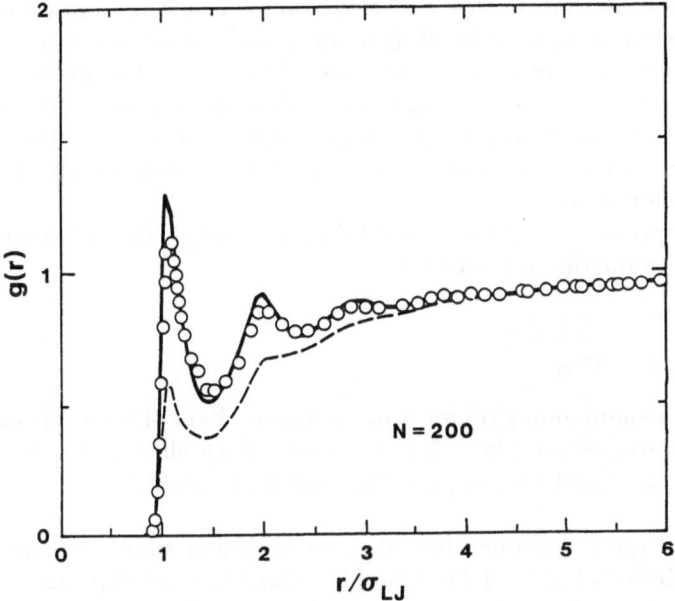

Fig. 3. Site-averaged intermolecular radial distribution function g(r) for N = 200. The *points* are the MD simulation of Kremer and Grest [27] (some points omitted for clarity). The *solid line* is the predicted result from the PRISM theory [27] using the simulated intramolecular $\hat{\omega}(k)$. The *dashed line* is the corresponding PRISM prediction using $\hat{\omega}(k)$ from a freely jointed chain model (with overlaps explicitly removed [14]). R_g is approximately 7.7 σ_{LJ} for the simulated chains

length is chosen to match the radius of gyration from simulation. Using this model intramolecular structure to compute g(r) also led to excellent agreement with the molecular dynamics simulations [24].

As the polymer fluid density decreases the agreement between PRISM and simulations becomes quantitatively poorer, even if the simulated $\hat{\omega}(k)$ is used in the PRISM calculation [29, 30]. Such a reduction of quantitative accuracy of the RISM approach also occurs for small, rigid molecule fluids and is a well-understood and documented trend [6].

3.3 Rotational Isomeric State Chains

The coarse grained models we have considered thus far are valuable for examining qualitative trends. However, in order to make comparisons directly with experimental data, more local structural details presumably need to be taken into account in the calculation of $\hat{\omega}(k)$. A realistic way of incorporating monomer structure is through the rotational isomeric state approximation, successfully employed [16] by Flory and others to describe isolated polymer chains in a theta solvent. In this description the continuous rotational potentials are replaced by discrete rotational states corresponding to the lowest vibrational state in the potential wells. Short range excluded volume resulting from rotational correlations between adjacent bonds (pentane effect) is routinely incorporated into the calculations. Lower order moments of the end-to-end distance can be computed for this model but the complete distribution function necessary for determining $\hat{\omega}(k)$ cannot be explicitly obtained. An expansion of $\omega(r)$ in terms of the moments of the distribution is possible, but unfortunately the convergence is very slow [16].

An alternative approach is to generate $\hat{\omega}(k)$ through a single chain, Monte Carlo simulation by evaluating the average

$$\hat{\omega}(k) = \frac{1}{N} \left\langle \sum_{\alpha,\gamma=1}^{N} \frac{\sin(kr_{\alpha\gamma})}{kr_{\alpha\gamma}} \right\rangle . \tag{3.7}$$

An example of such a simulation [31] for a polyethylene chain of 1001 repeat units is shown as a Kratky plot in Fig. 4. This result for a single chain could then be introduced into Eq. (3.10) to compute the interchain packing from the PRISM theory.

A convenient and quite accurate alternative to computer simulation has been developed by McCoy et al. [31]. Here $\hat{\omega}(k)$ is computed as an approximation by including the structural detail only on short length scales. This is accomplished by rewriting $\hat{\omega}(k)$ in the form

$$\hat{\omega}(k) = \frac{1}{N} \sum_{\alpha} \left[\sum_{|\alpha-\beta| \leqslant 5} \hat{\omega}_{\alpha\beta}(k) + \sum_{|\alpha-\beta| > 5} \hat{\omega}_{\alpha\beta}(k) \right] . \tag{3.8}$$

The first term for $|\alpha - \beta| \leqslant 5$ is evaluated by direct enumeration of rotational isomeric states, thereby including the pentane effect (longer range enumerations

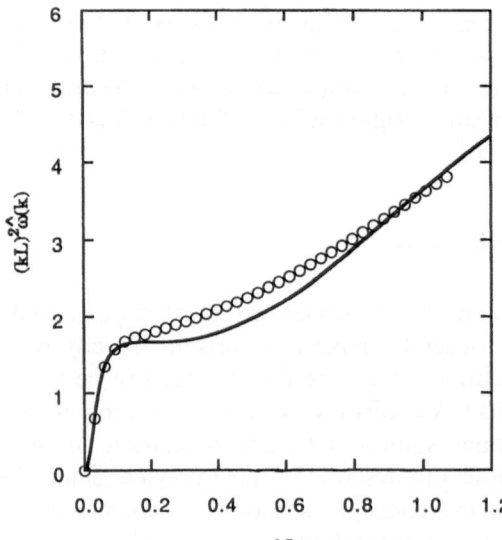

Fig. 4. A Kratky plot for a 1001 site rotational isomeric state polyethylene chain. The *circles* are from a Monte Carlo simulation [31] and the *curve* is the corresponding result using the approximate procedure described in the text. L is the carbon-carbon bond length

are also feasible). The long range contribution for $|\alpha - \beta| > 5$ is approximated by "splicing" on the Koyama distribution for a semiflexible chain [24, 25], where the second and fourth moments $\langle r_{\alpha\gamma}^2 \rangle$ and $\langle r_{\alpha\gamma}^4 \rangle$ are adjusted for each $|\alpha - \gamma|$ to match the corresponding moments from a rotational isomeric state computation. This approximate calculation is also shown in Fig. 4. The standard alkane rotational isomeric state parameters were used in both the Monte Carlo and the approximate calculation (bond length $\ell = 1.54$ Å, bond angles $\theta = 112°$, gauche states at $\phi = \pm 120°$, with energy of 500 cal/mol relative to trans, and 2000 cal/mol for the g^+g^- state) [16]. It is interesting to note that the approximate procedure yields a well defined intermediate scaling regime indicated by the plateau on the Kratky plot; no corresponding intermediate scaling regime is found for the Monte Carlo calculation. This issue and its experimental and theoretical implications are discussed in depth by McCoy and co-workers [31].

Honnell et al. computed [32, 33] the intermolecular packing correlations for a polyethylene melt (N = 6429) at 430 K from the PRISM theory using the above approximate procedure. The results for the structure factor are shown in Fig. 5. The static structure factor is defined as

$$\hat{S}(k) = \hat{\omega}(k) + \rho_m \hat{h}(k) \qquad (3.9)$$

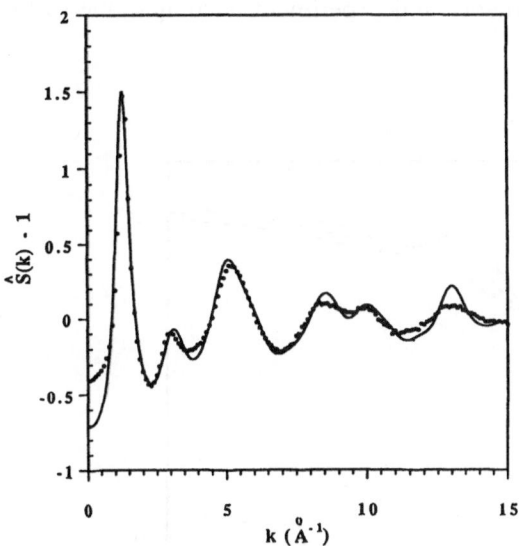

Fig. 5. Structure factor $\hat{S}(k)$ for polyethylene melt (N = 6429) as a function of wavevector k at T = 430 K. The *points* are experimental results [32, 33] of Narten and Habenschuss from X-ray scattering (the k ≈ 0 data is inaccurate due to sample preparation related scattering). The *solid curve* is the PRISM theory with the hard core diameter d = 3.90 Å. Use of a value of d = 3.7 Å results in roughly a 10% underestimate of the intensity of the amorphous halo feature. Disagreement between experiment and theory at large k is eliminated if thermal broadening, due to vibrational and torsional oscillations, is taken into account [33, 34]

and represents the Fourier transform of the total density fluctuation correlation function. The methylene groups along the chain were represented by overlapping hard sphere sites of diameter d. Also shown in this figure is the structure factor from experimental X-ray scattering measurements [32, 33] on the same polyethylene melt by Narten and Habenschuss. The hard core diameter d, adjusted in the PRISM theory to match the height of the first peak in the experimental $\hat{S}(k)$ curve, yielded a value of 3.90 Å. This is a very reasonable value [32] based on crystalline-packing-based geometrical estimates for the volume of a CH_2 unit. It can be seen from Fig. 5 that the agreement between the PRISM theory and X-ray scattering experiment is quite satisfactory. Analysis of the intramolecular and intermolecular contributions to the structure factor in Eq. (3.9) indicates that for $k > 6$ Å^{-1}, $\hat{S}(k)$ is almost entirely intramolecular in origin. The contributions to $\hat{S}(k)$ for $k < 6$ Å^{-1} are due to both intra- and intermolecular correlations. It is important to mention that good agreement with experiment was found even though we used hard core interactions between sites, and did not include attractions. This is not surprising since it is well known [5, 6] that the intermolecular radial distribution function and wide angle diffraction pattern are primarily controlled by repulsive interactions at high density. On the other hand, $\hat{S}(k = 0)$ is a thermodynamic property which is very sensitive to attractive interactions, and is generally not as accurately predicted by the RISM approach. Systematic PRISM calculations for the alkane series have also been carried out by Honnell et al. and compared with X-ray scattering measurements [34]. Agreement between theory and experiment is comparable to the polyethylene case.

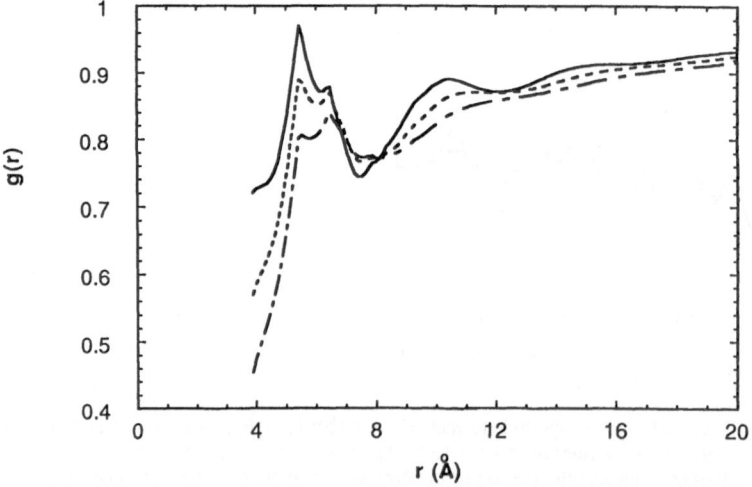

Fig. 6. Intermolecular radial distribution function g(r) computed for polyethylene (d = 3.90 Å, T = 430 K, N = 6429) using PRISM theory. $\rho_m d^3 = 2.2$ (*solid curve*), 2.0 (*short dashed*), 1.8 (*dash-dot*). At one atmosphere·pressure the experimental value of $\rho_m d^3$ is ~ 2.0

Figure 6 shows our PRISM theory predictions for g(r) of polyethylene at 430 K and three different densities using d = 3.90 Å for the hard core diameter. Note that g(r) for polyethylene is somewhat more complex than in the tangent hard sphere calculations in Figs. 2 due to the multiplicity of length scales (i.e. $d \neq \ell \neq \xi$). The first peak, which occurs at the hard core diameter for tangent hard spheres, is shifted out to $d + \ell$. The cusps that occur in g(r) arise because of the hard sphere and constant bond length constraints [6]. The correlation hole regime is also clearly evident beyond roughly 15–20 Å in which g(r) approaches unity on a scale of the radius of gyration ($R_g \sim 130$ Å) in a power law manner.

4 Equation-of-State

Based on comparison with Monte Carlo simulations in Fig. 3 and X-ray scattering experiments in Fig. 5, it would seem that PRISM theory is accurate for the intermolecular structure in polymer melts. Given the site-site radial distribution function g(r) it should, in principle, be possible to deduce the equation-of-state and other thermodynamic properties. Unlike g(r), which is sensitive only to the repulsive branch of the potential, one expects that the equation-of-state will also be a function of the range and magnitude of the attractions. The fact that g(r) is primarily controlled by repulsions at high density suggests that the effect of attractions on the pressure can be treated by perturbation theory [5, 6, 28, 35] about a hard core reference system. Such an expansion should be particularly valid at high temperatures or small $\beta\varepsilon$, where ε is the attractive interaction energy scale of, for example, a Lennard-Jones potential,

$$v(r) = 4\varepsilon[(\sigma_{LJ}/r)^{12} - (\sigma_{LJ}/r)^6] \,. \tag{4.1}$$

The effective hard core is optimally chosen such that a selected property of the continuous repulsive force fluid is accurately reproduced by the hard core reference system [5, 6, 28, 35].

In the Barker-Henderson [5, 35] perturbation theory the potential in Eq. (4.1) is divided into a repulsive branch $v_0(r)$ and an attractive branch $v_a(r)$

$$\begin{aligned}
v_0(r) &= v(r); \quad r \leqslant \sigma_{LJ} \\
&= 0; \quad r > \sigma_{LJ} \\
v_a(r) &= 0; \quad r \leqslant \sigma_{LJ} \\
&= v(r) \quad r > \sigma_{LJ} \,.
\end{aligned} \tag{4.2}$$

The optimum hard core diameter is then given by

$$d = \int_0^\infty \{1 - \exp[-\beta v_0(r)]\} dr \,. \tag{4.3}$$

Similarly in the perturbation theory of Weeks, Chandler and Andersen [5, 28] (WCA) the Lennard-Jones potential is decomposed according to

$$
\begin{aligned}
v_0(r) &= v(r) + \varepsilon, & r &\leqslant 2^{1/6}\sigma_{LJ} \\
&= 0, & r &> 2^{1/6}\sigma_{LJ} \\
v_a(r) &= \quad -\varepsilon, & r &\leqslant 2^{1/6}\sigma_{LJ} \\
&= v(r), & r &> 2^{1/6}\sigma_{LJ} .
\end{aligned}
\tag{4.4}
$$

Here the optimum hard core diameter d is chosen to satisfy

$$
\int_0^d r^2 \exp[-\beta v_0(r)]C_0(r; d)dr + \int_d^\infty r^2\{1 - \exp[-\beta v_0(r)]\}g_0(r; d)dr = 0
\tag{4.5}
$$

where C_0 and g_0 are the direct correlation function and radial distribution function, respectively, for the hard core reference system subject to the PY closure in Eq. (2.7).

Given a knowledge of $g_0(r)$, the pressure of the optimized hard core polymer liquid can be computed in several ways. The simplest route is merely to integrate the isothermal compressibility [5, 6] κ_T which is related to the zero wavelength structure factor $\rho_m k_B T \kappa_T = \hat{S}(0)$:

$$
\frac{\beta P}{\rho_m} = \frac{1}{\rho_m} \int_0^{\rho_m} \hat{S}^{-1}(0, \rho_m')d\rho_m' .
\tag{4.6}
$$

$\hat{S}(0)$ involves all length scales and presumably is relatively more sensitive to the long range part of $g_0(r)$. Another route to the pressure is through the Helmholtz free energy A. The free energy of a hard core system can be computed according to a "charging formula" given by [6]

$$
\frac{A - A_0}{V} = 2\pi\rho_m^2 d^3 k_B T \int_0^1 g^{(\lambda)}(\lambda d^+)\lambda^2 d\lambda
\tag{4.7}
$$

where the hard core diameter is gradually "turned on" as λ varies from 0 to 1. In Eq. (4.7), $g^{(\lambda)}(\lambda d^+)$ is the contact value of g(r) for a liquid comprised of interaction sites of diameter λd and A_0 is the free energy of an ideal gas. The pressure is then obtained from Eq. (4.7) by numerical differentiation with respect to volume. In the case of a simple hard sphere fluid, Eq. (4.7) can be shown [6] to reduce to the well known virial formula. For molecular fluids, however, one must explicitly carry out the integration over λ. Since the integrated quantity in Eq. (4.7) is the site-site radial distribution function at contact, the free energy route to the pressure is sensitive to the very short range character of the intermolecular packing. A third route to the pressure makes use of $g_w(r)$, the wall-fluid distribution function

$$
\frac{\beta P}{\rho_m} = g_w(0) .
\tag{4.8}
$$

The quantity $\rho_m g_w(0)$ is the density of molecular sites in contact with a hard wall. This method was employed recently by Dickman and Hall [36] to determine pressure from Monte Carlo simulations. Yethiraj and Hall [37] demonstrated that $g_w(r)$ can be approximately computed from PRISM theory by considering a mixture of polymer chains and spherical particles in the limit of zero concentration of particles with diameters approaching infinity.

If $g_0(r)$, $g^{(\lambda)}(r)$, and $g_w(r)$ are known exactly, then all three routes should yield the same pressure. Since liquid state integral equation theories are approximate descriptions of pair correlation functions, and not of the effective Hamiltonian or partition function, it is well known that they are thermodynamically inconsistent [5]. This is understandable since each route is sensitive to different parts of the radial distribution function. In particular, $g(r)$ in polymer fluids is controlled at large distance by the correlation hole which scales with the radius of gyration or $\sqrt{N}$. Thus it is perhaps surprising that the hard core equation-of-state computed from PRISM theory was recently found by Yethiraj et al. [38, 39] to become more thermodynamically inconsistent as N increases from the diatomic to polyethylene. The uncertainty in the pressure is manifested in Fig. 7 where the insert shows the equation-of-state of polyethylene computed [38] from PRISM theory for hard core interactions between sites. In this calculation, the hard core diameter d was fixed at 3.90 Å in order to maintain agreement with the experimental structure factor in Fig. 5.

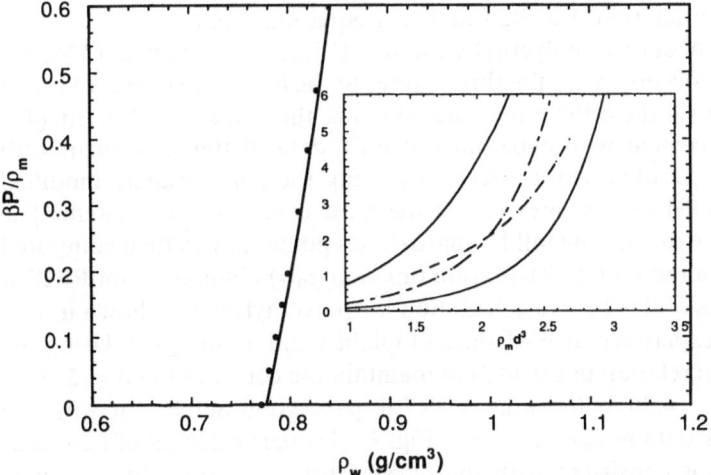

Fig. 7. Equation-of-state for polyethylene (d = 3.90 Å, T = 430 K, N = 6429). The *solid curve* was computed from the GFD hard sphere reference system [38] with the effect of attractions computed by Barker-Henderson perturbation theory [35] using g(r) obtained from PRISM theory. The points are experimental PVT data of Olabisi and Simha [41]. The *inset* shows the hard sphere equation-of-state computed by various routes: free energy (*upper solid*), compressibility (*lower solid*), wall (*dashed*) GFD (*long-dash-dot*).

It is difficult to assess the accuracy of the pressure obtained from the three routes since computer simulations of polyethylene hard core chains have not yet been performed at high density. In Fig. 7 we have also plotted the pressure predicted from a continuous space mean field equation-of-state, the generalized Flory dimer model [40] (GFD), appropriately generalized by Yethiraj et al. [38] to treat the overlapping site rotational isomeric state model. Prior detailed comparisons between the GFD predictions and off-lattice computer simulations suggest this equation-of-state is accurate for long chains at high density. It can be seen from the inset in Fig. 7 that the GFD equation-of-state is bounded by the compressibility and free energy routes. The wall route is reasonably close to the GFD result, but the isothermal compressibility is predicted to be too high.

The pressure of a realistic polymer liquid interacting with Lennard-Jones potential can now be approximated from the hard core reference system using thermodynamic perturbation theory:

$$A \cong A_{HS} + \tfrac{1}{2} \rho_m^2 \int g_0(r) v_a(r) d\vec{r} \ . \tag{4.9}$$

In Eq. (4.9) A_{HS} is the Helmholtz free energy of the hard core reference system with the optimized hard core diameter. The pressure of the full system can be found from Eq. (4.9) by differentiation with respect to volume. This pressure will be a function of the two Lennard-Jones parameters σ_{LJ} and ε.

In the case of polyethylene, we observed in the previous section that to be consistent with the experimental structure factor the hard core diameter should be fixed at d = 3.90 Å at 430 K. With this added constraint we therefore have only a single adjustable parameter since d is related to σ_{LJ} and ε through the Barker-Henderson relation, Eq. (4.3) or WCA equation, Eq. (4.5).

The equation-of-state of polyethylene at 430 K has been calculated [38, 39] using the above procedure with the three routes to the hard core reference liquid in Eqs. (4.6)–(4.8). Of these three reference systems, the "wall route" in Eq. (4.7) gave the best agreement with experimental PVT data. Better agreement with experiment could be obtained, however, by using the appropriately modified GFD approach [38] for the pressure of the hard core reference system. The pressure of the system with the full Lennard-Jones potential was then computed from Eq. (4.9) with the radial distribution function $g_0(r)$ obtained from PRISM theory. The results of this hybrid calculation for polyethylene are shown in Fig. 7 along with the experimental PVT data of Olabisi and Simha [41]. Using the Barker-Henderson relation in Eq. (4.3) to maintain the constraint of d = 3.90 Å, the values of σ_{LJ} = 4.36 Å and ε/k_B = 38.7 K gave good agreement with the experimental PVT data as can be seen in Fig. 7. This particular set of Lennard-Jones parameters is consistent with the recent Monte Carlo study of Lopez-Rodriguez and coworkers [42] on the second virial coefficients of hydrocarbon fluids.

The above considerations suggest that the hard core radial distribution function obtained for polyethylene was sufficiently accurate on short length scales to predict the perturbative contribution (Eq. (4.9)) to the attractive branch of the potential. The difficulty lies in computing the pressure of the hard core reference system. It is clear from the inset in Fig. 7 that there is a large uncertainty in the hard core pressure due to the thermodynamic inconsistency among the different routes. It is likely that this problem can be improved by using methods developed [5] for obtaining thermodynamically consistent results for atomic and small molecule fluids. Following these ideas, one adds a nonzero tail of some assumed functional form to the direct correlation function outside the hard core. The parameters in this function are then chosen in such a way as to force thermodynamic consistency constraints [43].

5 Polymer Crystallization

A number of years ago Flory developed [44] a mean field incompressible lattice model to describe the equilibrium freezing of polymers. Although the model has the virtue of simplicity, it suffers from at least three well known deficiencies. (1) It is an incompressible model that does not account for the density change associated with the first order freezing transition. This density change is not negligible for many polymers, for example, polyethylene exhibits approximately a 30% densification on crystallization from the melt. (2) Attractive intermolecular forces are ignored. (3) The Flory approach is a mean field theory that neglects intra- and intermolecular correlations. Nagle et al. [45] have argued that such approximations, plus the unrealistic nature of packing on a lattice, seriously affect the validity of the Flory crystallization theory.

More recently, density functional theory has been successfully applied to the freezing of monatomic [46, 47] and polyatomic [48–51] liquids. In a common application of density functional theory to freezing, the excess free energy of the solid is expanded about the structure of the liquid at density ρ_L. The free energy functional is customarily truncated at second order in the order parameter $\Delta\rho(r) = \rho(r) - \rho_L$ where $\rho(r)$ is the nonuniform, single particle density of the solid phase. It has been found that the errors introduced from the higher order terms can be compensated for, to some degree, by expanding the free energy difference between the system of interest and some ideal system which can be solved exactly. In this manner the higher order terms in the expansion of the ideal and true system tend to cancel, thereby providing a more accurate density functional.

McCoy and coworkers [52] have generalized this density functional scheme to describe the crystallization of polymers. In this approach the structure of the liquid is provided by PRISM theory. Three conditions are necessary in order to ensure equilibrium between the solid and liquid phases. (1) temperature equality, (2) equality of the chemical potentials μ, and (3) pressure equality. Because of these conditions it is more convenient to work with the Grand potential W rather than the more conventional Helmholtz free energy A. W is related to

A through a Legendre transform

$$W = A - \sum_i \int d\vec{r}\psi_i(\vec{r})\rho_i(\vec{r}) \tag{5.1}$$

where $\rho_i(\vec{r})$ is the density of sites of type i. In density functional theory it is envisaged that an external field $\psi_i(\vec{r})$ acts on sites of type i to give the non-uniform density profile $\rho_i(\vec{r})$. The Grand potential difference ΔW between the solid and liquid phase is defined as

$$\Delta W = W[T, V, \mu, \rho_i(\vec{r})] - W[T, V, \mu, \rho_L] \tag{5.2}$$

where μ is the chemical potential. Introducing an ideal system denoted by superscripts "o" and using PRISM theory for the structure of the liquid phase, the Grand potential functional ΔW can be written as [52]

$$\Delta W = \sum_i \int d\vec{r}[\psi_{L,i} - \psi_i(r)]\rho_i(\vec{r}) + \sum_i \int d\vec{r}[\psi_i^o(r) - \psi_{L,i}^o]\rho_i(\vec{r})$$

$$- \frac{1}{N}\sum_i \int d\vec{r}\Delta\rho_i(\vec{r}) - \frac{1}{2}\sum_{i,j} \iint d\vec{r}d\vec{r}' C_{ij}(\vec{r} - \vec{r}')\Delta\rho_i(\vec{r})\Delta\rho_j(\vec{r}')$$

$$+ \cdots \tag{5.3}$$

where $\psi_i(r)$ and $\psi_i^o(r)$ refer to the position dependent external fields for the real and ideal solid, and $\psi_{L,i}$ and $\psi_{L,i}^o$ are the constant external fields for the uniform real and ideal liquid phases. $C_{ij}(r)$ is the direct correlation function between sites i and j and is to be determined from PRISM theory.

A convenient ideal system for the liquid phase is a collection of chains each of which is described by the rotational isomeric state model [16]. For such a system we can write the external field $\psi_{L,i}^o$ as

$$\psi_{L,i}^o = \frac{1}{N}\ln\left(\frac{1}{N}\right)\sum_i \rho_{L,i} - \frac{(N-1)}{N}\ln\lambda \tag{5.4a}$$

where λ is determined [16] from the rotational isomeric state theory as

$$\lambda = \tfrac{1}{2}\{1 + s(1 + w) + \sqrt{[1 - s(1 + w)]^2 + 8s}\} \tag{5.4b}$$

and s and w are simply related to the gauche energy E_g and gauche$^+$gauche$^-$ energy E_{g+g-}:

$$s = \exp(-E_g/RT); \quad w = \exp(-E_{g+g-}/RT). \tag{5.4c}$$

For simplicity, torsional and other vibrations of the ideal crystalline solid are ignored, and it is assumed that the chains are in the all-trans configuration. In this case the ideal external field of the solid has the form

$$\psi^o = \frac{1}{N}\ln\left(\frac{1}{N}\right)\sum_i \rho_{L,i}. \tag{5.5}$$

For this ideal system, Eq. (5.3) can be simplified for the homopolymer case of identical sites to give [52]

$$\frac{\Delta W}{\rho_L V} = \int d\vec{r}\, \ln\lambda\rho(\vec{r}) - \tfrac{1}{2}\iint d\vec{r}d\vec{r}'\,C(|\vec{r}-\vec{r}'|)\Delta\rho(\vec{r})\Delta\rho(\vec{r}') . \qquad (5.6)$$

One proceeds numerically to minimize ΔW with respect to the liquid density ρ_L and a parametrized solid phase density profile $\rho(\vec{r})$. PRISM theory is used to compute the site-site direct correlation function $C(r)$. This procedure is repeated with different assumed solid profiles until one finds the profile such that $\Delta W_{min} = 0$. It can be demonstrated that this profile corresponds to the coexistence curve where the three conditions for equilibrium between liquid and solid are satisfied. Such a program has been carried out to describe the freezing of polyethylene (PE) and polytetrafluoroethylene (PTFE) melts [52] and the results are depicted in Fig. 8. In this work the orthorhombic crystal was assumed to be the most stable ordered phase.

In Fig. 8 the predicted temperature-density behavior of coexisting liquid and solid phases are shown along with experimental data for PE [53] and PTFE [54]. Based on the experimental equation-of-state [41] the melting temperature T_m for PE is predicted to be 427 K, in good agreement with experiment (415–420 K). Excellent agreement is also found for the liquid density at the coexistence point and some of the lattice parameters [52]. For PTFE,

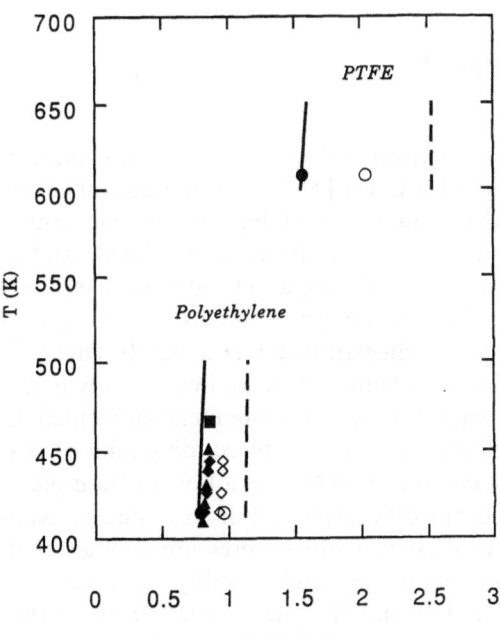

Fig. 8. Solid-liquid phase diagram for polyethylene (*lower*) and polytetrafluoroethylene (PTFE) (*upper*) computed from PRISM plus density functional theory [52]. The *solid* (*dashed*) *lines* are the theoretical liquid (solid) predictions. The *solid* (*open*) symbols are the experimental results [53, 54] for the coexistence of liquid (solid)

344 K.S. Schweizer and J.G. Curro

$T_m = 608$ K is predicted whereas the experimental values are given by 607 ± 10 K. Heuristic arguments suggest that PTFE should have a larger freezing temperature than PE because of the increased chain stiffness in PTFE. Following the Flory theory [44] of melting would imply that T_m/E_g should be approximately constant. Such reasoning would lead to the conclusion [52] that for PTFE, $T_m \sim 1000$ K, assuming that PE has a melting temperature of 420 K. Thus it appears that freezing of polymers is more subtle than a simple chain stiffness argument and presumably involves other factors such as chain packing and local chemical structure. It can also be observed from Fig. 8 that the predicted volume changes on crystallization from density functional theory are larger than observed experimentally. This trend is to be expected since it was assumed for calculation purposes that torsional oscillations, which would reduce the solid density, were not present. The removal of this simplification is certainly feasible and represents an important direction for future research.

From a more general perspective, the polymer density functional theory predicts that the location, or even existence, of a melting line is very sensitive to the Angstrom level chemical structure, volume change and attractive forces [52]. Neglect of any of these features can destroy the stability of the orthorhombic crystalline phase. For example, Gaussian and freely rotating chain models were found never to crystallize due to their excessive conformational entropy content in the disordered phase. Such behavior is consistent with the arguments of Nagle et al. [45], and the general understanding of a strongly first order phase transition as a highly nonuniversal process sensitive to many competing factors.

6 Polymer Blends: General Aspects

The structure and properties of phase-separating polymer alloys is a subject of considerable scientific and technological interest [55]. With the advent of the small angle neutron scattering (SANS) technique much information concerning concentration fluctuations in polymer alloys has been accumulated, and is commonly expressed in terms of an empirical "apparent chi-parameter", χ_s, which contains all non-ideal mixing information [56]. According to classical mean field Flory-Huggins ideas [1], the chi-parameter is a purely energetic quantity which is inversely proportional to temperature and completely determined by the local chemical interactions between monomers embedded in a structureless fluid. In practice, chi-parameters and phase diagrams of real polymer alloy materials are far more complex and exhibit a host of "non-mean field" features of both enthalpic and entropic origins. For example, χ_s often depends strongly on composition, local chain stiffness, pressure, global chain architecture, molecular weight, chain branching, and possibly wavevector of observation [57, 58]. The ability to understand the phase behavior of existing polymer mixtures, and ultimately to predict and control the miscibility of novel alloys, requires new theoretical methods that can account for the multitude of correlation effects.

The PRISM integral equation theory has been generalized for multicomponent polymer mixtures by Curro and Schweizer [23, 59–63]. Much formally exact analysis can be carried out based simply on the structure of the PRISM matrix integral equations without specifying a particular closure approximation [61]. In this section these aspects are summarized.

6.1 PRISM Formalism

A standard interaction site model of homopolymers is adopted where the pairwise decomposable site-site intermolecular potentials consist of a hard core repulsion plus a more slowly spatially varying tail.

$$u_{MM'}(r) = u_{MM'}^{(0)}(r) = \infty, \quad r < d_{MM'}$$

$$= v_{MM'}(r), \quad r \geq d_{MM'}. \tag{6.1}$$

Continuous repulsive potentials can easily be treated using the WCA perturbation approach [5, 6, 28] (see Eq. (4.5)) to map the problem of interest onto an effective hard core model. Generalization to the case of heteropolymers composed of more than one type of site is also straightforward. The PRISM matrix equations for the homopolymer mixture is given in Fourier-transform space by [59]

$$\hat{h}_{MM'}(k) = \hat{\omega}_M(k) \left[\hat{C}_{MM'}(k)\hat{\omega}_{M'}(k) + \sum_{M''} \hat{C}_{MM''}(k)\rho_{M''}\hat{h}_{M''M'}(k) \right] \tag{6.2}$$

where chain-end effects have been averaged over in the usual manner [13]. Here, ρ_M is the site number density of species M, $h_{MM'}(r) = g_{MM'}(r) - 1$, where $g_{MM'}(r)$ is the chain-averaged intermolecular pair correlation (or radial distribution) function between interaction sites of species M and M', $C_{MM'}(r)$ is the corresponding intermolecular site-site direct correlation function, and $\hat{\omega}_M(k)$ is the intramolecular structure factor of species M.

The partial density-density collective structure factors are given by

$$\hat{S}_{MM'}(k) = \rho_M \hat{\omega}_M(k)\delta_{M,M'} + \rho_M\rho_{M'}\hat{h}_{MM'}(k) \tag{6.3}$$

We concentrate here on a binary homopolymer blend under constant volume conditions; generalizations to ternary or more complex mixtures is straightforward. Substituting the PRISM Eq. (6.2) in Eq. (6.3) yields

$$\hat{S}_{AA}(k) = \rho_A\hat{\omega}_A(k)[1 - \rho_B\hat{\omega}_B(k)\hat{C}_{BB}(k)]/\Lambda(k) \tag{6.4a}$$

$$\hat{S}_{BB}(k) = \rho_B\hat{\omega}_B(k)[1 - \rho_A\hat{\omega}_A(k)\hat{C}_{AA}(k)]/\Lambda(k) \tag{6.4b}$$

$$\hat{S}_{AB}(k) = \rho_A\rho_B\hat{\omega}_A(k)\hat{\omega}_B(k)\hat{C}_{AB}(k)/\Lambda(k) \tag{6.4c}$$

$$\Lambda(k) = 1 - \rho_A \hat{\omega}_A(k)\hat{C}_{AA}(k) - \rho_B \hat{\omega}_B(k)\hat{C}_{BB}(k) + \rho_A \rho_B \hat{\omega}_A(k)\hat{\omega}_B(k)$$
$$\times [\hat{C}_{AA}(k)\hat{C}_{BB}(k) - \hat{C}_{AB}^2(k)] . \tag{6.5}$$

The direct correlation functions contain the fundamental microscopic information regarding interactions and correlations in blends. In general there are three independent functions for a binary homopolymer mixture, which enter the scattering functions in a nonlinear fashion. On the relatively long length scales relevant to small angle scattering measurements the approximation $\hat{C}_{MM'}(k) \cong \hat{C}_{MM'}(0) \equiv C_{MM'}$ is appropriate. The spinodal condition is given by the simultaneous divergence at $k = 0$ of all the partial structure factors

$$0 = 1 - \rho_A N_A C_{AA} - \rho_B N_B C_{BB} + \rho_A \rho_B N_A N_B [C_{AA}C_{BB} - C_{AB}^2] . \tag{6.6}$$

The total density fluctuations are characterized by the isothermal compressibility, κ_T, which is given by [63, 64]:

$$\kappa_T^{-1} = k_B T \sum_{M, M'} \rho_M \rho_{M'} \hat{S}_{MM'}^{-1}(0)$$
$$= k_B T \left\{ \frac{\rho_A}{N_A} + \frac{\rho_B}{N_B} - [\rho_A^2 \hat{C}_{AA}(0) + \rho_B^2 \hat{C}_{BB}(0) + 2\rho_A \rho_B \hat{C}_{AB}(0)] \right\} \tag{6.7}$$

where N_M is the number of interaction sites comprising a molecule of species M. The inverse isothermal compressibility generally remains nonzero at the spinodal defined by Eq. (6.6), but will vanish at a liquid-gas-like transition. As recently emphasized in the simple liquid integral equation context [65], liquid-liquid phase separation is a coupled density and concentration fluctuation process and thus compressibility effects generally play an important role.

6.2 Connections with IRPA and SANS Chi-Parameter

The incompressible random phase approximation (IRPA) is routinely used by experimentalists to analyze small angle scattering data from polymer alloys. A common approach to empirically defining an apparent SANS chi-parameter, χ_S, is based on the total scattering intensity extrapolated to $k = 0$ as [56, 57]:

$$\frac{k_N}{\hat{S}_E(0)} \equiv \frac{\left(\dfrac{b_A}{V_A} - \dfrac{b_B}{V_B}\right)^2}{\displaystyle\sum_{M, M'} b_M b_{M'} \hat{S}_{MM'}(0)} = \frac{1}{\phi_A N_A V_A} + \frac{1}{\phi_B N_B V_B} - 2\frac{\chi_S}{V_0} . \tag{6.8}$$

Here, $\phi_M = \rho_M V_M/\eta$ is the volume fraction of sites of type M, V_M is the volume of a site of type M, V_0 is a "reference volume", and b_M is the total neutron scattering length of a site of species M. Note that this chi-parameter will generally diverge as $\phi_M \to 0$ due to the unrealistic incompressibility assumption. The SANS chi is, by construction, equivalent to the incompressible Flory value at the spinodal.

Substitution of Eq. (6.4) into Eq. (6.8) yields a general expression for the SANS chi-parameter which can be employed in PRISM studies to make direct contact with experiments on specific blends. Such an approach to comparing theory and experiment has been emphasized by Dudowicz, Freed, and co-workers [2] as the most appropriate procedure. The resultant empirical chi-parameter contains not only microscopic information concerning intermolecular interactions in blends (i.e. some combination of the independent $C_{MM'}$), but also reflects all the errors made by the neglect of compressibility effects which depend on N_M, T, ϕ_M, etc. [2].

The general, wavevector-dependent IRPA scattering function for pure concentration fluctuation scattering is given by [57, 61]

$$\hat{S}_c^{-1}(k) = r^{-1/2}[\phi_A \hat{\omega}_A(k)]^{-1} + r^{1/2}[\phi_B \hat{\omega}_B(k)]^{-1} - 2\hat{\chi}_s(k) \tag{6.9}$$

where $r = V_A/V_B$ and $\eta = \rho_A V_A + \rho_B V_B$ is the total site packing fraction. The corresponding IRPA spinodal condition is

$$2\chi_s = r^{-1/2}(N_A \phi_A)^{-1} + r^{1/2}(N_B(1 - \phi_A))^{-1} . \tag{6.10}$$

Note that in the hypothetical incompressible limit the form of both the scattering functions and spinodal condition are much simpler than the rigorous expressions of Eqs. (6.4)–(6.6). Although the IRPA can usually be fitted to low wavevector experimental scattering data, and an "apparent chi-parameter" thereby extracted, the *literal* use of the IRPA for the calculation of thermodynamic properties and phase stability is generally expected to represent a poor approximation due to the importance of density-fluctuation-induced "compressibility" or "equation-of-state" effects [2, 65, 66]. The latter are non-universal, and are expected to increase in importance as the structural and/or intermolecular potential asymmetries characteristic of the blend molecules increase.

The conditions required for accuracy of an incompressibility assumption at the level of the scattering functions and spinodal condition are easily derived within the PRISM formalism [67]. From Eq. (6.7) a small isothermal compressibility implies that $- \rho \hat{C}_{MM'}(0) \gg 1$, which is generally true for any dense fluid. If the related wavevector-dependent condition

$$- \rho_M \hat{\omega}_M(k) \hat{C}_{MM}(k) \gg 1 \tag{6.11}$$

applies then considerable simplification of Eqs. (6.4) occurs [67]:

$$\hat{S}_{AA}^{-1}(k) \cong \frac{\tilde{\Lambda}(k)}{- \rho_A \rho_B \hat{\omega}_A(k) \hat{\omega}_B(k) \hat{C}_{BB}(k)}$$

$$\cong (\rho_A \hat{\omega}_A(k))^{-1} + (\rho_B \hat{\omega}_B(k))^{-1} \frac{\hat{C}_{AA}(k)}{\hat{C}_{BB}(k)}$$

$$- \hat{C}_{BB}^{-1}(k)\{\hat{C}_{AA}(k)\hat{C}_{BB}(k) - \hat{C}_{AB}^2(k)\}$$

$$\hat{S}_{BB}^{-1}(k) \cong \frac{\hat{C}_{BB}(k)}{\hat{C}_{AA}(k)} \hat{S}_{AA}^{-1}(k) \quad \hat{S}_{AB}^{-1}(k) \cong - \frac{\hat{C}_{BB}(k)}{\hat{C}_{AB}(k)} \hat{S}_{AA}^{-1}(k) . \tag{6.12}$$

Equation (6.11) can be viewed as an *"effective incompressibility condition"*, which is expected to be accurate on long length scales for dense, high polymer blends. However, the interrelationship between the three partial structure factors is not precisely given by the literal incompressibility relations: $\hat{S}_{AA}(k) = \hat{S}_{BB}(k) = -\hat{S}_{AB}(k)$, and the form of $\hat{S}_{AA}(k)$ is not the same as the IRPA-like expression of Eqs. (6.9). Most significantly, the analog of the effective chi-parameter is not of the simple linear arithmetic difference form (see Eqs. (6.16) below), but is fundamentally nonlinear in the direct correlation functions. These differences between the "effective" and "literal" incompressible approximations are generally very important [67]. Similar general conclusions have been drawn by Freed and co-workers [2].

An alternative approach to extracting an apparent chi-parameter from SANS data is to fit the scattering curves to an IRPA form over the entire measured wavevector range. Using Eqs. (6.12) one obtains for the scattering profile in the "effectively incompressible" approximation:

$$\hat{S}_c(k) \equiv \rho^{-1} \sum_{M, M'} b_M b_{M'} \hat{S}_{MM'}(k) = F_A \tilde{S}_c(k) \tag{6.13a}$$

where the "amplitude" factor, F_A, and dimensionless concentration fluctuation scattering function, $\hat{S}_C(k)$, are given by

$$F_A \equiv b_A^2 \sqrt{\frac{C_{BB}}{C_{AA}}} + b_B^2 \sqrt{\frac{C_{AA}}{C_{BB}}} - 2b_A b_B \frac{C_{AB}}{\sqrt{C_{AA} C_{BB}}} \tag{6.13b}$$

$$\tilde{S}_c^{-1}(k) \equiv \frac{\rho}{\rho_A \hat{\omega}_A(k)} \sqrt{\frac{C_{BB}}{C_{AA}}} + \frac{\rho}{\rho_B \hat{\omega}_B(k)} \sqrt{\frac{C_{AA}}{C_{BB}}} - \rho \frac{C_{AA} C_{BB} - C_{AB}^2}{\sqrt{C_{AA} C_{BB}}}. \tag{6.13c}$$

Here, ρ is the total site number density, and the small angle scattering approximation $\hat{C}_{MM'}(k) \cong \hat{C}_{MM'}(0) \equiv C_{MM'}$ has been employed. The wavevector dependence is contained in $\hat{S}_C(k)$ which is of the same general mathematical form as the empirical IRPA expression. To make direct contact with the experimental data analysis $\hat{S}_C(k)$ can be equated to the IRPA form of Eq. (6.9). The result of this procedure is an explicit expression for the apparent SANS chi-parameter. For the simpler case of equal A and B site volumes (generalizations are straightforward) one obtains [67]:

$$\hat{\chi}_s(k) = \frac{\rho}{2\sqrt{C_{AA} C_{BB}}} (C_{AA} C_{BB} - C_{AB}^2)$$

$$+ \frac{1 - \sqrt{C_{BB}/C_{AA}}}{2\phi_A \hat{\omega}_A(k)} + \frac{1 - \sqrt{C_{AA}/C_{BB}}}{2(1 - \phi_A) \hat{\omega}_B(k)}. \tag{6.14}$$

Note that the apparent SANS chi-parameter can acquire a k-dependence even in the small angle regime via a "cross-term" between the intramolecular and intermolecular correlation contributions. Thus, the wavevector dependence may be viewed as an "artifact" of the incompressibility approximation. It may be particularly important for SANS experiments on strongly structurally asymmetric mixtures for which the quantity $|1 - C_{BB}/C_{AA}|$ is expected to be non-negligible. Such an "anomalous" k-dependent chi-parameter has been experimentally observed recently by Brereton et al. [57] in a binary blend of polymers with significantly different aspect ratios (polystyrene and polytetramethyl carbonate). For most amorphous blends composed of relatively flexible chains, experimental SANS data can apparently be adequately fitted using the IRPA and a single, wavevector-independent chi-parameter [56–58]. The latter can be identified with the k = 0 limit of Eq. (6.14):

$$\hat{\chi}_s(k) = \frac{\rho}{2\sqrt{C_{AA}C_{BB}}}(C_{AA}C_{BB} - C_{AB}^2)$$

$$+ \frac{1 - \sqrt{C_{BB}/C_{AA}}}{\phi_A N_A} + \frac{1 - \sqrt{C_{AA}/C_{BB}}}{(1 - \phi_A)N_B}. \tag{6.15}$$

The results of PRISM theory can be substituted in this relation to make contact with SANS experiments which extract an apparent chi-parameter via the k-dependent fitting procedure. Note, however, that in general the predictions of an apparent SANS chi-parameter using Eq. (6.15) will not be the same as the extrapolated zero angle intensity approach of Eq. (6.8). This point re-emphasizes the phenomenological nature of the single, effective chi-parameter approach.

In our initial PRISM studies of polymer blends a literal connection with the IRPA was derived [59, 61] and utilized in model calculations of the effective chi-parameter [59–63]. The incompressibility assumption is fundamentally incompatible with the PRISM approach which naturally includes pure density, pure concentration, and coupled density-concentration fluctuations on all length scales. However, a "literal incompressibility constraint", i.e. $\rho_A(r) + \rho_A(r) = \rho$, can be enforced on the PRISM theory in a post facto manner by employing a functional Taylor expansion of the free energy [59, 61]. This procedure results in a single scattering function of precisely the IRPA form of Eq. (6.9), but the apparent chi-parameter is a wavevector dependent correlation function which we denote here as $\hat{\chi}_{INC}(k)$. In the long wavelength limit, $\hat{\chi}_{INC}(k) \cong \chi_{INC}$ and is given by [61]

$$2\chi_{INC} = \rho[\phi_A r^{-1/2} + \phi_B r^{1/2}]^{-1}$$

$$\times [r^{-1}\hat{C}_{AA}(0) + r\hat{C}_{BB}(0) - 2\hat{C}_{AB}(0)\}. \tag{6.16}$$

If the blend possesses segmental volume and degree of polymerization symmetries, i.e., r = 1 and $N_A = N_B = N$, then Eqs. (6.16) and (6.10) become particularly simple:

$$2\chi_{INC} = \rho[C_{AA} + C_{BB} - 2C_{AB}] \tag{6.17}$$

$$2\chi_{INC}\phi(1 - \phi)N = 1 \tag{6.18}$$

where Eq. (6.18) applies at the spinodal. Note the analogy here with Flory-Huggins theory. χ_{INC} is a generalization of the incompressible mean field chi-parameter where the $k = 0$ component of the direct correlation functions replace the "bare" attractive tail potentials. These direct correlation functions contain the many body correlation information neglected by the simple mean field approximation.

Most prior PRISM predictions [23, 59–63] for the effective SANS chi-parameter were based on the literal incompressible form of Eq. (6.16), and not the more experimentally appropriate relation of Eq. (6.8) or (6.15). Thus, it is instructive to inquire whether there are any special cases and/or well-defined additional approximation for which Eqs. (6.6) and (6.12) reduce to the literal IRPA forms. As discussed in Sect. 8 and elsewhere [61, 67, 68], the one special case corresponds to the theoretically much studied, but experimentally unrealizable, "*symmetric polymer blend*". More generally, the additional approximation required to recover the IRPA forms corresponds in integral equation language to the $k = 0$ statement [67]:

$$C_{MM'} \cong C_0 - \beta v_{MM'}, \quad \text{with} \ -C_0 \rightarrow \infty \ . \tag{6.19}$$

The motivation for this approximation is that the integrated strength of the direct correlation functions (often viewed as "excluded volume parameters") consists of a purely repulsive part plus a weak attractive contribution. If the former is viewed as the bare singular hard core potential (i.e., a literal $\rho \rightarrow \infty$ incompressibility approximation for the direct correlation function), then Eq. (6.19) is obtained. Substitution of Eq. (6.19) in Eqs. (6.12) yields the literal IRPA results, with an effective *enthalpic* chi-parameter given by the simple arithmetic difference Flory form of Eq. (6.17) expressed solely in terms of the "weak attractive" contributions $\beta v_{MM'}$. The new molecular closures [68–70] discussed in Sect. 8 do possess a mathematical structure very similar to Eq. (6.19), but with two crucial differences. First, the large repulsive force contributions are not infinite since real liquids are compressible, and second, in general they depend on the two interacting sites (i.e. M and M′ labels) since the local interchain correlations are sensitive to molecular structure. Even in the hypothetical $\rho \rightarrow \infty$ limit, the repulsive interaction contribution to $C_{MM'}$ is generally species-dependent which destroys the reduction of Eqs. (6.6) and (6.12) to the literal IRPA forms.

6.3 Other General Aspects

In this paper we focus solely on the spinodal as a measure of the phase behavior. The corresponding binodal can be derived by (numerically) integrating the compressibility relations. Alternatively, the phase behavior could be deduced directly via the so-called "free-energy", or charging parameter route discussed in Sect. 4. Implementation of the "free energy" route is generally much more demanding computationally and has only very recently [70b] been numerically studied within the blend PRISM formalism. Constant volume, not constant pressure, conditions are assumed. The latter requires a theory for the equation-of-state of blends which has not yet been fully developed. This restriction motivates consideration primarily of upper critical solution temperature (UCST) phase transitions, although some PRISM work on the lower critical solution temperature (LCST) phenomenon has recently been done [67, 71].

Specific results require two additional pieces of information. First, the single chain polymer structure as contained in the species-dependent structure factors must be known. In principle, these correlation functions should be self-consistently computed along with the intermolecular pair correlations. This most rigorous approach is now feasible and is briefly discussed in Sect. 10. However, all published PRISM work on blends has employed the "Flory ideality ansatz" discussed in Sect. 2, i.e., the required $\hat{\omega}_M(k)$ functions are presumed given and effectively independent of blend composition and proximity to the spinodal. To date, published analytical and numerical work for alloys has employed coarse-grained models such as the Gaussian, freely-jointed, and semi-flexible chain. Second, a closure approximation is required which relates the intermolecular pair and direct site-site correlation functions outside the hard core diameter. Both traditional "atomic" and novel "molecular" closures have been studied.

7 Athermal Blends

The model "athermal blend" is defined [59, 62] as the hypothetical limit of vanishing interchain attractive potentials relative to the thermal energy, i.e., $\beta v_{MM'}(r) = 0$. For this situation the atomic site-site Percus-Yevick closure approximation of Eq. (2.7) is employed where the subscripts now refer to the species type. The constant volume athermal blend is of theoretical interest since it isolates the purely entropic packing effects. However, as emphasized by several workers [2, 62, 63, 67], the athermal reference blend is not an adequate model of any real phase separating system. Its primary importance is as a "reference" system for the theories of thermally-induced phase separation discussed in Sect. 8.

A number of numerical PRISM studies [59, 61, 62] of model athermal blends have performed at various levels of single chain description. These mixtures are characterized by structural asymmetries between the two components of either a local nature (size and/or stiffness differences) or global nature (chain length difference, blend of rings and chains). When Gaussian models are employed, numerical studies to date show that these blends are completely miscible in the sense that no spinodal instability is found. However, for dense mixtures of rods and coils [71] and semiflexible chains [72] very recent work has found entropy-driven phase separation. For very stiff polymers, the constant volume assumption, and the neglect by RISM of significant orientational correlations and nematic phase formation [26], may make constant volume PRISM less appropriate.

Typical results [62] for the intermolecular concentration correlations of amorphous Gaussian chain athermal blends are shown in Fig. 9. The amplitude of the structural fluctuations is relatively small. The stiffness/size asymmetric mixture exhibits primarily local deviations from random mixing which decay to zero monotonically with increasing spatial separation. On the other hand, the bimodal chain length blend displays nonmonotonic fluctuations which change sign and peak on a length scale characteristic of macromolecule size. The amplitude of the structural fluctuations in athermal semiflexible polymer mixtures increases significantly as the magnitude of the aspect ratio and/or the difference in persistence lengths of the components increase [71, 72], and phase separation can eventually occur.

The incompressible chi-parameter defined in Eq. (6.16) has also been extensively studied. Many of the numerical results for site volume and/or statistical segment length asymmetric athermal Gaussian chain blends are adequately reproduced at a qualitative level by the analytic thread model discussed in Sect. 2. For an athermal stiffness blend of very long Gaussian threads the $k = 0$ direct correlation functions are [23, 62]:

$$C_{BB}^{(0)} = \gamma^4 C_{AA}^{(0)} \quad \text{and} \quad C_{AB}^{(0)} = \gamma^2 C_{AA}^{(0)} \tag{7.1}$$

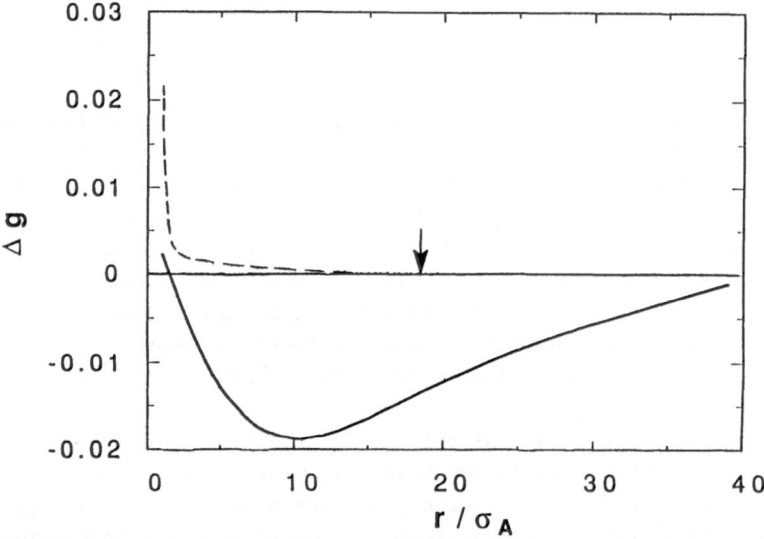

Fig. 9. Intermolecular fluctuation correlation function, $\Delta g(r) = g_{AA}(r) + g_{BB}(r) - 2g_{AB}(r)$, as a function of interchain site separation normalized by the statistical segment length of the A-chain ρ_A (which equals the site hard core diameter). This function is one measure of the length-scale-dependent non-random packing in the mixture. A Gaussian chain model is employed and the volume fraction of A and B chain segments are equal [62]. For completely random mixing $\Delta g(r) = 0$. The *dashed curve* is for a stiffness/size asymmetric case of $\gamma = \sigma_B/\sigma_A = 1.2$ with $N = 2000$ and total packing fraction of 0.45. The *solid curve* is for the bimodal chain length blend with $N_A = 2000$, $N_B = 200$ and a total packing fraction of 0.5. The *arrow* denotes the radius-of-gyration of a 2000 unit ideal Gaussian chain

$$\sigma_A^{-3} C_{AA}^{(0)} = -\frac{\pi^2}{108} \rho \sigma_A^3 [\phi_A + \gamma^2 (1 - \phi_A)]$$

$$= -\frac{\pi}{18} \eta \Gamma^3 [\phi_A + \gamma^2 (1 - \phi_A)] \qquad (7.2)$$

where $\gamma = \sigma_B/\sigma_A$ is the stiffness (or statistical segment length) asymmetry ratio, $\Gamma = \sigma_A/d$ is an aspect ratio, and the superscript "0" emphasizes the results are for athermal blend. Substituting these expressions in Eq. (6.16) yields a negative incompressible chi-parameter:

$$\chi_{INC}^{(0)} = -\frac{\eta^2 \Gamma^6}{6} (\gamma^2 - 1)^2 [\phi_A + \gamma^2 (1 - \phi_A)] . \qquad (7.3)$$

This result implies that stiffness asymmetry stabilizes the constant volume athermal blend in the sense that pure concentration fluctuations are reduced relative to the $\gamma = 1$ melt. Moreover, within a literal incompressible description it is tempting to conclude from Eq. (7.3) that statistical segment length asymmetry will always decrease the effective chi-parameter and stabilize the miscible phase of real blends. Such a conclusion is generally incorrect with regards to real phase-separating mixtures since it ignores compressibility effects which destroy a rigorous separation of enthalpic and entropic contributions to the thermodynamics of mixing. The irrelevance of a negative chi-parameter for the model athermal blend to the question of miscibility in real polymer mixtures is not a failing of the PRISM theory, but rather only the RPA-like definition of an incompressible chi-parameter in Eq. (6.16). Although perhaps not emphasized enough in early papers, we have explicitly demonstrated that an incompressible idealization, and hence a single chi-parameter, is a poor approximation for a thermal blends [62, 63, 67, 72].

8 Thermally Phase-Separating Blends

A natural foundation for developing an integral equation theory for phase-separating polymer alloys would be an understanding of atomic and small molecule fluid mixtures. Unfortunately, very little work has been done on such systems, and only recently have atomic mixture calculations based on various closure approximations begun to be systematically done and compared with computer simulations [65]. Even less work has been performed for small molecule mixtures using the RISM formalism. From a fundamental theoretical perspective, an accurate treatment of fluctuations induced by the attractive branch of the intermolecular potential is difficult even for simple atomic fluids [5, 6]. These considerations suggest that integral equation theories of polymer alloys should initially focus on the simplest possible mixture in order to test the unavoidable closure approximation.

8.1 Symmetric Blends and Predictions of Atomic Closures

A "symmetric" binary blend is defined as consisting of two types of homo-polymer chains, A and B, which are structurally identical in every respect. Hence, the corresponding single chain structure factors obey the identity $\omega_A(r) = \omega_B(r) \equiv \omega(r)$. The pair potentials between sites are:

$$u_{AA}(r) = u_{BB}(r) = u_0(r) + v_{AA}(r)$$

$$u_{AB}(r) = u_0(r) + v_{AB}(r) \tag{8.1}$$

where $u_0(r)$ is a hard core potential with diameter d, and the $v_{MM'}(r)$ potentials are defined for $r > d$. For the "fully symmetric" case of equal concentration of A and B chains there is complete symmetry between A and B, and long wavelength concentration fluctuations are expected to be accurately described by the IRPA-like formula [61, 67]:

$$\hat{S}_c^{-1}(k) = [\hat{\omega}(k)[\phi(1 - \phi)]]^{-1} - 2\hat{\chi}_{INC}(k) \tag{8.2}$$

where

$$\hat{\chi}_{INC}(k) = \rho\Delta\hat{C}(k) \equiv \rho[\hat{C}_{AA}(k) - \hat{C}_{AB}(k)] \tag{8.3}$$

and $\phi = 1/2$ for the fully symmetric blend. The "incompressible chi-parameter" is given by the zero wavevector limit of Eq. (8.3), i.e. $\chi_{INC} = \hat{\chi}_{INC}(k = 0)$. The corresponding "mean field" or "bare" chi-parameter, denoted χ_0, is a purely energetic quantity given by:

$$\chi_0 = \beta\rho \int d\vec{r}\{v_{AB}(r) - v_{AA}(r)\} \equiv \beta\rho\{\hat{v}_{AB}(0) - \hat{v}_{AA}(0)\} \tag{8.4}$$

which is the "classical" Flory-Huggins result for the off-lattice situation. Physically, Eq. (8.4) assumes that the intermolecular interactions and chain connectivity induce neither density nor concentration fluctuations in the blends, i.e. a literal random mixing assumption. The quantity χ_{INC}/χ_0 will be referred to as the "renormalization ratio" which serves as a simple scalar measure of the influence of intra- and intermolecular correlations.

Curro and Schweizer have carried out numerical [60, 61] and analytical [23] studies of the symmetric blend using the Mean Spherical Approximation (MSA) closure successfully employed for atomic, colloidal, and small molecule fluids [5, 6]. This closure corresponds to the approximation:

$$C_{MM'}(r) \cong -\beta v_{MM'}(r), \quad r > d. \tag{8.5}$$

It is important to emphasize that the PRISM theory with "atomic site-site closures" assumes that all the consequences of chemical bonding are adequately accounted for via the structure of the integral equations alone. As depicted in Fig. 10, even at the two molecule level there are "indirect" pathways by which

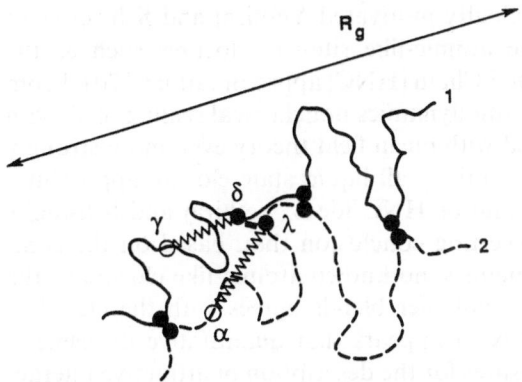

Fig. 10. Schematic representation of two-polymer molecule correlation pathways. The *wiggly bonds* denote long range chemical bonding pair correlations due to chain connectivity, while the *short solid line* denotes the spatially local direct correlation function. There are of order $a + bN^\theta$ pair "contacts" between two interpenetrating polymer chains of degree of polymerization N which are embedded in a correlation volume of linear dimension $R_g \propto N^{1/d_F}$. Here d_F is the mass fractal dimension of the polymer, $\theta = 2 - (D/d_F)$, D is the spatial dimension, and a and b are numerical constants of order unity

two particular sites α and γ on different molecules may interact. As a consequence, RISM is never exact even in the "weak coupling" regime where the "tail" potential obeys the inequality $\beta v_{MM'}(r) \ll 1$.

The most provocative prediction of the PRISM-MSA theory was a non-classical relation between the critical temperature for phase separation, T_c, and degree of polymerization, N, of the form $T_c \propto N^{1/2}$. This unexpected result corresponds to a massive stabilization of the mixed phase (via a long range concentration fluctuation process) relative to Flory-Huggins mean field theory which predicts $T_c \propto N$. Moreover, in the high temperature limit, $\chi_{INC}/\chi_0 \propto N^{-1/2}$, thereby implying that the Flory-Huggins form is not recovered even in the perturbative or weak coupling limit of $\chi_0 \to 0$. This fact establishes the important conclusion that the origin of the massive renormalization effect is a two-molecule correlation process which depends on both spatial and polymer fractal dimensionalities [23]. Although there are a number of experiments [58] on more chemically complex alloys which display significant deviations from Flory-Huggins scaling and exhibit N-dependent apparent chi-parameters in the direction predicted by PRISM-MSA, definitive work on simple systems has been lacking.

Very recently, Deutsch and Binder [73] carried out a large scale lattice Monte Carlo simulation on a model symmetric polymer mixture, and Gehlsen et al. [74] performed a SANS study on a family of specially-designed high molecular weight isotopic blends. The classical Flory-Huggins scaling law was found to a high degree of accuracy. This strongly suggests that the PRISM theory with the atomic-like MSA closure is in qualitative error, and our prior PRISM calculations [60, 61, 63] on thermally phase-separating model blends are not generally reliable. This difficulty motivated Yethiraj and Schweizer to numerically investigate alternative atomic-like site-site closures such as the Percus-Yevick (PY) and Hypernetted Chain (HNC) approximations [70]. From the compressibility route to the thermodynamics nonclassical scaling of T_c with N was again found which disagreed with mean field theory even more strongly than the MSA prediction. Since essentially all liquid state closure approximations are based on the PY, MSA and/or HNC ideas, Yethiraj and Schweizer [68, 70] were forced to the disconcerting conclusion that, based on the compressibility route to the thermodynamics, no known atomic-like closure to the PRISM equations for random coil polymer blends agrees with the classified Flory-Huggins scaling of T_c with N. It appears that quantitative deficiencies associated with atomic site-site closures for the description of attractive interactions in small molecule fluids become grossly amplified into qualitative errors for phase-separating macromolecular systems.

One possible alternative theoretical approach is to calculate the blend thermodynamics and phase diagram not from the compressibility and spinodal divergence, but from the so-called free energy route [5, 6]. As emphasized very recently by Chandler [75], only the spatially local consequences of the attractive potentials on the interchain correlations enter the free energy approach. Thus, long wavelength difficulties of integral equation theories will be "cut off" and a classical $T_c \propto N$ scaling law is obtained [75]. However, there remain severe conceptual difficulties since PRISM-MSA theory is characterized by a massive (N-dependent) thermodynamic inconsistency problem. Moreover, the description of concentration fluctuations on the macromolecular and longer length scales in the one-phase region, which are of prime importance in SANS measurements on blends and microphase separation of diblock copolymers, will still be treated in a qualitatively incorrect fashion at all temperatures [68–70].

8.2 Molecular Closures and Predictions for Symmetric Blends

The theoretical difficulties described in the preceding section have lead Yethiraj and Schweizer to reconsider the question of the closure approximation for interaction-site molecular and polymer fluids [68–70]. Their approach to formulating new "molecular" closures is strongly motivated by the observation that the fundamental error incurred by the atomic-like closures for long wavelength correlations appears in the weak coupling limit [the $\rho \rightarrow 0$ two-molecule level for one-component fluids, and the $T \rightarrow \infty$ high temperature regime in blends], and is associated with the influence on the site-site direct correlation functions of the number of contacts between a pair of interacting macromolecules. The construction of the new molecular closures is not based on the individual site-site correlation functions, but rather on their full two-molecule counterpart.

Technical development of the new closures has been guided by three general considerations. (1) Separate approximations are employed to treat the consequences of the hard core and (generally attractive) tail parts of the potential.

This strategy is common in atomic fluid theory at low to moderate densities, and for Coulombic systems, and corresponds to the "reference" idea ubiquitous in liquid state theory [5]. The purely hard core problem is treated using the accurate [27] PY closure. (2) The construction of a closure approximation for the tail part of the potential is subject to the constraint of exactly describing the weak coupling limit. In physical terms, for fractal-like interpenetrating molecules these indirect processes may strongly couple the direct correlation functions associated with those pairs of sites which are in simultaneous contact. The number of such two-molecule pair contacts, N_c, scales with N as [23]:

$$N_c \propto 1 + cN^2/R_g^D \propto 1 + c'N^{2-\left(\frac{D}{d_f}\right)} \tag{8.6}$$

where c and c' are numerical constants of the order of unity, R_g is the radius-of-gyration, and D and d_f are the spatial and mass fractal dimensionalities, respectively. For ideal coils the number of contacts grows as $\sqrt{N}$ and this geometrical factor is the source of the massive "renormalization" predicted by the PRISM-MSA theory. (3) The relation between the intermolecular attractive potentials and the "direct" part of the interchain correlation processes is estimated based on our present understanding of the analogous problem for simple atomic fluids [5, 6].

The simplest molecular closure based on the above ideas is one that builds in the hard core reference behavior and correctly treats the longer ranged attractive potentials in the weak coupling limit. It is called the "Reference Molecular Mean Spherical Approximation" (RMMSA) and is given in real space for a homopolymer blend by [68–70]

$$\omega_M * C_{MM'} * \omega_{M'}(r) \cong \omega_M * C_{MM'}^{(0)} * \omega_{M'}(r) - \omega_M * \beta v_{MM'} * \omega_{M'}(r), \quad r > d_{MM'}. \tag{8.7}$$

The "reference" direct correlation functions associated with the athermal blend are denoted by $C_{MM'}^{(0)}$, and are computed separately using the Percus-Yevick approximation. There are two distinguishing features of this closure. (1) Even outside the hard core, the direct correlation functions between different pairs of sites on two molecules are explicitly coupled. (2) The site-site direct correlations inside the hard core are intimately coupled to their behavior outside the core. These two features are in strong contrast to the atomic closures, and represent the influence of the indirect, chemical-bonding-mediated processes between two molecules.

A more general formulation consistent with the three considerations enumerated above is given for a *homopolymer* blend by [68]:

$$\omega_M * C_{MM'} * \omega_{M'}(r) \cong \omega_M * C_{MM'}^{(0)} * \omega_{M'}(r) + \omega_M * \Delta C_{MM'} * \omega_{M'}(r), \quad r > d_{MM'}. \tag{8.8}$$

where $\Delta C_{MM'}(r)$ is an approximate atomic site-site closure relation for the attractive branch of the potential. For relatively short ranged attractions, the Percus-Yevick closure is quite accurate [5, 6]. This suggests the approximation [68]:

$$\Delta C_{MM'}(r) \cong [1 - \exp(\beta v_{MM'}(r))] g_{MM'}(r), \quad r > d_{MM'} \tag{8.9}$$

which is non-perturbative in the strength of the tail potentials. The combination of Eqs. (8.8) and (8.9) is called the "Reference-Molecular Percus-Yevick (R-MPY)" closure. The qualitative physical content of this approximation is that the "MSA part" of the closure is corrected by the multiplicative factor of the species-dependent radial distribution function in a manner analogous to the calculation of an internal energy or enthalpy. The influence of local fluctuations on the site-site direct correlation functions are then self-consistently determined via solution of the coupled PRISM equations and closure. A useful simplification of the R-MPY closure corresponds to introducing a high temperature approximation (HTA) into Eq. (8.9):

$$\Delta C_{MM'}(r) \cong - \beta v_{MM'}(r) g_{MM'}^{(0)}(r), \quad r > d_{MM'} \tag{8.10}$$

where the superscript "0" refers to the hard core reference blend. This closure approximation is called the R-MPY/HTA and is conceptually of the R-MMSA form but is expected to be more accurate since the influence of reference system correlations on the attractive force component of the direct correlation functions is included.

Extensive studies of the predictions of the new molecular closure to the blend PRISM theory for the symmetric binary blend have been carried out by Yethiraj and Schweizer [68–70]. Here, a few of their major results are summarized, beginning with the numerical studies. The PRISM equations with the molecular closures can be solved using standard Picard iteration methods and the fast Fourier transform [5, 70].

The reduced critical temperature, $T_c^* = k_B T_c / |\epsilon|$ where ϵ is the energy parameter of the repulsive Lennard-Jones tail potential between A and B sites, is plotted vs N in Fig. 11 for the R-MMSA and R-MPY closures and two values of total packing fraction [69, 70]. The latter is defined for the tangent semiflexible chain [24] as $\eta = \pi \rho d^3 / 6$. A linear scaling law is found with a nonuniversal prefactor that decreases with density. As seen in Fig. 11 and Table 1, the R-MPY predicts a significant reduction of the critical temperature relative to both R-MMSA and Flory-Huggins mean field theory which becomes increasingly significant as the polymer density is lowered [70]. This prediction is easily understood as a consequence of the local correlation hole (at lower densities) on the length scale of the tail potential which drives phase separation. Representative results [70] for the composition dependence of the effective incompressible chi-parameter defined in Eq. (6.17) are shown in Fig. 12. The R-MMSA predicts (not shown) virtually no composition dependence, while the R-MPY exhibits [70a] a parabolic-like, concave-upwards behavior the amplitude of which is a strong function of polymer density and stiffness, but weakly dependent on degree of polymerization over the range of $N \leqslant 200$. However, in the large N limit this composition dependence disappears [70b], consistent with thermodynamic perturbation theory arguments [75]. Note that the R-RMSA closure appears to be nearly the integral equation realization of Flory-Huggins theory for the *symmetric* blend, while the R-MPY exhibits a host of fluctuation corrections associated with local correlations in the blend which are in excellent

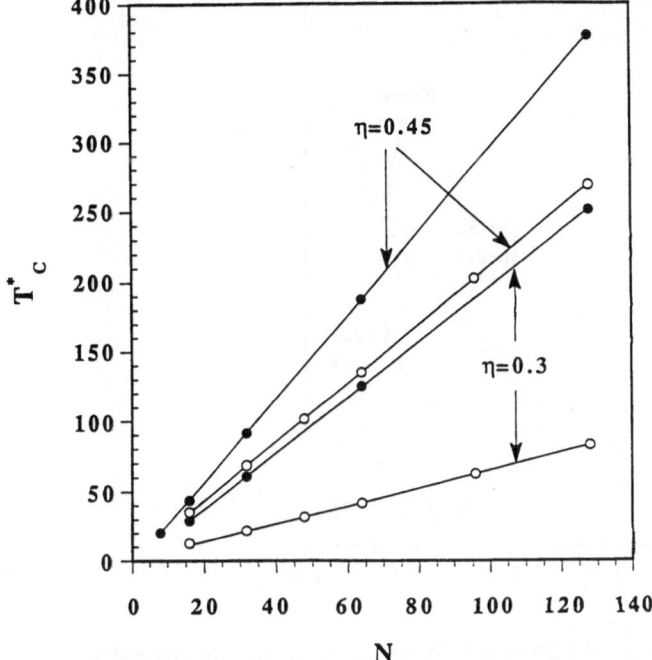

Fig. 11. Variation of the reduced critical temperature, $T_c^* = T_c/\varepsilon$, with degree of polymerization of symmetric binary blends composed of semiflexible chains with zero bending energy [70]. ε is the repulsive AB Lennard-Jones interaction parameter, $v_{AA}(r) = v_{BB}(r) = 0$, and $\phi = 0.5$. Results for the R-MMSA (*solid circles*) and R-MPY (*open circles*) are shown for two values of packing fraction. The *straight lines* are linear fits to the theoretical results for the four highest chain lengths

Table 1. Variation of the ratio of the reduced critical temperature, T_c, predicted by PRISM theory with the R-MPY closure relative to its corresponding mean field Flory-Huggins value, T_0, with packing fraction, η, for a $N = 32$ unit tangent semiflexible chain model [24] (zero bending energy). The interchain tail potentials obey the symmetric restriction with the AA and BB potentials set equal to zero [70]

η	T_c/T_0
0.20	0.126
0.30	0.339
0.45	0.712
0.55	0.930

agreement [70] with recent simulations [73] (with the exception that the critical divergence aspect is not Ising-like).

The density and composition dependent correlation effects present in the R-MPY theory arise from non-random mixing intermolecular packing and

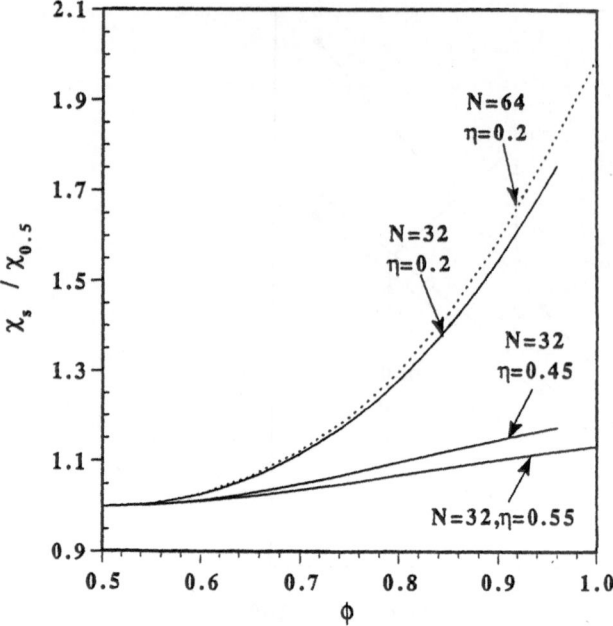

Fig. 12. Numerical predictions of PRISM/R-MPY for the variation with composition of the symmetric blend incompressible chi-parameter (normalized by its values for $\phi = 0.5$) [70]

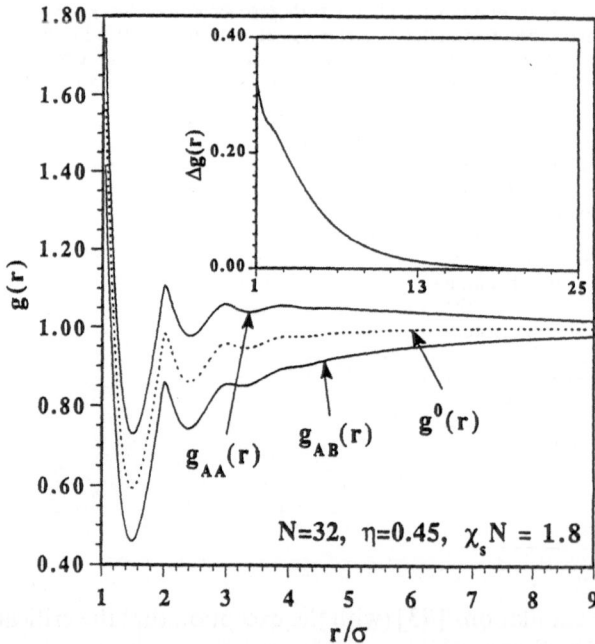

Fig. 13. Intermolecular structure of the $\phi = 0.5$ symmetric blend close to the critical temperature [70]. The *curve labeled* $g^0(r)$ is the corresponding homopolymer melt (infinite temperature) result. The *inset* depicts a measure of the non-random packing expressed as $\Delta g(r) = g_{AA}(r) - g_{AB}(r)$. The behavior of the latter function is qualitatively different when using the atomic MSA closure of the PRISM equations [23, 61, 68]. The radius-of-gyration is 2.91 for $N = 32$

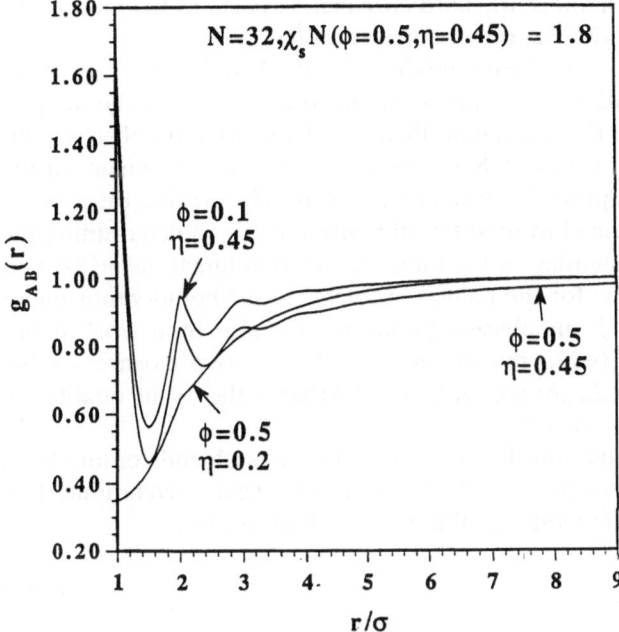

Fig. 14. Selected predictions of PRISM/R-MPY theory for the dependence of the symmetric blend $g_{AB}(r)$ on composition and total packing fraction at fixed temperature [70]

typical results [70] are shown in Figs. 13 and 14. As expected from the discussion of Sect. 8.1, the post facto IRPA simplification is found to be an excellent approximation for the small k scattering functions, effective chi-parameter, and spinodal boundary of the idealized symmetric blend model. The relationship of these calculations for symmetric model blends to experiments on simple isotopic polymer mixtures is discussed elsewhere [67, 68, 70].

Extensive analytic results for the symmetric thread blend have also been derived [68, 70b]. In the thread-polymer limit the hard core condition becomes "irrelevant" for the molecular closure relations. In particular, for the R-MMSA and R-MPY/HTA approximations the $\hat{C}_{MM'}(k)$ functions are fully specified by the closure relations, and their k = 0 values are given in general by

$$C_{MM'} = C_{MM'}^{(0)} - \beta \tilde{H}_{MM'}, \quad d_{MM'} \to 0$$

$$\tilde{H}_{MM'} \equiv v_{MM'}, \qquad \text{R-MMSA}$$

$$\equiv \int d\vec{r} \, v_{MM'}(r) g_{MM'}^{(0)}(r), \quad \text{R-MPY/HTA} . \qquad (8.11)$$

For the symmetric blend the reference system is the homopolymer melt and hence the reference correlation functions are independent of species label (M, M′). It is important to emphasize that Eq. (8.11) is not valid for polymers of nonzero hard core thickness. Thus, the "thread" idealization is a very special limit characterized by a unique simplification of the integral equation theory. These equations have the analytical structure of a "high temperature and/or mean field" approximation about the hard core blend system at the level of the "effective potentials", i.e., site-site direct correlation functions.

For the symmetric thread blend model, the R-MMSA theory predicts $\chi_{INC} = \chi_0$ for all compositions, densities, etc., and thus represents the integral equation realization of the Flory-Huggins theory with regard to the effective chi-parameter and critical temperature. Such simplicities are a consequence of the "weak coupling" or "asymptotic" nature of a MSA-like closure, i.e., the $r \rightarrow \infty$ limit of $g_{MM'}(r) = 1$ is assumed to hold for all r with regards to determining the effects of the attractive potential on the direct correlation functions. MSA and R-MMSA analytical results for the related problems of the liquid-vapor phase transition and polymer-solvent phase separation have also been worked out [68]. Stark differences between the predictions of these two closures of the PRISM equations are again found, with the R-MMSA theory in qualitative agreement with mean field ideas.

The R-MPY/HTA predictions for the symmetric thread blend contain local fluctuation corrections associated with the reference blend correlations. For example, the renormalization ratio in the $N \rightarrow \infty$ limit is [68]:

$$\frac{\chi_{INC}}{\chi_0} = \frac{a}{a + \xi_\rho} \tag{8.12}$$

where a is the spatial range of the (screened Coulomb) tail potential and ξ_ρ is the collective density fluctuation length scale of the reference homopolymer melt. The latter length scale decreases monotonically with increasing polymer volume fraction (see, for example, Eq. (3.4a)). Thus the renormalization ratio is predicted to decrease strongly as the polymer density (or spatial range parameter) decreases. However, the renormalization effect vanishes, i.e., $\chi_{INC} = \chi_0$, in the hypothetical $\rho \rightarrow \infty$ incompressible limit. The R-MPY/HTA approximation predicts a composition-independent chi-parameter, in disagreement with the full R-MPY numerical results [70] and simulations for the finite N mixtures studied [73]. Thus, the origin of the composition dependence is finite temperature local concentration fluctuations which become monotonically stronger as the spinodal boundary is approached. Perturbation arguments and explicit calculations suggest the HTA should be exact for symmetric mixtures as $N \rightarrow \infty$ [5, 70b, 75]. Thus, the "intrinsic" composition-dependence of the effective chi-parameter as computed from the incompressible definition of Eq. (6.17) would vanish for very large N, although compressibility effects could still introduce an apparent composition-dependence in the experimentally relevant SANS chi-parameter as defined in Eqs. (6.8) or (6.15). From a general perspective, the influence of temperature-dependent structural changes on phase behavior is expected to be of considerable importance for polymers of moderate molecular weight, and may be the genesis of the poorly understood lower critical solution temperature (LCST) phenomena commonly observed in polymer blends.

Summarizing, to recover the classic Flory mean field scaling law for the critical temperature within the original compressibility-route PRISM framework requires a reformulation of the closure approximations. The fundamental quantity susceptible to "simple" approximation appears not to be the site-site direct correlation functions, but its collective two-molecule counterpart, $\omega*C*\omega(r)$. Despite the detailed differences between the R-MMSA, R-MPY and R-MPY/HTA closures, it is important to stress that they are conceptually similar in the sense that globally mean field predictions are obtained in the long chain limit ($T_c \propto N$). The new molecular closures proposed by Yethiraj and Schweizer [68–70] are certainly not unique, and the search for the "most accurate and computationally convenient" closure remains an important problem. The ability to obtain analytical results in the thread limit is a major advantage of the molecular closures described above. We believe the re-formulated PRISM theory of alloys can now be employed to investigate reliably the multitude of physical effects which are beyond mean field and/or lattice approaches.

An alternative approach to circumventing the atomic closure difficulties has very recently been developed by Melenkevitz and Curro [76]. They generalized the "optimized cluster" theory of Chandler and Andersen [77] and Lupkowski and Monson [78] to polymer melts and blends using the diagramatically well-founded RISM formalism [26, 79]. Numerical results for the symmetric blend give the correct mean field scaling of T_C with N, and the effective chi-parameter is found to be weakly composition-dependent.

8.3 Structural and Interaction Asymmetry Effects

"Symmetric polymer blends do not exist in reality. A host of "asymmetries" are present in real chemical alloys of interest. These include attractive potential asymmetries (present even for isotopic blends) and specific interactions, molecular weight asymmetries and polydispersity, and single chain structural differences between the blend components (e.g., monomer shape and volume, backbone stiffness, and tacticity). Realistic accounting for most of these effects would seem to require an off-lattice description which includes local interchain density and concentration correlations, and "compressibility" effects [1, 2, 63, 66, 67, 80].

Detailed analytical and numerical studies of the above questions are in progress, and a very rich and nonadditive dependence of the phase behavior on the precise nature of the attractive potentials, single chain architecture, and thermodynamic state is found [67, 72]. A full understanding of these issues would provide a scientific basis for the rational "molecular design" of polymeric alloys. The influence of asymmetries on the spinodal phase boundary of simple model polymer alloys using analytic PRISM theory with molecular closures has been derived by Schweizer [67]. In this section a few of these results are briefly discussed.

For a binary blend of thread homopolymers the structure of each chain is specified by its statistical segment length, σ_M, segmental hard core diameter, $d_M \equiv d$, degree of polymerization, N_M, and the attractive "tail" potentials, $v_{MM'}(r)$. The structural asymmetry on an equal volume basis is given by the ratio of statistical segment lengths: $\gamma = \sigma_B/\sigma_A$, which, in general, is temperature dependent. The attractive intermolecular potentials between nonpolar molecules are of the van der Waals or London dispersion form. As a first approximation the corresponding Lennard-Jones energy parameters obey "Berthelot scaling relations":

$$\varepsilon_{BB} = \lambda^2 \varepsilon_{AA}, \quad \varepsilon_{AB} = \lambda \varepsilon_{AA} \tag{8.13}$$

where λ is a positive constant proportional to molecular polarizability and ionization potential ratios [5, 81]. Such scaling relations are also often employed for the integrated strength of the entire attractive potential, $\hat{v}_{MM'}(0)$, or for the internal or cohesive energies in a multicomponent fluid, $\int d\vec{r} \, v_{MM'}(r) \, g_{MM'}(r)$. The latter is the basis of the empirical "solubility" or "Hildebrand" parameter of regular solution theory [82]. These two interpretations correspond to a Berthelot scaling in the analytic thread closures of Eq. (8.11) of the form

$$\tilde{H}_{BB} = \lambda^2 \tilde{H}_{AA}, \quad \tilde{H}_{AB} = \lambda \tilde{H}_{AA} \,. \tag{8.14}$$

The rigorous spinodal boundary is given by Eq. (6.6) and generally strongly differs from that predicted by the literal incompressible RPA approximation of Eqs. (6.10) or (6.18). Analytical results for the spinodal temperature can be derived based on Eqs. (6.6), (7.1), (7.2), (8.11), and (8.14). Other properties such as the critical composition, SANS chi-parameter, free energy of mixing, etc. can also be obtained as discussed in depth elsewhere [67].

For simplicity let $N_A = N_B = N$. The predicted spinodal is [67]:

$$k_B T_S = \frac{\rho |\tilde{H}_{AA}|}{\phi + \gamma^4(1 - \phi)} \left\{ N(\lambda - \gamma^2)^2 \, \phi(1 - \phi) + \frac{\phi + \lambda^2(1 - \phi)}{-\rho C_{AA}^{(0)}} \right\} . \tag{8.15}$$

Here the "attractive energy scale" variable is

$$|\tilde{H}_{AA}| = |\hat{v}_{AA}(0)|, \quad \text{R-MMSA}$$

$$= |\hat{v}_{AA}(0)| \frac{a}{a + \xi_{EFF}}, \quad \text{R-MPY/HTA} \tag{8.16}$$

where ϕ is the volume fraction of A segments, and ξ_{EFF} is a temperature-independent, but density, composition and aspect ratio dependent, effective density-density screening length in the reference athermal thread blend [67]. The location of the critical composition, and shape of the predicted spinodal envelope, are generally not of classical form due to both stiffness asymmetry and explicit compressibility corrections. For example, the R-MMSA closure predicts the critical composition in the long chain limit is $\phi_c = 1/(1 + \gamma^{-2})$. The dependence of phase separation temperature on statistical segment length asymmetry given by Eq. (8.15) is in excellent agreement [67] with recent experiments by Bates et al. [83] on polyolefin blends and diblock copolymers. The fundamental driving force for phase separation is predicted by compressible PRISM theory to be enthalpic in nature but with multiple correlation corrections. This aspect is in strong contrast with the recent phenomenological field theory for stiffness asymmetric blends [84] based on a "nematic correlation" process and incompressibility.

The second, "explicit compressibility correction" term in the braces of Eq. (8.15) is essentially N-independent. Since it is positive definite, explicit compressibility effects always destabilize the blend. Its dependence on polymer density is in general very different than the leading "concentration fluctuation" contribution. The latter monotonically increases with polymer density, while the former strongly decreases with density.

The "explicit compressibility contribution" is unimportant if the inequality $- N\rho C_{AA}^{(0)}(\lambda - \gamma^2)^2 \phi(1 - \phi) \gg 1$ is obeyed. Adopting the latter condition can be viewed as enforcing an "effective incompressibility" constraint in a thermodynamically post facto manner It differs enormously from the spinodal predicted based on the literal IRPA approach of Eqs. (6.17) and (6.18) which is given by [67]

$$k_B T_{S,\,INC} = \rho |\tilde{H}_{AA}|(\lambda - 1)^2 \, N\phi(1 - \phi)\{1 + (\eta^2 \Gamma^6/6)(\gamma^2 - 1)^2$$
$$\times N\phi(1 - \phi)[\phi + \gamma^2(1 - \phi)]\}^{-1}. \tag{8.17}$$

Equation (8.17) incorrectly predicts that increasing stiffness asymmetry monotonically stabilizes the mixture. The basic error incurred by the literal IRPA is the implicit assumption that "enthalpic" and (packing) "entropic" contributions to the non-ideal free energy of mixing are independent and additive.

For the large N and high densities of primary interest in polymer alloy materials, the inequality $- N\rho C_{AA}^{(0)}(\lambda - \gamma^2)^2 \phi(1 - \phi) \gg 1$ will be violated only for the special case of $\lambda \approx \gamma^2$. However, if such "cancellation" or "compensation" of the attractive potential and stiffness asymmetry factors occurs, then strong stabilization of the blend is predicted since the spinodal temperature obeys the law $T_S \propto N^0$. This suggests an interesting and novel "strategy" for molecular engineering miscible polymer blends or increasing the interfacial region in phase-separated alloys. Equation (8.15) is also consistent with the physical expectation that blends of chains of greatly disparate aspect ratios (e.g., "rods and coils") will phase separate at high temperatures, low N, and/or low total polymer densities, due to packing-induced "frustration". Moreover, if the energetic and structural asymmetries are comparable, then their consequences are never separable since the "cross terms" are always significant and can either stabilize or destabilize the blend depending on the sign of the factor $(\gamma^2 - 1)(1 - \lambda)$. A dramatic example is when $\gamma = 1$, and hence $\chi_0 = 0$, corresponding to an "effectively" athermal case in the mean field sense, but not the literal $\beta v_{MM'}(r) = 0$ athermal situation considered in Sect. 7. Equation (8.15) obviously still predicts phase separation at a temperature which grows strongly with stiffness asymmetry, while the IRPA analysis based on Eqs. (6.17) and (6.18) incorrectly predicts the effective chi-parameter is negative and the blend is completely miscible!

For most of the λ, γ parameter space, correlation effects result in a critical temperature higher than predicted by Flory-Huggins theory [67]. However, there are "windows" of parameter space where stiffness asymmetry stabilizes the blend. Whether the theory predicts relative stabilization or destabilization due to an increase of the stiffness depends crucially on the sign of the quantity $\gamma - \sqrt{\lambda}$.

The apparent SANS chi-parameter is also easily determined analytically for stiffness asymmetric Berthelot thread model with the R-MMSA or R-MPY/HTA closure approximations. For algebraic simplicity we consider the neutron data analysis approach which leads to Eq. (6.15). In the "effectively incompressible" regime, defined here as $|C_{MM'}^{(0)}| \gg |\beta\tilde{H}_{MM'}|$ in Eq. (8.11), one easily obtains the result [67]

$$2\chi_S = \rho\gamma^{-2}\beta|\tilde{H}_{AA}|(\gamma^2 - \lambda)^2 + \frac{1 - \gamma^2}{N\phi} + \frac{1 - \gamma^2}{N(1 - \phi)}. \tag{8.18}$$

This prediction is of the general form found in many SANS experiments, i.e., $\chi_S = A + (B/T)$ where "A" and "B" are often empirically interpretated as "entropic" and "enthalpic" contributions, respectively. Note that for the present idealized thread model, "B" is always positive, but the molecular weight dependent A-factor can in general be positive or negative depending on the precise values of the stiffness asymmetry ratio and blend composition.

In summary, the predictions of analytic PRISM theory [67] for the phase behavior of asymmetric thread polymer blends display a very rich dependence on the single chain structural asymmetry variables, the interchain attractive potential asymmetries, the ratio of attractive and repulsive interaction potential length scales, a/d, and the thermodynamic state variables η and ϕ. Moreover, these dependences are intimately coupled, which mathematically arises within the compressible PRISM theory from "cross terms" between the repulsive (athermal) and attractive potential contributions to the $k = 0$ direct correlations in the spinodal condition of Eq. (6.6). The nonuniversality and nonadditivity of the consequences of molecular structural and interaction potential asymmetries on phase stability can be viewed as a virtue in the sense that a great variety of phase behaviors are possible by rational chemical structure modification. Finally, the relationship between the analytic thread model predictions and numerical PRISM calculations for more realistic nonzero hard core diameter models remains to be fully established, but preliminary results suggest the thread model predictions are qualitatively reliable for thermal demixing [72, 85].

9 Block Copolymers and Other Polymer Alloy Problems

PRISM theory based on the new molecular closures has recently been generalized by David and Schweizer [86] to treat periodic block copolymers. For simplicity, consider "2M-block copolymers", $(A_xB_y)_M$, which consist of alternating sequences of A and B segments of length x and y, respectively. The following variables are defined: concentration of A-site $f = x/(x + y)$, the total degree of polymerization $N = M(x + y)$, $N_A = xN$, $N_B = yN$, and ρ_{cp} = number density of copolymer molecules. As for the homopolymer case, a tractable theory requires ignoring explicit chain end effects at the level of the direct correlation functions. In addition, an extra zeroth order approximation enters corresponding to neglecting "junction effects". That is, the direct correlation functions associated with a pair of sites of type M and M' are assumed not to depend on the precise location of the segments within their respective blocks. The resulting matrix PRISM equations in Fourier space are given by [86]:

$$\hat{\underline{H}}(k) = \hat{\underline{\Omega}}(k)\hat{\underline{C}}(k)[\hat{\underline{\Omega}}(k) + \hat{\underline{H}}(k)] \tag{9.1}$$

where the intermolecular site-site pair correlation function matrix is defined as $\hat{H}_{MM'}(k) \equiv \rho_M\rho_{M'} \hat{h}_{MM'}(k)$, $\hat{C}_{MM'}(k)$ is the site-site direct correlation function, and the intramolecular partial structure factor matrix is

$$\hat{\Omega}_{MM'}(k) \equiv \rho_{cp} \sum_{\alpha=1}^{N_M} \sum_{\gamma=1}^{N_{M'}} \hat{\omega}_{\alpha M \gamma M'}(k) \tag{9.2}$$

where $\hat{\omega}_{\alpha M \gamma M'}(r)$ is the normalized probability distribution function where a site α of type M is a distance r from site γ of type M'. It is convenient to introduce a related set of intramolecular correlation functions defined as

$$\hat{\omega}_{MM}(k) \equiv N_M^{-1} \sum_{\alpha,\gamma=1}^{N_M} \hat{\omega}_{\alpha M \gamma M}(k),$$

$$\hat{\omega}_{AB}(k) \equiv (N_A + N_B)^{-1} \sum_{\alpha=1}^{N_A} \sum_{\gamma=1}^{N_B} \hat{\omega}_{\alpha A \gamma B}(k) . \tag{9.3}$$

The collective partial structure factors, $\hat{S}_{MM'}(k)$, are easily written down, and in an approximate theory finite length scale spinodal instabilites *may* be present corresponding to $\hat{S}_{MM'}(k = k^*) = \infty$ which is equivalent to [86]:

$$0 = 1 - f\rho_S\hat{\omega}_{AA}(k^*)\hat{C}_{AA}(k^*) - (1 - f)\rho_S\hat{\omega}_{BB}(k^*)\hat{C}_{BB}(k^*)$$

$$- 2\rho_S\hat{\omega}_{AB}(k^*)\hat{C}_{AB}(k^*) + f(1 - f)\rho_S^2 \delta\hat{\omega}(k^*)\delta\hat{C}(k^*) \tag{9.4}$$

$$\delta\hat{\omega}(k) \equiv \hat{\omega}_{AA}(k)\hat{\omega}_{BB}(k) - f^{-1}(1 - f)^{-1}\hat{\omega}_{AB}^2(k),$$

$$\delta\hat{C}(k) \equiv \hat{C}_{AA}(k)\hat{C}_{BB}(k) - \hat{C}_{AB}^2(k)$$

where k^* is the most unstable wavevector and $\rho_S = N\rho_{cp}$. Since microphase separation is a finite length scale ordering phenomena akin to crystallization, a critical point is not generally expected to exist in physical reality.

The mean field (Landau) theory of block copolymers developed by Leibler [87] is based on an IRPA treatment of the liquid correlations. Enforcing the latter constraint on PRISM theory in a post facto manner yields for an AB block copolymer for which the A and B site volumes are equal [86]

$$\hat{S}_{RPA}^{-1}(k) = \frac{f\hat{\omega}_{AA}(k) + (1-f)\hat{\omega}_{BB}(k) + 2\hat{\omega}_{AB}(k)}{f(1-f)\hat{\omega}_{AA}(k)\hat{\omega}_{BB}(k) - \hat{\omega}_{AB}^2(k)} - 2\hat{\chi}_{INC}(k) \ . \tag{9.5}$$

The effective chi-parameter is given by the k-dependent generalization of Eq. (6.17). If a small angle approximation where $\hat{\chi}_{INC}(k)$ is constant is invoked, then Eq. (9.5) is identical in form with Leibler's RPA result [87] and the spinodal condition is far simpler than Eq. (9.4). However, since the true "chi-parameter" is a wavevector-dependent correlation function, not a phenomenological number, it is functionally related to all the other intramolecular and intermolecular pair correlations in the system. This non-mean-field feature has many important consequences such as the fact that k* is influenced by many chain correlations [86, 88]. It must be emphasized that although Eq. (9.5) should be accurate for the hypothetical symmetric block copolymer model, since it does not properly treat compressibility effects it is expected to be inadequate for most real copolymer systems.

The same molecular closures proposed for homopolymer blends [68–70] apply to copolymers but the intramolecular structure factor matrix is now non-diagonal. Equation (8.8) becomes [86]

$$[\underline{\Omega} * \underline{C} * \underline{\Omega}(r)]_{MM'} \cong [\underline{\Omega} * \underline{C^{(0)}} * \underline{\Omega}(r)]_{MM'}$$

$$+ [\underline{\Omega} * \underline{\Delta C} * \underline{\Omega}(r)]_{MM'}, \quad r > d_{MM'} \ . \tag{9.6}$$

For the analytically tractable thread polymer model, and the R-MMSA or R-MPY/HTA closure approximations, $k = 0$ values of the direct correlation functions are precisely the same in the long chain limit as found for polymer blends in Sect. 8. In particular, for the symmetric block copolymer, the R-MMSA closure yields [67, 86] the mean field result $\chi_{INC} = \chi_0$. Thus, within the symmetric thread idealization and the incompressible approximation of Eq. (9.5), PRISM/R-MMSA theory reduces to Leibler theory for all compositions and block architectures [67, 86].

For the more interesting thread model case with statistical segment asymmetry and a Berthelot attractive potential model, Schweizer has derived the following expression for the spinodal temperature [67]

$$\frac{k_B T_S}{\rho_S |\tilde{H}_{AA}|} = f(1-f)N(\gamma^2 - \lambda)^2 \tilde{F}(k^*, f, \gamma)$$

$$+ [-\rho_S C_{AA}^{(0)}]^{-1} \tilde{G}(k^*, f, \gamma, \lambda) \tag{9.7}$$

where the block architecture-dependent (x and y variables) functions F and G are given by

$$\tilde{F} \equiv N^{-1} \frac{\hat{\omega}_{AA}(k^*)\hat{\omega}_{BB}(k^*) - [f(1-f)]^{-1}\hat{\omega}_{AB}^2(k^*)}{f\hat{\omega}_{AA}(k^*) + \gamma^4(1-f)\hat{\omega}_{BB}(k^*) + 2\gamma^2\hat{\omega}_{AB}(k^*)}$$

$$\tilde{G} \equiv \frac{f\hat{\omega}_{AA}(k^*) + \lambda^2(1-f)\hat{\omega}_{BB}(k^*) + 2\lambda\hat{\omega}_{AB}(k^*)}{f\hat{\omega}_{AA}(k^*) + \gamma^4(1-f)\hat{\omega}_{BB}(k^*) + 2\gamma^2\hat{\omega}_{AB}(k^*)} \ . \tag{9.8}$$

The functions F and G are both independent of N in the long chain limit. For a stiffness symmetric $f = 1/2$ diblock copolymer in the $N \gg 1$ limit, the quantity $F = 2/10.495$. The ordering wavevector, k^*, is (numerically) determined by maximizing the right hand side of Eq. (9.7), and is a function of all the variables in the problem. Note that the mathematical structure of Eq. (9.7) is qualitatively similiar to the blend case of Eq. (8.15).

As discussed in Sect. 8.3 and elsewhere [67, 68, 86], for the R-MMSA and R-MPY/HTA closures, taking the thread limit results in a theory where the hard core exclusion condition is not relevant (except for determining the reference fluid $C_{MM'}^{(0)}$). This great simplification does not occur for any non-zero hard core diameter. Calculations by David and Schweizer [86] for symmetric diblocks have shown that PRISM with the R-MMSA closure for non-thread symmetric diblocks does not exhibit a critical point or spinodal instabilities if either the temperature or inverse degree of polymerization are nonzero. The destruction of the critical point appears to be a "finite size" fluctuation effect, as in the Brazovski-Fredrickson-Helfand phenomenological field theory [89], but the physical origin of the nonlinear feedback mechanism is very different.

An example of the intermolecular concentration correlations and scattering intensity for a symmetric diblock melt [86] composed of ideal freely-jointed chains are shown in Fig. 15. Note that as the melt is cooled, strong local and

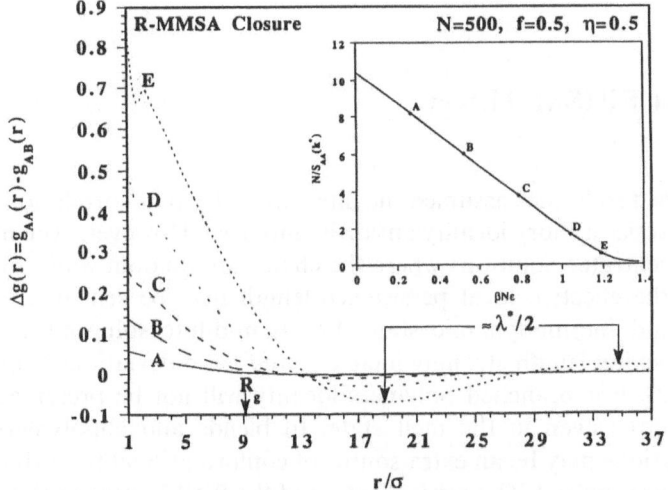

Fig. 15. Temperature dependence of $\Delta g(r) = g_{AA}(r) - g_{AB}(r)$ for the $f = 0.5$ symmetric diblock copolymer using the R-MMSA closure [86]. An ideal freely-jointed single chain model is employed and the choice of tail potentials is identical to the symmetric blend discussed in Sects. 6 and 8. Note that upon cooling strong local correlations emerge. Eventually a weaker, long range oscillatory feature appears associated with correlations on the domain size length scale ($\lambda^* \cong 2\pi/k^*$). The inset shows the corresponding points (A–E) in the inverse scattering peak intensity $N/S_{AA}^{(k)}$ as a function of dimensionless, scaled inverse temperature $\beta N\epsilon$. Note the linear "mean field" behavior at high temperatures but the nonlinear stabilization at the low temperatures where the long range oscillation in $\Delta g(r)$ emerges

long wavelength correlations emerge, the detailed nature of which depend on copolymer composition, N, single chain structure, density, etc. Such local fluctuations may be very important for understanding thermodynamic properties and experimental measurements of short and intermediate length scale structure in the diblock melt (e.g., NMR, dielectric spectroscopy, wide angle scattering).

There remains the controversial question of whether "fluctuation phenomena" in diblock copolymers are a finite size effect [89], or an intrinsic process which survives in the $N \to \infty$ limit as recently suggested based on both simulations [88] and non-perturbative theoretical arguments [90]. Numerical PRISM calculations based on the most sophisticated R-MPY closure favor an approximately intrinsic process for symmetric diblocks of $N \leqslant 500$; however, the true asymptotic behavior appears to be a finite size effect [86].

Several other polymer alloy problems have just begun to the investigated using the PRISM approach. These include (a) LCST phase behavior in binary blends where "specific interactions" are present [67, 71]; (b) segregation of blends near a surface and in confined spaces [91a] using the wall-PRISM theory of Yethiraj and Hall [37]; (c) mixtures of statistically random copolymers [67]; and (d) polymer-colloid mixtures [91b]. The application of existing PRISM theory to treat more complex mixtures such as a binary blend plus solvent, or a block copolymer/homopolymer mixture, are straightforward. All these problems will be even richer and more complex with regards to their sensitivity to system-specific structural and interaction potential asymmetries.

10 Self-Consistent PRISM Theory

All the theory described so far has assumed the intramolecular pair correlations are known, and in practice a Flory ideality ansatz is employed. However, even in dense melts and concentrated solutions where the chains are random walks on large length scales, the effective local persistence length may be sensitive to chemical structure and thermodynamic state. For semi-dilute solutions the excluded volume screening length strongly increases and large deviations from ideality occur [3a, 22]. For branched polymers ideality will not be preserved near the branch point(s) even in the melt state. In blends and copolymers concentration fluctuations may be an extra source of conformational perturbations and are poorly understood. The generalization of the PRISM approach to self-consistently calculate intramolecular and intermolecular pair correlations functions has been initiated [92] but is still in its early stages. Here, we sketch the basic ideas, briefly describe one application, and discuss ongoing research.

For simplicity, consider a solution of homopolymers where the "solvent" is treated as "free volume" and thus an effectively one-component description applies. The single polymer effective potential energy in the condensed phase, $W(\underline{R})$, contains three distinct terms [14, 92]

$$W(\underline{R}) = U_0 + U_E + \Delta\mu \tag{10.1}$$

where $\underline{R}$ denotes a complete set of coordinates which specifies a particular polymer configuration. The "ideal" contribution, U_0, contains all the short range interactions such as bond length and angle constraints, torsional potentials, etc. The "long range" intramolecular potential, U_E, is taken to be pairwise decomposable, and in specific applications to date has been chosen to be a hard core or soft repulsion. The "medium-induced" contribution, $\Delta\mu$, is the excess chemical potential for the polymer constrained to a particular conformation $\underline{R}$ due to its interactions with the surrounding molecules. This object is extremely complex since it involves many body correlations.

Chandler and co-workers constructed a tractable medium-induced potential in the context of the solvated electron problem [93, 94]. Their "self-consistent pair" approximation also applies to real polymeric fluids, and for homopolymers with neglect of explicit chain end effects approximation one has [92–94]

$$\Delta\mu \cong \sum_{\alpha,\gamma} w(|\vec{r}_\alpha - \vec{r}_\gamma|), \quad w(r) = -\beta^{-1}C*\rho\,S*C(r) . \tag{10.2}$$

This result can be deduced from a number of perspectives such as renormalized perturbation theory [93], polymer density functional theory [95, 96], and Percus functional expansion methods [97]. The medium-induced pair potential is determined by the direct correlation function and collective density fluctuations which are both functionally related to the intramolecular pair correlations via the PRISM equation. Hence, a coupled intramolecular/intermolecular theory is obtained.

Implementation of this "self-consistent" PRISM theory is non-trivial and there are two general classes of approach. First, approximate theories can be constructed based on a tractable reference system description of single chain correlations and an approximate free energy. The initial work along this line by Schweizer, Honnell and Curro [92] used an optimized perturbation scheme for dense melts which predicted chain dimensions and intermolecular pair correlations in very good agreement with molecular dynamics simulations [27]. However, this approach is rather limited in the choice of reference system, and does not correctly describe stiff polymers nor long chain semi-dilute solutions [30]. More general variational approaches have been recently developed by several workers [97, 98] which remove these limitations. They have been extensively applied to predict (both analytically and numerically) chain dimensions as a function of polymer concentration, molecular weight, aspect ratio, and global architecture. The best approximation probably depends on the particular system and thermodynamic conditions of interest, and theory development remains an active area of present research.

The second approach is to use Monte Carlo simulation to solve exactly the nonlocal single chain problem (defined by Eqs. (10.1) and (10.2)) self-consistently with the PRISM theory of many chain correlations. Although more computationally demanding, the introduction of an approximate free energy expression and reference system are avoided. The first implementation of this "PRISM/Monte Carlo" (PMC) scheme has been performed by Melenkevitz et al. [96] for the case of homopolymer solutions composed of semiflexible chains [24] (with zero local bending energy). A more extensive PMC study of the same model over the entire range of density from dilute solution to the dense melt using a more sophisticated Monte Carlo algorithm has also been performed [99]. The latter results are summarized in Table 2. The PMC theory with the solvent-induced potential of Eq. (10.2) accurately predicts both the magnitude and density dependence of chain dimensions in semi-dilute and weakly concentrated solutions [96, 99]. However, for concentrated solutions and melts the chains tend to "collapse" locally (not in the $\langle R^2 \rangle \propto N^{2/3}$ global sense), and the physically expected ideal random coil behavior is not recovered [96, 99]. This subtle high density problem is still under active study, but the work of Grayce and co-workers [97, 99] suggests the difficulty lies with the approximate medium-induced potential of Eq. (10.2) which predicts too strong a compressive force at high densities. Both a new solvent-induced pair potential, and a criterion for a priori accessing the accuracy of approximate solvation potentials, have been formulated [97]. Preliminary results are encouraging since PMC with the new solvation potential correctly predicts the qualitative conformational behavior of flexible polymers over the entire density range [99].

Table 2. Mean-square end-to-end distance, $\langle R^2 \rangle$, in units of the hard core diameter, as a function of packing fraction and N. $\langle R^2 \rangle_{PMC}$ refers to the results of [99] using self-consistent PRISM/Monte Carlo based on Eq. (10.2) for a hard core tangent semiflexible chain model [24]. $\langle R^2 \rangle_{MC}$ are the results of many chain "exact" Monte Carlo simulations of Yethiraj and Hall [29]. The corresponding density-independent self-avoiding walk (ideal walk) values of $\langle R^2 \rangle$ are: 50.78 (30.79), 152.03 (80.8), and 348.51 (164.12) for N = 20, 50, and 100, respectively

N	η	$\langle R^2 \rangle_{PMC}$	$\langle R^2 \rangle_{MC}$
20	0.10	45.24 ± 0.41	43.01 ± 1.79
20	0.20	41.92 ± 0.09	37.37 ± 1.32
20	0.30	38.43 ± 0.48	34.95 ± 2.81
20	0.35	35.69 ± 0.49	32.23 ± 4.30
20	0.40	32.39 ± 0.35	
20	0.45	26.56 ± 0.30	
20	0.50	17.76 ± 0.64	
50	0.20	131.57 ± 1.90	118.70 ± 4.83
50	0.30	119.40 ± 1.19	106.84 ± 6.05
100	0.20	299.57 ± 2.55	242.51 ± 6.53
100	0.30	276.67 ± 5.54	220.09 ± 5.08
100	0.40	224.03 ± 5.67	
100	0.45	149.83 ± 15.4	

Formally, the self-consistent PRISM theory is easily generalized to treat polymer blends and copolymers [92] where significant non-ideal conformational effects may occur which intensify as phase separation is approached.

11 Future Directions

The full development and application of many aspects of the PRISM theory discussed in this paper remain to be done. In both melts and alloys, the construction of a "thermodynamically self-consistent" theory is an important task in order to compute accurately the equation-of-state and thereby allow constant pressure calculations to be carried out. This direction may also be important for understanding at a molecular-level LCST phase transitions. Construction of the binodal curve is also an important technical direction, as are careful studies of the free energy-based route to the phase diagram [70b]. Further fundamental research concerning the closure difficulties encountered in phase-separating polymer alloys is an ongoing topic, as is continual testing of the accuracy of PRISM theory against carefully designed off-lattice computer simulations and experimental measurements on model systems.

From the point of view of atomistic modeling, the application of PRISM theory with chemically realistic single chain models, such as the RIS description, is a major thrust of present and planned research particularly in the area of multiphase alloys. The influence of microstructural features such as chain branching (e.g., olefins) [108], monomer shape, and tacticity [71] on packing and phase stability can thus be unambiguously investigated. A priori, one might expect that thermodynamic properties and phase diagrams are extremely sensitive to local chemical structure and packing. However, many theoretical approaches and/or models rely on the hope that for flexible macromolecules a significant amount of "self-averaging" of the Angstrom-level chemical details occurs, and thus intermediate-level models of chain structure are useful. Systematic PRISM studies of polymer models of increasing chemical complexity will allow this subtle and very important question to be addressed.

The combination of polymeric density functional methods and PRISM theory for the liquid correlations allow a wide range of closure and inhomogeneous material problems to be studied [109]. Present research involves using this approach to treat at an atomistic level the crystallization of the entire alkane series, and the structure of hydrocarbon fluids near surfaces and interfaces [109]. An alternative, purely integral equation approach to the latter problem is to employ the wall-PRISM theory of Yethiraj and Hall [37].

The further development of the self-consistent version of PRISM theory will be particularly important in two areas: (i) liquids of flexible conjugated polymers where the electron delocalization length and interchain dispersion forces are strongly coupled to chain conformation [100], and (ii) polymer alloys where nonuniversal conformational perturbations due to both density and concentration fluctuations are intimately coupled and may significantly influence the location, and perhaps the nature, of the phase boundaries.

The PRISM theory has recently been generalized and applied by Grayce and Schweizer [101] to treat star-branched polymer solutions and melts. For such systems the equivalent site approximation must be abandoned, and non-ideal conformational effects become particularly important near the crowded central branch point. Generalization of their new methods to treat complex fluids composed polymer-coated colloidal particles is also feasible and presently under study.

Other important unsolved general problems within the PRISM formalism include the following. (a) The treatment of quenched randomness, e.g., sequence disorder in random copolymers. (b) The description of first order microphase separation transitions in block copolymers. Polymeric density functional methods, such as recently employed by Melenkevitz and Muthukumar [102], present an attractive strategy. (c) Macromolecular liquids with strong and/or directional attractive forces, such as ionomers and polyelectrolytes. Nonideal conformational perturbations are likely to be particularly important in these systems, and the appropriate closure approximation for Coulombic interactions in macromolecular fluids remains an open question. (d) Polymer liquid crystals, and molecular "composites" composed of rod-like and flexible species. The proper description of strong orientational correlations and/or the nematic phase represent major challenges for site-site integral equation approaches. All the equilibrium structural information calculable from PRISM theory will also find additional applications in microscopic theories of macromolecular dynamics such as the polymeric mode-coupling approach [103].

Finally, we mention that very recently three other integral equation approaches to treating polymer systems have been proposed. Chiew [104] has used the "particle-particle" perspective to develop theories of the intermolecular structure and thermodynamics of short chain fluids and mixtures. Lipson [105] has employed the Born-Green-Yvon (BGY) integral equation approach with the Kirkwood superposition approximation to treat compressible fluids and blends. Initial work with the BGY-based theory has considered lattice models and only thermodynamics, but in principle this approach can be applied to compute structural properties and treat continuum fluid models. Most recently, Gan and Eu employed a Kirkwood hierarchy approximation to construct a self-consistent integral equation theory of intramolecular and intermolecular correlations [106]. There are many differences between these integral equation approaches and PRISM theory which will be discussed in a future review [107].

Acknowledgement. The work described in this paper was done with many excellent collaborators. We gratefully acknowledge the important theoretical contributions of K.G. Honnell, J.D. McCoy, J. Melenkevitz, A. Yethiraj, C.J. Grayce, and E.F. David. We would also like to thank G.S. Grest, K. Kremer and J.D. Honeycutt for fruitful theory/computer simulation collaborations, and A.H. Narten and A. Habenshuss for theory/experiment collaborations. Helpful discussions and/or correspondence with many scientists are also appreciated with a special thanks extended to D. Chandler. We are grateful to C.J. Grayce, G. Szamel, A. Yethiraj, and E.F. David for their critical reading of this manuscript. The work described in this paper has been done over a period of several years. At Sandia National Laboratories support came from the U.S. Department of Energy under contract DE-AC047DP00789, and also the Basic Energy Sciences/Division of Materials Science Program. Support at the University of Illinois is from the Division of Materials Sciences, Office of Basic Energy Sciences, U.S. Department of Energy in cooperation with the UIUC Frederick Seitz Materials Research Laboratory (via grant number DEFG02-91ER45439) and Oak Ridge National Laboratory. Support from the central computing facilities of the UIUC-MRL are also gratefully acknowledged.

References

1. Flory PJ (1953) Principles of polymer chemistry. Cornell University Press, Ithaca
2. Dudowicz J, Freed MS, Freed KF (1991) Macromolecules 24: 5096; Freed KF, Dudowicz J (1992) Theoretica, Chimica Acta 82; Dudowicz J, Freed KF (1993) Macromolecules, 26: 213; Freed K, Dudowicz J (1992) J Chem Phys 97: 2105
3. (a) For polymer solutions: Doi M, Edwards SF (1986) Theory of polymer dynamics. Oxford Press, Oxford; (b) For block copolymers: Bates FS, Fredrickson GH (1990) Ann Rev Phys Chem 41: 525
4. Roe RJ (ed) (1991) Computer simulations of polymers. Prentice Hall, Englewood Cliffs, N.J.; Colburn EA (ed) (1992) Computer simulations of polymers. Longman, Harlow; Binder K (1993) Advances in Polymer Science, in press
5 Hansen JP, McDonald IR (1986) Theory of simple liquids, 2nd edn. Academic, London
6. Chandler D (1982) In: Montroll EW, Lebowitz L (eds) Studies in statistical mechanics, vol. VIII. North-Holland, Amsterdam, p. 274 and references cited therein
7. Percus JK (1964) In: Frisch HL, Lebowitz KL (eds) Classical fluids. Wiley, New York
8. Chandler D, Andersen HC (1972) J Chem Phys 57: 1930
9. Lowden LJ, Chandler D (1974) J Chem Phys 61: 5228; (1973) 59: 6587; (1975) 62: 4246
10. Chandler D, Hsu CS, Streett WB (1977) J Chem Phys 66: 5231; Sandler SI, Narten AH (1976) Mol Phys 32: 1543; Narten AH (1977) J Chem Phys 67: 2102; Hsu CS, Chandler D (1978) Mol Phys 36: 215 ; Mol Phys 37: 299 (1979)
11. Schweizer KS, Curro JG (1987) Phys Rev Lett 58: 246
12. Curro JG, Schweizer KS (1987) Macromolecules 20: 1928
13. Curro JG, Schweizer KS (1987) J Chem Phys 87: 1842
14. Schweizer KS, Curro JG (1988) Macromolecules 21: 3070
15. Schweizer KS, Curro JG (1988) Macromolecules 21: 3082
16. Volkenstein MV (1963) Configurational statistics of polymer chains. Interscience, New York; Flory PJ (1969) Statistical mechanics of chain molecules. Interscience, New York
17. Flory PJ (1949) J Chem Phys 17: 203
18. Curro JG (1976) J Chem Phys 64: 2496; (1979) Macromolecules 12: 463; Vacatello M, Avitabile G, Corradini P, Tuzi A (1980) J Chem Phys 73: 543
19. Ballard DG, Schelton J, Wignall GD (1973) Eur. Polymer Journal, 9: 965; Cotton JP, Decker D, Benoit H, Farnoux B, Higgins J, Jannick G, Ober R, Picot C, des Cloizeaux J (1974) Macromolecules 7: 863
20. Lue L, Blanckschtein D (1992) J Phys Chem 96: 8582
21. Elliot JR, Kanetar US (1990) Mol Phys 71: 871 and 883
22. deGennes PG (1979) Scaling concepts in polymer physics. Cornell University Press, Ithaca
23. Schweizer KS, Curro JG (1990) Chemical Physics, 149: 105; Schweizer KS, Curro JG (1991) J Chem Phys 94: 3986
24. Honnell KG, Curro JG, Schweizer KS (1990) Macromolecules, 23: 3496
25. Koyama R (1973) J Phys Soc Japan, 22: 1029; Mansfield ML (1986) Macromolecules, 19: 854
26. Chandler D, Silbey RS, Ladanyi BM (1982) Mol Phys 46: 1335; Richardson DM, Chandler D (1984) J Chem Phys 80: 4484
27. Curro JG, Schweizer KS, Grest GS, Kremer K (1989) J Chem Phys 91: 1357; Kremer K, Grest GS (1990) J Chem Phys 92: 5057
28. Andersen HC, Weeks JD, Chandler D (1971) Phys Rev A 4: 1597; Weeks JD, Chandler D, Andersen HC (1971) J Chem Phys 54: 5237

29. Yethiraj A, Hall CK (1991) J Chem Phys 93: 4453; (1992) 96: 797
30. Yethiraj A, Schweizer KS (1992) J Chem Phys 97: 1455
31. McCoy JD, Honnell KG, Curro JG, Schweizer KS, Honeycutt JD, Macromolecules (1992) 25: 4905
32. Honnell KG, McCoy JD, Curro JG, Schweizer KS, Narten AH, Habenschuss A (1991) J Chem Phys 94: 4659
33. Narten AH, Habenschuss A, Honnell KG, McCoy JD, Curro JG, Schweizer KS (1992) J Chem Soc Faraday Trans 88: 1791
34. Honnell KG, McCoy JD, Curro JG, Schweizer KS, Narten AH, Habenshuss A (1991) Bull Am Phys Soc 36(3): 481, and paper in preparation
35. Barker JA, Henderson D (1967) Chem Phys 47: 4714; (1972); Ann Rev Phys Chem 23: 439; (1976) Rev. Mod Phys 48: 587
36. Dickman R, Hall CK (1988) J Chem Phys 89: 3168
37. Yethiraj A, Hall CK (1991) J Chem Phys 95: 3749
38. Yethiraj A, Curro JG, Schweizer KS, McCoy JD (1993) J Chem Phys 98: 1635
39. Curro JG, Yethiraj A, Schweizer KS, McCoy JD, Honnell KG (1993) Macromolecules 26: 2655
40. Dickman R, Hall CK (1986) J Chem Phys 85: 4108; Honnell KG, Hall CK (1989) J Chem Phys 90: 1841
41. Olabisi O, Simha R (1975) Macromolecules 8: 206
42. Lopez-Rodriguez A, Vega C, Freire JJ, Lago S (1991) Mol Phys 73: 691
43. Martynov GA, Vompe AG (1993) Phys Rev E 47: 1012
44. Flory PJ (1956) Proc Roy Soc A 234: 60
45. Nagle JF, Gujrati PD, Goldstein M (1984) J Phys Chem 88: 4599
46. Ramakrishnan TV, Yussouff M (1979) Phys Rev B 19: 2775
47. Haymet ADJ, Oxtoby DW (1981) J Chem Phys 74: 2559; Laird BB, McCoy JD, Haymet ADJ (1987) J Chem Phys 87: 5451
48. Chandler D, McCoy JD, Singer SJ (1986) J Chem Phys 85: 5977; McCoy JD, Singer SJ, Chandler D (1987) J Chem Phys 87: 4953
49. McCoy JD, Rick SW, Haymet ADJ (1989) J Chem Phys 90: 4622; (1990) 92: 3034; Rick SW, McCoy JD, Haymet ADJ (1990) J Chem Phys 92: 3040
50. Ding K, Chandler D, Smithline SJ, Haymet ADJ (1987) Phys Rev Lett 59: 1698
51. McMullen WE, Freed KF (1990) J Chem Phys 92: 1413
52. McCoy JD, Honnell KG, Schweizer KS, Curro JG (1991) Chem Phys Lett 179: 374; J Chem Phys 95: 9348
53. Wunderlich B, Czornj G (1977) Macromolecules 10, 906
54. Starkweather HW, Zoller P, Jones GA, Vega AJ (1982) J Polym Sci., Polym Phys 20: 751
55. See, for example, Bates FS (1991) Science 251: 898; Sanchez IC (1983) Ann Rev Mater Sci 13: 387; Solc K (ed) (1981) Polymer compatibility and incompatibility. Midland, Michigan
56. Wignall GD (1987) in Encyclopedia of polymer science and engineering, second edition. Wiley, New York, vol. 12, p. 112
57. See, for e.g., Jung WG, Fischer EW (1988) Makromol, Chem Makromol Symp 16: 281; Brereton MG, Fischer EW, Herkt-Maetzky C, Mortensen K (1987) J Chem Phys 87: 6114; Han CC, Bauer BJ, Clark JC, Moroga Y, Matsushita Y, Okada M, Tran-cong Q, Chang T, Sanchez IC (1988) Polymer 29: 2002; Bates FS, Muthukumar M, Wignall GD, Fetters LJ (1988) J Chem Phys 89: 535
58. Marie P, Selb J, Rameau A, Gallot Y (1988) Makromol Chem Makromol Symp 16: 301; Jung WG, Fischer EW (1988) ibid 16: 281; Hashimoto T, Ijichi Y, Fetters LJ (1988) J Chem Phys 89: 2463; Ijichi Y, Hashimoto T, Fetters LJ (1989) Macromolecules 22: 2817; Tanaka H, Hashimoto T (1991) Macromolecules 24: 5398
59. Schweizer KS, Curro JG (1988) Phys Rev Lett 60: 809
60. Curro JG, Schweizer KS (1988) J Chem Phys 88: 7242
61. Schweizer KS, Curro JG (1989) J Chem Phys 91: 5059
62. Curro JG, Schweizer KS (1990) Macromolecules 23: 1402
63. Curro JG, Schweizer KS (1991) Macromolecules 24: 6736
64. Kirkwood JG, Buff FP (1951) J Chem Phys 19: 774
65. Chen XS, Forstmann F (1992) J Chem Phys 97: 3696; Malescio G (1992) J Chem Phys 96: 648, and references cited therein; Arrieta E, Jedrzejek C, Marsh KN (1991) ibid 95: 6806 and 6838
66. Sanchez IC (1991) Macromolecules 24: 908
67. Schweizer KS (1993) Macromolecules 26: 6033 and 6050
68. Schweizer KS, Yethiraj A (1993) J Chem Phys 98: 9053

69. Yethiraj A, Schweizer KS (1992) J Chem Phys 97: 5927
70. (a) Yethiraj A, Schweizer KS (1993) J Chem Phys 98: 9080; (b) Singh C, Schweizer KS, Yethiraj A (1994) J Chem Phys, submitted
71. Honeycutt JD (1992) ACS Polymer Preprints 33(1): 529; (1992) Proc. of CAMSE'92, Yokohama, Japan, in press; private communication
72. Singh C, Schweizer KS (1994) J Chem Phys, submitted
73. Deutsch H-P, Binder K (1992) Europhysics Lett 17: 697; (1993) Macromolecules 25: 6214; (1993) J Phys II France 3: 1049, see also, Sariban A, Binder K (1988) Macromolecules 21: 711
74. Gehlsen MP, Rosedale JH, Bates FS, Wignall GD, Hansen L, Almdal K (1992) Phys Rev Lett 68: 2452
75. Chandler D (1993) Phys Rev E 48: 2898
76. Melenkevitz J, Curro JG (1994) in preparation
77. Andersen HC, Chandler D (1972) J Chem Phys 57: 1918, 1930
78. Lupkowski M, Monson PA (1987) J Chem Phys 87: 3618
79. Ladanyi BM, Chandler D (1975) J Chem Phys 62: 4308
80. Dudowicz J, Freed KF (1990) Macromolecules 23: 1519; Tang H, Freed KF (1991) Macromolecules 24: 958
81. Rowlinson JS, Swinton FL (1982) Liquids and Liquid Mixtures, Butterworth Scientific, London
82. Hildebrand J, Scott R (1949) The Solubility of Nonelectrolytes, 3rd Edition, Reinhold, New York
83. Bates FS, Schulz MF, Rosedale JH (1992) Macromolecules 25: 5547
84. Liu AJ, Fredrickson GH (1992) Macromolecules 25: 5551
85. Yethiraj A, Schweizer KS (1993) Bull Am Phys Soc 38(1), 485; David EF, Schweizer KS (1994) in preparation
86. David EF, Schweizer KS (1994) J Chem Phys 100: May 15
87. Leibler L (1980) Macromolecules 13: 1602
88. Fried H, Binder K (1991) J Chem Phys 94: 8349
89. Fredrickson GH, Helfand E (1987) J Chem Phys 87: 697; Brazovski SA (1975) Sov Phys JETP 41: 85
90. Tang H, Freed KF (1992) J Chem Phys 96: 862
91. (a) Yethiraj A, Kumar S, Hariharan A, Schweizer KS (1994) J Chem Phys 100: 4691; (b) Yethiraj A, Hall CK, Dickman R (1992) J Colloid and Interface Sci 151: 102
92. Schweizer KS, Honnell KG, Curro JG (1992) J Chem Phys 96: 3211
93. Chandler D, Singh Y, Richardson DM (1984) J Chem Phys 81: 1975; Nichols AL, Chandler D, Singh Y, Richardson DM (1984) ibid 81, 5109; Laria D, Wu D, Chandler D (1991) ibid, 95: 4444
94. Chandler D (1987) Chem Phys Lett 139: 108
95. Singh Y (1987) J Phys A-Math Gen 20: 3949
96. Melenkevitz J, Schweizer KS, Curro JG (1993) Macromolecules 26: 6190
97. Grayce CJ, Schweizer KS (1994) J Chem Phys 100: May 1
98. Melenkevitz J, Curro JG, Schweizer KS (1993) J Chem Phys 99: 5571
99. Grayce CJ, Yethiraj A, Schweizer KS (1994) J Chem Phys 100: May 1
100. Schweizer KS (1986) J Chem Phys 85: 1156, 1176; Synthetic Metals 28: C565 (1989)
101. Grayce CJ, Schweizer KS (1993) Bull Am Phys Soc 38(1), 485; Macromolecules to be submitted.
102. Melenkevitz J, Muthukumar M (1991) Macromolecules 24: 4199
103. Schweizer KS (1989) J Chem Phys 91: 5802 and 5822; J Non-Cryst Sol 131–133, 643 (1991); Physica Scripta (1993) T49: 99
104. Chiew YC (1990) Mol Phys 70: 129 and 73: 359 (1991); J Chem Phys 93: 5067 (1990).
105. Lipson JEG (1991) Macromolecules 24: 1334; Lipson JEG, Andrews AA (1992) J Chem Phys 96: 1426
106. Gan HH, Eu BC (1993) J Chem Phys 99: 4084, 4103
107. Schweizer KS, Curro JG, Ann Rev Phys Chem, in preparation
108. Curro JG (1994) Macromolecules, in press
109. Donley JP, Curro JG, McCoy JD (1994) J Chem Phys, in press; Sen S, Cohen JM, McCoy JD, Curro JG (1994) J Chem Phys, submitted

Received July 1993

Author Index Volumes 101-116

Subject Index

Erratum to Volume 112

Contr.: Conventional and Living Carbocationic Polymerizations United.

The right order of authors is:

J. Majoros, A. Nagy, J.P. Kennedy.

Advances in Polymer Science, Vol. 112
© Springer-Verlag Berlin Heidelberg 1994